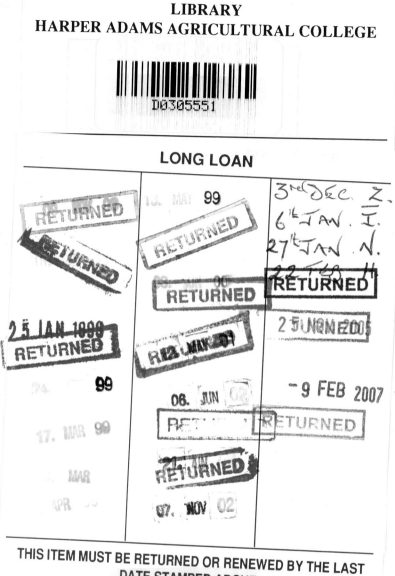

# The Transformation of the Agrifood System in Central and Eastern Europe and the New Independent States

*For Pam, Anna and Ian*

# The Transformation of the Agrifood System in Central and Eastern Europe and the New Independent States

Jill E. Hobbs

William A. Kerr

*and*

James D. Gaisford

CAB INTERNATIONAL

CAB INTERNATIONAL
Wallingford
Oxon OX10 8DE
UK

Tel: +44 (0)1491 832111
Fax: +44 (0)1491 833508
E-mail: cabi@cabi.org

CAB INTERNATIONAL
198 Madison Avenue
New York, NY 10016-4314
USA

Tel: +1 212 726 6490
Fax: +1 212 686 7993
E-mail: cabi-nao@cabi.org

A catalogue record for this book is available from the
British Library, London, UK
A catalogue record for this book is available from the
Library of Congress, Washington DC, USA

ISBN 0 85199 176 9

Typeset in Melior by Wyvern 21 Limited
Printed and bound in the UK by Biddles Ltd,
Guildford and King's Lynn

# Contents

# *Preface*

The motivation for writing this book came, not surprisingly perhaps, from our frustration with what has been written before. The transition from an economy based on planning and command to one based on markets had no precedent prior to the earth-shattering political events in Central and Eastern Europe and the Soviet Union after the fall of the Berlin Wall in November 1989. Hence, those who were to become directly involved in providing advice had nothing to go on, much less those within the former command economies who were responsible for shepherding the process of transition. There certainly was a model of the endpoint of the transition process, but no roadmap on how to get there. When we first went to the former command economies, it seemed that we had little to offer. While we could explain how businesses and markets functioned in the West, we were at a loss when it came to providing suggestions as to how to proceed to achieve the goal of a modern market economy. This was true whether teaching those directly involved in agribusiness in Poland, helping to devise university curricula in Slovakia, teaching prospective Russian agribusiness instructors or in numerous discussions with individuals charged with or involved in the transition process in the agrifood system. These latter conversations took place in a wide variety of venues ranging from the World Bank headquarters in Washington DC to the Whisky Trail in Scotland to private homes in Poland to the Internet. The experience of others in our situation was much the same. Helpful information simply did not exist.

Of course, since the early days of economic liberalization, a mass of material has been written on the transition process, some of it by us. In agriculture, a great deal has been written on the process of transforming former collective and state farms. On the other hand, little has been written about the other facets of the agrifood system: agroinputs, processing, retailing, international commercial relations, etc. Even less has been written about how the various components of the agrifood sector interact. We became convinced that these interactions should be the focal point for the analysis of the transition of the agrifood sector.

The approach we have taken is to look at the agrifood sector as a system. The systems perspective is particularly important for the agrifood sector for two reasons:

**1.** For many agricultural commodities, it is important to ensure that movement through the various stages of production, processing and distribution takes place in a continuous and timely fashion for reasons of perishability and food safety.
**2.** A breakdown in the system can affect food security and impose considerable hardship on individuals.

Waste, and hence reduced efficiency, is the result of the former. The latter can explain much of the political reluctance to deregulate the agrifood sector and why the sector has not been in the forefront of transition. The poor performance of the agrifood system since liberalization is also an important contributing factor to the resurgence in popularity of reformed communist parties in almost all countries.

One of the other deficiencies with much of what is available to read is that it includes very little economic theory which is applicable to transition. Hence, while there are theoretical discussions of market economies, many descriptions of various aspects of the reform process and some blow by blow accounts of the difficulties encountered in fostering transition, these have seldom been put within a broader theoretical discussion and, hence, lack a focus. We have tried to provide a theoretical framework for the problems associated with the process of transition and then to consistently use that framework to draw insights for the agrifood system. Hence, this is a book of ideas and not a chronology of transition nor a catalogue of liberalization initiatives. Many excellent chronologies and catalogues now exist, but in the rapidly changing economies of Central and Eastern Europe and the former Soviet Union, they are often dated before they go to press. Readers interested in these aspects of transition should look elsewhere. We hope this book is more insightful and will have a longer shelf-life.

The book is organized into five sections. Section I provides the introduction and a chapter describing the agrifood system in a command economy. The three chapters comprising Section II lay out the theoretical ideas that underpin the rest of the book. While our theoretical discussions are not entirely new – some of them are very old – they have not been explicitly put into the context of the transformation of agribusinesses in Central and Eastern Europe and the former Soviet

Union before. This section is written for readers with a basic understanding of economics. Where more formal theoretical models are important, they are available in the appendix to the chapter. Section III, which is the core of the book, is organized to move along the agrifood system from agroinputs through to retail. Special chapters are provided on perishables, international trade and transnational firms. Section IV provides suggestions for fostering the market institutions which are necessary for the creation of a market based agrifood sector. In Section V, which concludes the volume, we attempt to look into the future.

This book was conceived through a number of discussions among the three of us. We formally agreed to write the book at Westmead Farm, Elcombe in Wiltshire. Jim Gaisford signed on there after a long flight from Canada and may have regretted it ever since. The book was written in a wide range of venues including The University of Calgary, The University of Wales–Aberystwyth, San Diego State University and Lincoln University in New Zealand. We would like to thank all of those institutions for their support. We would also like to thank the Agricultural and Rural Economics Department, Scottish Agricultural College, Aberdeen, and the Economics Division, Department of Agriculture, University of Aberdeen, who gave us our first chances to participate in the transition process. In particular, in Aberdeen we would like to thank Professor Ken Thomson, Graham Dalton, Jacquie Middleton, Philip Leat and Garth Entwhistle. A special thanks to Simon Davies and Al McNeil for their continuing insights.

A number of people are inevitably involved in a book. Low-tech author Kerr would like to thank Laurie Stephens and Moira Jensen for efficiently typing and retyping his chapters. Jim Gaisford would like to thank Pam Bradley, Don Gaisford, Annette Hester and Leslie Tang for their input on various chapters. Robin Poitras and the Department of Geography at the University of Calgary are thanked for providing us with the maps. Tim Hardwick at CAB INTERNATIONAL is thanked for his support and for shepherding the project. Finally, we would like to thank all our friends from the economies in transition without whom this book would not have been possible.

J.E.H.
W.A.K.
J.D.G.

Calgary,
September 1996

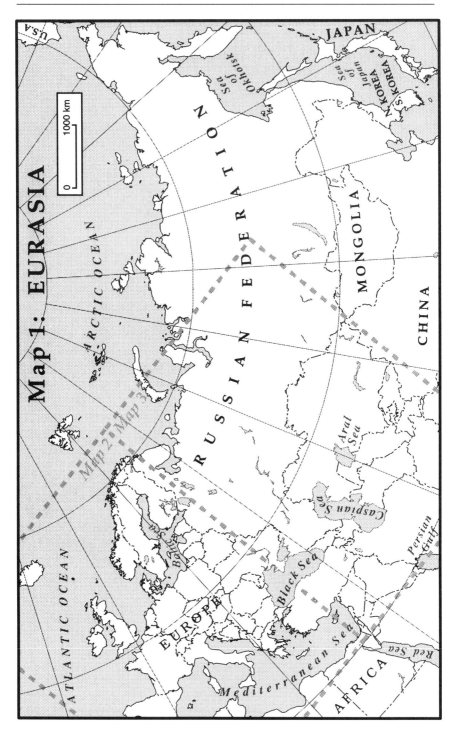

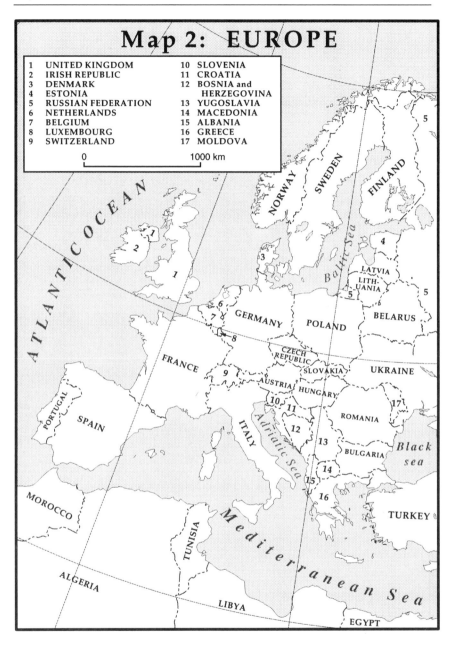

# Map 2: EUROPE

| | | | |
|---|---|---|---|
| 1 | UNITED KINGDOM | 10 | SLOVENIA |
| 2 | IRISH REPUBLIC | 11 | CROATIA |
| 3 | DENMARK | 12 | BOSNIA and |
| 4 | ESTONIA | | HERZEGOVINA |
| 5 | RUSSIAN FEDERATION | 13 | YUGOSLAVIA |
| 6 | NETHERLANDS | 14 | MACEDONIA |
| 7 | BELGIUM | 15 | ALBANIA |
| 8 | LUXEMBOURG | 16 | GREECE |
| 9 | SWITZERLAND | 17 | MOLDOVA |

0                    1000 km

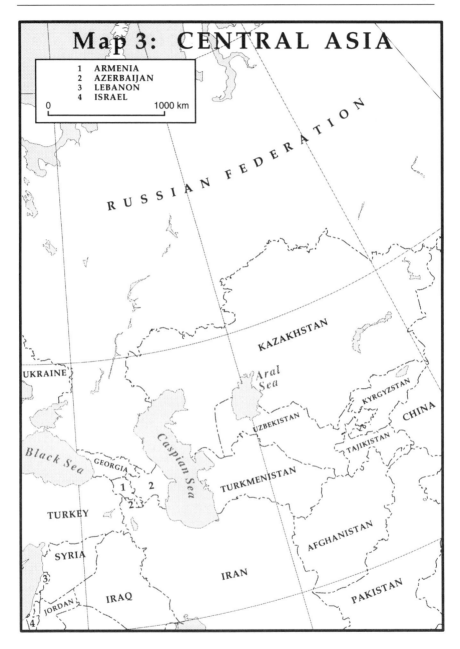

# Map 3:  CENTRAL ASIA

| 1 | ARMENIA |
|---|---------|
| 2 | AZERBAIJAN |
| 3 | LEBANON |
| 4 | ISRAEL |

0                    1000 km

RUSSIAN FEDERATION

KAZAKHSTAN

UKRAINE

*Aral Sea*

KYRGYZSTAN

UZBEKISTAN

CHINA

*Black Sea*

GEORGIA

*Caspian Sea*

TAJIKISTAN

1   2

TURKMENISTAN

2

TURKEY

AFGHANISTAN

SYRIA

IRAN

3

IRAQ

PAKISTAN

JORDAN

4

# 1 Background to Economic Transformation

The opening of the Berlin Wall in November 1989 marked a momentous historic watershed. In June of that year, the Communist Party had been forced to share power with the once-outlawed Solidarity trade union in Poland. By the end of the year, the mantle of 45 years of Soviet domination had been swept from Central and Eastern Europe in a wave of mainly peaceful popular uprisings.[1] By mid-1990, there had been free elections in East Germany, Romania, Czechoslovakia and Bulgaria, and, in the autumn, Solidarity leader Walesa was elected as president of Poland. While Germany was re-unified in the autumn of 1990, Czechoslovakia had split into the Czech and Slovak Republics by the end of 1992.

More gradual political change had been underway in the Soviet Union since the rise to power of Gorbachev in 1985. Glasnost (or openness) and perestroika (or restructuring) had become the order of the day. The collapse of Soviet hegemony in Central and Eastern Europe coupled with domestic political and labour unrest, led to an attempted coup in August 1991. The coup was successfully resisted by Russian President Yeltsin with the popular support of Muscovites. Thereafter, the pace of change speeded up. While Gorbachev was briefly reinstalled as leader of the Soviet Union, that union was dissolved at the end of 1991.

Both for residents of the countries of Central and Eastern Europe and the former Soviet Union and for onlookers in the West, the new political era was initially greeted with euphoria and exhilaration. In addition to political democracy, economic liberalization was on the agenda. While the move towards a market-based economic system seemed to hold promise of much greater efficiency and prosperity than

1

the ponderous system of central planning, any hope for a smooth and easy economic transformation proved to be naive. All of the former command economies were beset by economic depression in the early 1990s. In turn, the economic collapse has precipitated                    resurgence of the Communist Party in several countries, incl           ssia. It is now apparent that the economic transition from a                 to a market system is a very arduous process. This book is a              oblems of the transition in the agrifood sectors of the Centr              stern European Countries (CEECs) and the New Independent S          ...Ss) of the former Soviet Union.

## 1.1. The Central and Eastern European Countries

There are seven countries in the group that we refer to as the Central and Eastern European Countries (CEECs). Extending in a band from the Baltic Sea in the northwest to the Black Sea in the southeast are: Poland, the Czech Republic, the Slovak Republic (Slovakia), Hungary, Romania and Bulgaria. The seventh CEEC is the Mediterranean state of Albania which lies on the Adriatic coast to the north of Greece.

Two important exclusions from our list of CEECs should be noted. First, we do not include the former German Democratic Republic (East Germany) because German re-unification has resulted in a dramatically different context for economic transition. Second, we also exclude the former Yugoslav Republics from our CEEC group. Here, complications to any eventual economic transformation have arisen both directly from the hostilities that erupted first in Croatia and then swept Bosnia-Hertzegovina, and indirectly from the resulting international sanctions that were imposed against Serbia (and Montenegro). Nevertheless, much of our discussion is applicable to Slovenia which escaped almost entirely unscathed from the turmoil in the former Yugoslavia.

At the end of World War II, Poland, Czechoslovakia, Hungary, Romania, Bulgaria and what was to become East Germany were occupied by the Soviet Red Army. Eventually, Communist governments were installed throughout the region. The CEECs' economies were reorganized on the basis of central planning and command using the Soviet model. Whatever the merits and demerits of the command-based economic system, the political yoke of Moscow's authority was never completely accepted in those countries that fell under Soviet domination. Popular uprisings in Hungary in 1956 and Czechoslovakia in 1968 were brutally suppressed by the Soviet military, while in Poland the Communist authorities proclaimed marshal law and outlawed the trade union Solidarity in 1981 under pressure from Moscow.

By western European standards, per-capita Gross National Products (GNPs) of the CEECs are low. Table 1.1 shows that in 1994 the per-capita GNPs of Hungary and the Czech Republic, at $3840 and $3210 respectively, were significantly higher than those of the remaining CEECs and all of the NISs. By contrast, Albania's per-capita GNP of $360 was extremely low. From an agricultural perspective, the CEECs are highly varied; climatic conditions range from cooler temperate climates in the northern and central countries to warmer more Mediterranean climates in Albania and Bulgaria. Table 1.2 shows that in all the CEECs except for the Czech Republic, over 30% of the population was still located in rural areas in 1996. Further, agriculture generated over 15% of employment in 1991 in all of the CEECs except the Czech and Slovak Republics. The fact that labour's share of employment exceeded its share of output in all of the CEECs (see Table 1.2), however, reveals that the average productivity of labour is lower in agricultural than non-agricultural production. Table 1.3 shows that each of the CEECs (for which information is available) was a net food exporter in 1989. Nevertheless, for all CEECs except Hungary, there has been a decline in net exports of agrifood products between 1989 and 1993 as a result of the general economic collapse and the reduction in trade with the NISs. Four of the seven CEECs were actually net importers in 1993. The region as a whole has faced difficulties in trying to increase exports to western Europe and other modern market economies where agriculture is highly protected.

The CEECs participated in both military and economic alliances with the Soviet Union and, thus, can be regarded as part of the former Soviet bloc. The Warsaw Pact or Warsaw Treaty Organization was the Cold-War counterpoint to the North Atlantic Treaty Organization or NATO. Its membership included: East Germany, the Soviet Union and all of the CEECs. Albania, however, withdrew from the Warsaw Pact in 1958. Although a Communist himself, Tito managed to keep Yugoslavia from being drawn into the Soviet political orbit. Thus, Yugoslavia remained outside the Warsaw Pact as a non-aligned country.[2] The Warsaw Pact crumbled with the political emancipation of the CEECs and met its official end in 1991.

The Council for Mutual Economic Assistance (CMEA) or COMECON was set up to promote economic cooperation and coordination among the Communist countries. Bulgaria, Czechoslovakia, Hungary, Poland, Romania and the Soviet Union were the initial CMEA countries. These countries were joined almost immediately by Albania, but it withdrew from the CMEA in 1961 at the time of the Sino-Soviet split. Over the years East Germany, Mongolia, Cuba and Vietnam joined the CMEA and Yugoslavia participated as an associate member. By the 1980s, any move towards greater integration within the CMEA was blocked by

**Table 1.1.** General economic indicators in the New Independent States (NISs), the Central and Eastern European Countries (CEECs) and Slovenia.

| | Land area (mi²) | Population (mid 1996) | Per-capita GNP (1994, $US) | 4-Year real GDP growth (1990–1993, %) | 2-Year real GDP growth (1994–1995, %) | Unemployment rate (1993, %) | Inflation rate (1993, %) |
|---|---|---|---|---|---|---|---|
| Russia | 6592.82 | 147.7 | 2650 (b) | −28.5 | −19.1 | 1.1 | 915 |
| Baltic NISs | | | | | | | |
| Estonia | 16.32 | 1.5 | 2820 | −47.0 | 10.2 | 2.0 | 35.6 |
| Latvia | 23.96 | 2.5 | 2290 | −49.8 | 1.0 | 6.6 | 32.0 |
| Lithuania | 25.17 | 3.7 | 1350 | −57.3 | 5.1 | 3.8 | 188.7 |
| Western NISs | | | | | | | |
| Belarus | 80.15 | 10.3 | 2160 | −21.7 | −30.9 | 1.3 | 1190.0 |
| Moldova | 12.73 | 4.3 | 870 | −43.8 | −20.5 | 0.7 | 1183.0 |
| Ukraine | 233.10 | 51.1 | 1570 | −31.3 | −33.4 | 0.3 | 4735.0 |
| Caucasus NISs | | | | | | | |
| Armenia | 10.89 | 3.8 | 670 | −65.6 | 10.2 | 6.2 | 1820.0 |
| Azerbaijan | 33.44 | 7.6 | 500 | −56.2 | −35.2 | 0.7 | 1113.0 |
| Georgia | 26.91 | 5.4 | 580 | −72.4 | −31.2 | 2.0 | 1278.0 |
| Central Asian NISs | | | | | | | |
| Kazakhstan | 1049.15 | 16.5 | 1110 | −34.6 | −31.8 | 0.6 | 1663.0 |
| Kyrgyzstan | 76.64 | 4.6 | 610 | −31.5 | −30.9 | 0.2 | 1194.0 |
| Tajikistan | 54.29 | 5.9 | 350 | −54.8 | −25.2 | 1.1 | 2136.0 |
| Turkmenistan | 188.50 | 4.6 | 1380† | −9.1* | −5.0** | 2.4‡(c) | 1631.0 |
| Uzbekistan | 172.74 | 23.3 | 950 | −11.9 | −6.4 | 0.1 | 1232.0 |

| | | | | | | | |
|---|---|---|---|---|---|---|---|
| **CEECs** | | | | | | | |
| Albania | 10.58 | 3.3 | 360 | −29.6 | 13.8 | 22.3 | 85.0 |
| Bulgaria | 42.68 | 8.4 | 1160 | −27.8 | 3.0 | 16.1 | 72.9 |
| Czech Republic | 29.84 | 10.3 | 3210 | −21.1 | 7.7 | 3.4 | 20.8 |
| Hungary | 35.65 | 10.2 | 3840 | −15.6 | 4.6 | 12.1 | 22.5 |
| Poland | 117.55 | 38.6 | 2470 | −12.4 | 12.9 | 15.7 | 37.0 |
| Romania | 88.93 | 22.6 | 1230 | −28.2 | 9.6 | 10.4 | 256.0 |
| Slovakia | 18.56 | 5.4 | 2230 | −25.1 | 12.1 | 10.4 | 23.2 |
| **Former Yugoslavia** | | | | | | | |
| Slovenia | 7.71 | 2.0 | 7140 | −16.1 (a) | 9.2 | 14.5 | 32.3 |

* 3-year real GDP growth for 1990–1993; ** 1-year real GDP growth for 1995; † 1993; ‡ 1991.

Sources: Population Reference Bureau (1996) – data in columns 1–3; World Bank (1996) – data used in the calculation of columns 4 and 5; UN Economic Commission for Europe (1995) – data in columns 6 and 7. Where necessary, alternative data from the following sources were used: (a) UN Economic Commission for Europe (1994, 1995); (b) World Bank (1996); (c) IMF (1994a).

**Table 1.2.** Agrifood indicators in the New Independent States (NISs), the Central and Eastern European Countries (CEECs) and Slovenia.

| | Rural vs total population (mid 1996, %) | Ag. vs total employment (1991, %) | Ag. value added vs GDP (1991, %) | 4-Yr gross ag. output growth (1990–1993, %) | Gross ag. output growth (1994, %) | 4-Year grain output growth (1990–1993, %) | 4-Year meat output growth (1990–1993, %) |
|---|---|---|---|---|---|---|---|
| Russia | 27 | 13.4 | 12.3 | −18.7 | −9 | −30.3* | −32.1* |
| Baltic NISs | | | | | | | |
| Estonia | 30 | 12.0 | 17.1 | −54.8 | −9 | −16.1 | −63.3 |
| Latvia | 31 | 16.2 (a) | 22.5 | −43.5 | −20 | −22.7 | −42.0 |
| Lithuania | 32 | 17.0 | 19.6 | −39.0 | −22 | −19.7 | −48.4 |
| Western NISs | | | | | | | |
| Belarus | 31 | 19.1 | 21.6 | −17.8 | −14 | −13.3* | −37.3* |
| Moldova | 53 | 32.5 | 34.8 | −29.5 | −28 | −6.5§ (a) | −34.3† (a) |
| Ukraine | 32 | 19.5 | 26.2 | −21.6 | −17 | −30.3* | −40.3* |
| Caucasus NISs | | | | | | | |
| Armenia | 31 | 18.9 | 19.8 | 6.1 | 3 | 16.6* (a) | 50.1* (a) |
| Azerbaijan | 47 | 32.9 | 31.5 | −36.3** | −13 | 49.2† (a) | −39.4† (a) |
| Georgia | 45 | 26.7 (b) | 27.3 | −48.5** | −10 | −7.0† (b) | −3.6† (b) |
| Central Asian NISs | | | | | | | |
| Kazakhstan | 44 | 16.5 | 28.5 | −7.8 | −17 | −42.3* | −32.7* |
| Kyrgyzstan | 65 | 32.8 | 43.2 | −22.0 | −15 | −3.2† (b) | −8.7† (b) |
| Tajikistan | 72 | 42.9 | 33.3 | −30.8 | −25 | −13.7 (a) | −40.3 (a) |
| Turkmenistan | 55 | 42.4 | 32.4 | 8.4 | 2 | 94.5† (a) | −3.8‡ (b) |
| Uzbekistan | 61 | 42.0 (b) | 37.1 | −0.1 | −1 | 18.9‡ (a) | −2.3‡ (a) |

| CEECs | | | | | | | |
|---|---|---|---|---|---|---|---|
| Albania | 63 | 47.0 | 30.7 | n.a. | 15.0‡‡ | −29.5 | 20.0 |
| Bulgaria | 32 | 19.0 | 16.1 | −32.5 | −1.1 | −38.9 | −35.5 |
| Czech Republic | 25 | 8.2 | 5.0 (c) | −31.4 | −7.4 | −15.3 | −14.6 |
| Hungary | 36 | 15.8 | 8.6 | −32.8 | 5.0 | −46.3 | −32.0 |
| Poland | 38 | 27.6 | 6.9 | −14.8 | −10.0 | −13.1 | −6.6 |
| Romania | 45 | 27.9‡‡ | 19.9 | −4.3 | 0.2 | −15.8 | 1.7 |
| Slovakia | 43 | 10.7 | 5.7 | −21.9 | 5.5†† | −25.8 | −22.7 |
| Former Yugoslavia | | | | | | | |
| Slovenia | 50 | 11.8 | 5.0 (d) | −14.6 | 1.6 | −14.5 | −19.1 |

* 4-year cumulative growth, 1991–1994; † 3-year cumulative growth, 1990–1992; ** 2-year cumulative growth, 1992–1994; ‡ 2-year cumulative growth, 1991–1992; § 2-year cumulative growth, 1990–1991; †† 1993; ‡‡ 1989; n.a., data not available.

Sources: Population Reference Bureau (1996) – data in column 1; OECD (1995a) – data for the calculation of columns 6 and 7; World Bank (1995b) – data used in column 3; UN Economic Commission for Europe (1993) – data in column 2; UN Economic Commission for Europe (1994, 1995) – data in column 5 and data for the calculation of column 4. Where necessary, alternative data from the following sources were used: (a) World Bank (1993a,b,c; 1994a,b,c; 1995a); (b) IMF (1993a,c 1994a,b,); (c) OECD (1995b); (d) UN Economic Commission for Europe (1995).

**Table 1.3.** External indicators in the New Independent States (NISs), the Central and Eastern European Countries (CEECs) and Slovenia.

| | Agrifood net exports, 1989 ($US millions) | Agrifood net exports, 1993 ($US millions) | Exports vs GDP 1993 (%) | Export growth, 1994 (%) | Current acct deficit vs GDP (1993, %) | Gross invest. vs GDP (1993, %) | Net debt vs GDP (1993, %) |
|---|---|---|---|---|---|---|---|
| Russia | n.a. | −4809.1 (c) | 13.5* | 8.4* | −0.8* | 36.0 | 23.9 |
| Baltic NISs | | | | | | | |
| Estonia | n.a. | 56.0 | 19.1 | 64.6 | −0.8 | 26.2 | −5.9 |
| Latvia | n.a. | 67.4 | 20.4 | −7.3 | −4.3 | 10.5 | −2.2 |
| Lithuania | n.a. | −203.0 | 46.7 | 9.1 | −1.9 | 18.1 | −4.6 |
| Western NISs | | | | | | | |
| Belarus | n.a. | n.a. | 2.5* | 35.3* | −2.8* | 35.0 | 6.0 (T,E,e) |
| Moldova | n.a. | n.a. | 4.1* | −18.9* | 3.6 | 7.0 | 4.6 |
| Ukraine | n.a. | n.a. | 2.9* | 46.8* | 0.8* | 8.4 | 3.3 (T,E,e) |
| Caucasus NISs | | | | | | | |
| Armenia | n.a. | n.a. | 34.4* | 44.9* | 1.8* (e) | 14.4 | 6.4 (T,E,e) |
| Azerbaijan | n.a. | n.a. | 23.5* (E) | 7.0* | −8.6*† (e) | 14.1‡ | 0.7 (T,E,e) |
| Georgia | n.a. | n.a. | 16.0* (E) | n.a. | 6.3* (E,e) | 32.3† | 19.0 (T,E,e) |
| Central Asian NISs | | | | | | | |
| Kazakhstan | n.a. | −40.5† (b) | 5.1* (E) | −13.9* | 5.0*† (e) | 31.3‡ | 6.6 (T,E,e) |
| Kyrgyzstan | n.a. | n.a. | 2.8* (E) | 3.8* | 2.3*† (e) | 25.4‡ | 7.8 (T,E,e) |
| Tajikistan | n.a. | n.a. | 4.4* (E) | 21.7* | n.a. | 17.7‡ | 1.1 (T,E,e) |
| Turkmenistan | 5.6 (a) | n.a. | 17.6*† | −63.6* | −18.0*† (e) | 46.0‡ | n.a. |
| Uzbekistan | n.a. | −12.0† (b) | 3.5 | 33.6* | 1.8*† (e) | 29.4§ (e) | 0.0 (T,E,e) |
| Former Soviet Union | −14541.1 (c) | n.a. | n.a. | n.a. | n.a. | n.a. | n.a. |

| | | | | | | | |
|---|---|---|---|---|---|---|---|
| **CEECs** | | | | | | | |
| Albania | 9.1 | −134.3 | 11.2† | 27.8 | 4.6† | 10.4† | 90.2† |
| Bulgaria | 1360.4 | 355.7 | 37.8 | 11.2 | 12.1 | 20.1 | 120.6 |
| Czech Republic | n.a. | 36.0 | 33.2* (d) | 8.3 | −0.9* (d) | 23 (a) | 15.0 (d) |
| Hungary | 1556.7 | 1174.3 | 23.4 | 20.0 | 9.1 | 19.7 | 47.0 |
| Poland | 448.0 | −610.0 | 15.4 | 20.5 | 2.7 | 15.6 | 50.4 |
| Romania | 160.0 | −637.9 | 17.2 | 22.6 | 4.8 | 27.1 | 12.3 |
| Slovakia | n.a. | −212.0 | 28.3* | 24.0 | 6.3* | 21 (a) | 28.9 |
| **Former Yugoslavia** | | | | | | | |
| Slovenia | n.a. | −288.0 | 29.4 (d) | −15.5 | −0.9 (d) | 22 (a) | 6.3** (d) |

* Does not include trade within the former Soviet Union by the NISs or trade within the former Czechoslovakia by the Czech and Slovak Republics; ** includes only the debt of the former Yugoslavia that has been allocated to date; † 1992; ‡ 1991; § 1990; T, total rather than net debt; E, estimates in the underlying data; n.a., data not available.

Sources: OECD (1995a) – data in columns 1 and 2; World Bank (1995b) – data in column 6, data for the calculation of columns 3, 5 and 7; UN Economic Commission for Europe (1995) – data in column 4, data for the calculation of columns 3, 5 and 7. Where necessary, alternative data from the following sources were used: (a) UN Economic Commission for Europe (1995); (b) IMF (1994a, 1993b); (c) FAO (1994); (d) Underlying GDPs from IMF (1995); (e) All data for entry from World Bank (1995b).

growing east–west trade coupled with a gradual economic liberalization
in Hungary and the traditional demands for autonomy by Romania. Despite a desire for movement in the direction of becoming a market-based
trade area in some quarters, the CMEA was officially disbanded in 1991.

## 1.2. The New Independent States

The 15 New Independent Countries were formerly the Soviet Socialist
Republics (SSRs) that comprised the Union of Soviet Socialist Republics
(USSR) or Soviet Union. It was hoped that the Commonwealth of Inde-
pendent States (CIS), which includes all the NISs except for the Baltic
States of Latvia, Lithuania and Estonia, would prevent economic disin-
tegration from accompanying the political dissolution of the Soviet
Union. The CIS, however, has not been very effective and there has
been a marked decline in trade among the NISs.

The Russian Federation (Russia) contains the bulk of the territory
and population of the former Soviet Union. Extending from the Baltic
Sea to the Pacific and from the Arctic Ocean to the Black and Caspian
Seas, Russia is easily the largest country in the world. Table 1.1 shows
that in 1994, Russia was a middle-income country and had a higher
per-capita income than any of the other NISs aside from Estonia. Obvi-
ously, Russia is enormously varied in terms of its climate, soils and
agricultural production. In the Soviet era, Russia was a significant net
food importer and this has remained the case during transition (see
Table 1.3). Table 1.2 shows that, in comparison with the other NISs and
indeed most of the CEECs, agriculture in Russia accounts for a small
proportion of both output and employment. Russia is also more urban-
ized than the other NISs and all of the CEECs except the Czech Republic. It is noteworthy that Russia does have enormous oil, mineral and
forest resources.

From North to South along the Baltic coast are: Estonia, Latvia,
Lithuania and the Kaliningrad Oblast which is an enclave of Russia that
was seized from Germany after World War II. During the Soviet era,
living standards were higher in the Baltics than elsewhere but a drastic
decline in output had moved Latvia's and, especially, Lithuania's per-
capita GNP below that of Russia by 1994. Although the Baltic region is
predominantly industrial, it has traditionally exported dairy products,
meat, fish, and processed foods. The Baltics are also an important transshipment area for the other NISs. In particular, they play a significant
role in the importation of wheat to Russia.

Moving south from the Baltics along the western border of Russia
and adjacent to the CEECs are: Belarus, the Ukraine, and Moldova. Of

these, both Belarus and Moldova are land-locked while the Ukraine has access to the Black Sea.[3] Belarus' per-capita income is somewhat lower than that of Russia and the per-capita income of the Ukraine is somewhat lower still. In Moldova, strife between ethnic Romanians and Russians has complicated the process of transition and contributed to a very dramatic drop in per-capita income. Belarus is predominantly industrial and, in the Soviet era, it was a major producer of fertilizer and agricultural equipment. Despite the fact that the Ukraine is a major grain producer – it has traditionally been viewed as the bread-basket of Europe – environmental problems such as the legacy of contamination from the Chernobyl nuclear meltdown impinge on agriculture. Moldova is more reliant on agriculture than the other two countries and over 50% of its population was classified as being rural in 1996 (see Table 1.2). Moldova has traditionally exported fruit and wine as well as meat.

Between the Caspian Sea and Black Sea in the Caucasus Mountains are: Georgia, Armenia and Azerbaijan. Hostilities have broken out throughout the entire Caucasus region and have contributed to drastic declines in income and widespread poverty (see Table 1.1). Georgia has been torn by civil war, and Armenia and Azerbaijan have been at war with each other. More recently, Russia has become mired in an attempt to suppress secessionist forces in Chechnya which borders on the Caucasus region.[4] Georgia and Azerbaijan are generally fertile countries, and Georgia, in particular, produces wine and citrus fruit. Armenia has traditionally been a net food importer. Nevertheless, it has had to increase its food production (see Table 1.2) despite a drastic decline in overall economic activity because of war and political and economic isolation. Azerbaijan is endowed with oil but Armenia and Georgia are not.

To the east of the Caspian Sea are the five Central Asian NISs. Kazakhstan, which is the largest and most populous of the five, borders on Russia, but the others do not. Most of Kazakhstan is in a semi-arid or steppe climatic zone. While this climate is suitable for grain production, the yields vary widely depending on the amount of annual precipitation. Kazakhstan is also rich in oil and many minerals. Lying to the south of Kazakhstan and moving west from the Caspian Sea are: Turkmenistan, Uzbekistan, Tajikistan, and the Kyrgyz Republic (Kyrgyzstan). Although Turkmenistan and Uzbekistan are mainly covered by desert, fertile river valleys provide the basis for significant food and cotton production. Uzbekistan has significant mineral and gas deposits and Turkmenistan produces both oil and gas. While Kyrgyzstan and Tajikistan are both mountainous regions, the former has more fertile soil and a somewhat more diversified agriculture. Table 1.1 shows that with 1994 per-capita GNPs in the $1000 range, Kazakhstan, Uzbekistan and Turkmenistan were significantly more prosperous than their mountainous neighbours. The rural proportion of the population

is high in all five of the Central Asian NISs but agriculture is less significant in terms of employment and output in Kazakhstan than the
others. A comparison of agriculture's contribution to both output and
employment (see Table 1.2) reveals that the average product of labour
is much higher in agricultural versus non-agricultural activities in
Kazakhstan and Kyrgyzstan but less productive in agriculture in the
other three countries.

Except for the Baltic states which were absorbed into the Soviet
Union at the beginning of World War II, the NISs have a much longer
history under the command-based economic system than the CEECs.
Indeed, Russian domination of the region predates the Russian Revolution which began during the autumn of 1917 in the midst of World War
I. Under Communist rule, millions died during the 1930s as Stalin
forced the collectivization of agriculture and repeatedly purged the
Communist Party. Other than in Poland, where agriculture remained
predominantly based on peasant farms, large-scale collectivized agriculture was to eventually become the norm throughout the Soviet bloc.
After World War II and into the 1960s, the Soviet economy exhibited
rapid industrial growth based on forced saving and capital accumulation. By 1949, the Soviets had developed nuclear weapons and they
were the first to launch a man into space with Sputnik in 1957. Rapid
economic growth, however, proved to be only transitory. Military occu-
pation and guerrilla war in Afghanistan coupled with an intensification
of the arms race with the US contributed to economic stagnation in the
Soviet Union during the late 1970s and early 1980s. This, in turn, generated pressure for economic and political reform to which Gorbachev
responded with glasnost and perestroika.

## 1.3. The Economic Transition: From Elation to Despair

Ironically, the economic determinism of Marx provides one of the most
compelling lenses for viewing and understanding the anti-Communist
revolutions in Central and Eastern Europe and the Soviet Union.[5] The
1970s and especially the 1980s were decades of economic stagnation
where, to paraphrase Marx (1859), the command-based relations of production had increasingly become a fetter on the forces of production.
Technological innovation (i.e. the application of new scientific
discoveries) and organizational innovation were increasingly constrained by the planning bureaucracy and contradictory with it. The
mode of production based on central planning had entered a phase of
crisis and collapse.

History, according to Marx, only poses those questions that it is capable of answering. While the reformulation of the relations of production would undoubtedly take some time, the transformation should unleash the forces of production and soon lead to a phase of rapid economic development. Indeed, many economists and business executives were of the view that the market-based reform opened immediate and enormous opportunities. Many believed that there would be a domestic and foreign investment boom that would lead to rapid growth and a convergence in per-capita outputs with those in the modern market economies.[6] For example, many expected a strong convergence between the CEECs and their immediate neighbours in western Europe.[7]

Unfortunately for the CEECs and NISs, the much vaunted economic miracle has not happened. The initial loud and enthusiastic rhetoric from businesses and the economics profession was soon silenced by events. By and large the potential foreign investors slipped away and sought more profitable opportunities elsewhere while the economists were left to ponder an increasingly bleak reality. Rather than a gradual move into a boom, the initial liberalization towards market systems signalled the onset of a major economic depression. Table 1.1 shows that without exception, the CEECs and NISs posted very major declines in Gross Domestic Product (GDP) during the four years from 1990 through 1993.[8] In 1994 and 1995, the CEECs had started to regain lost ground but the depression continued to worsen in the NISs. Some observers have attributed the earlier recovery in the CEECs to more thorough ongoing reforms (World Bank, 1996). While this is probably true in part, the CEECs had many other advantages over the NISs including an earlier start to the transition process.

The picture in the agrifood sector was similar to that in the economy in general. Table 1.2 shows that significant declines in food production have accompanied the overall economic collapse in almost all of the CEECs and NISs during the transition period of the 1990s. The recent turnaround in agriculture in the CEECs has been less dramatic than in the economy at large. Although Hungary, Slovakia and Albania managed to post gains in agricultural production in 1994, there were significant further declines in Poland and the Czech Republic. In 1994 almost all of the NISs experienced further significant declines in agricultural output.

The CEECs, with the exception of the Czech Republic, have experienced high unemployment rates as they have embarked on the transition towards a market economy. The official unemployment rates for the NISs that are reported in Table 1.1 dramatically understate the degree of underutilization of labour in the NISs. There are definitional problems as well as data problems. For example, the unemployment rate for Russia would be 5.5% rather than 1.1% when International Labour Organization definitions are used.[9] In addition, there is widespread underemployment, and involuntary leavers and reduced hours have become increasingly common.

While inflation has been a problem in all of the transitional economies, Table 1.1 shows it has been much less dramatic in the CEECs than the NISs. By 1995 Albania, the Czech Republic and Slovakia posted inflation rates in the 10% range or below (World Bank, 1996). In Romania inflation has also fallen dramatically, and in Poland and Bulgaria the trend remains in the downward direction. In Hungary, however, inflation actually rose modestly in 1995. As a general rule, the data in Table 1.1 catch the peak of the hyper-inflationary bursts in the NISs. By 1995, the inflation rate was below 50% in five of the NISs and only above 500% in three countries. The sources of extreme inflationary pressure in the CEECs and NISs are not hard to find. Even with the decline in government involvement that the transition presupposes, it has been difficult for the nascent market system to generate anything close to enough tax revenue to finance government expenditures. Further, the controlled prices that were a hallmark of the command system generated a monetary overhang since households were unable to consume at the desired level. With the liberalization of prices in the initial stages of the transition, the liquidation of these balances contributed to a burst of inflation in many of the CEECs and NISs.

Table 1.3 shows that the CEECs are somewhat more open to international trade than the NISs, but the degree of openness to international trade varies widely within each group. Further, the degree of openness of each NIS is somewhat understated because trade with other NISs, which is typically conducted on a non-hard-currency basis, is excluded. Given this exclusion, it is not surprising that the geographically peripheral NISs in the Baltic and Caucasus have exhibited the highest degree of openness.[10] By 1994, all CEECs were experiencing a major recovery in exports to the rest of the world. Although general economic conditions continued to deteriorate in the NISs, Table 1.3 shows that many of them began to experience a significant recovery in their exports in 1994.

The CEECs and NISs ran current account deficits in 1993 and, indeed, throughout the early 1990s (see Table 1.3). Thus, the CEECs and NISs are typically becoming more indebted to the rest of the world. Gross investment, however, exceeds current account deficits in all cases. Since gross investment is not adjusted for depreciation and since it includes any unintended inventory accumulation that occurred as the economic situation worsened in many of the transitional economies in 1993, the margin of net indebted investment over external borrowing would have been considerably smaller. The extent of net external debt in convertible currencies varies widely as shown in the last column of Table 1.3. Relative to income, debt is highest in Bulgaria and Albania.

Nevertheless, Poland and Hungary also continue to carry significant debts that were amassed in the 1970s and 1980s but have been reduced since that time.[11]

## 1.4. What is to be Done?

Of course the CEECs and NISs are all very different. Although they all fall into the broad classification of middle-income countries, we have seen that there is evidence of very significant differences in standards of living. Table 1.1 shows that the per-capita GNP of Hungary was over ten times that of Tajikistan in 1994. Per-capita GNP tends to be higher in the CEECs than the NISs although there is much diversity within both groups. As a rule of thumb for the entire region, per-capita income tends to be lower in the more easterly and southerly countries. The types of economic activity are equally varied. Within the agrifood sector, we have seen that there are dramatic variations in the types of agricultural activity.

In spite of the enormous variations among these countries, the command system has bequeathed them a set of common structural problems. These structural problems, and what is done about them, are fundamental to the success or failure of the transition to a market system. It is the effect of these structural problems on the agrifood sector that is the subject of this book. Our theme is a simple one. Economies in general and market economies in particular do not operate frictionlessly; rather transaction costs are important. The sudden dissolution of the command system has inevitably created very high transaction costs. These costs act as both a direct and indirect constraint on the transformation process. Clearly, transaction costs directly reduce net benefits and thereby inhibit economic activity. The high transaction costs also act as an impediment to competition. By exacerbating the problem of market power, economic activity is stifled to a further degree.

While economic decline is certainly not a permanent state of affairs for the CEECs and NISs, there is no immediate or painless solution to the problem of high transaction costs for either their economies in general or their agrifood sectors in particular. How fast the transitional economies reverse the state of decline and how fast they grow thereafter are clearly crucial matters. Fortunately, all is not doom and gloom on this front. There is much that can be done at present in the CEECs and NISs to build their market institutions. By so doing, they will gradually reduce transaction costs and improve their medium and long term prospects. Even with a very solid foundation, however, any convergence of the CEECs and NISs with seemingly comparable modern market economies is likely to be slow at best.[12]

# 1.5. Conclusions

The initial euphoria which accompanied the fall of communism is now long gone. In transition, life is much riskier for most citizens of the CEECs and the NISs than it was in the command era. While some individuals enjoy the challenges (and possible rewards) which come with increased risk, many do not. The transition to a market economy is fundamentally about providing opportunities for those willing to accept some risk while providing increased security for those who are not. If entrepreneurship is overly constrained by the economic environment, then employment opportunities and tax financed improvements to health and social welfare will not be forthcoming for those who desire increased security. They will look to reformed communists and others offering a partial return to the past. The agrifood sectors are particularly important for the transition process given their relative importance in most CEECs and NISs and because food security is fundamental to individuals' welfare.

The optimism regarding economic transition, including in the agrifood sector, which accompanied the political changes in the late 1980s and early 1990s was based on the assumption that once prices were liberalized and privatization was accomplished, the institutions which underpin modern market economies would arise spontaneously and quickly. They have not. As a result, there has been a grudging realization that transition will be difficult. Understanding the role of market institutions is the key to transition. The agrifood sectors of modern market economies have many specialized market institutions which have evolved to facilitate the smooth functioning of farms and other agribusiness firms who must deal with the often unique characteristics of agricultural production and food processing and distribution. If transition of the agrifood sector of the CEECs and NISs is to be accomplished, it will be necessary to foster market institutions. Understanding these institutions, as well as the constraints imposed by the legacy of the command system, is the first step.

# Notes

1. In Romania, however, the initial armed repression of protesters led to the eventual execution of the Communist dictator, Ceauçescu and his wife in December 1989. Meanwhile, the non-Soviet Communist regime in Albania temporarily clung to power.
2. Although Cuba also remained outside the Warsaw Pact, it was politically closely aligned with Moscow in spite of its official non-aligned status.
3. Moldova is encircled by the Ukraine and Romania.
4. In the Soviet era, Checheno-Ingushetia was an Autonomous Soviet Socialist Republic

(ASSR) within the Russian Soviet Federative Socialist Republic. Thus, upon the dissolution of the Soviet Union, Chechnya was part of the Russian Federation.

5. In fairness, it should be acknowledged that the command-based mode of production may have been rather different from Marx's notion of either socialism or communism.

6. Astute observers would have cautioned that convergence in standards of living would take longer because of the amassed foreign debt of CEECs such as Poland and Hungary.

7. The prospects of the various CEECs were, of course, expected to differ. While Hungary might approach the Austrian level, Bulgaria might aspire to that of Greece.

8. The figures in Table 1.1 probably overstate the decline in GDP to some extent. No doubt, the move towards a market-based economy with turnover taxes and the like has driven some economic activity underground.

9. See UN Economic Commission for Europe (1995) Table 3.4.4, Memorandum item g. For a further commentary on the problems of measuring unemployment in the NISs, see Oxenstierna (1993).

10. We have seen that the location of the Baltic NISs invites transit trade between Russia and the other NISs and the rest of the world. Similarly, the Caucases and, to a certain extent, Turkmenistan, are well situated for trade with Iran and Turkey. With regard to Table 1.3, it should be observed that the dramatic decline in GDP in both the Baltic and Caucasus countries has probably pushed up the export to GDP ratios and overstated the degree of openness.

11. Some assessments of the economic transition (e.g. Köves, 1992) have attached great important to debt issues, but the data in Table 1.3 suggests that this is not an acute problem.

12. Even within the highly favourable circumstances of a re-unified Germany, the transition has been slow and painful for many former East Germans.

# References

FAO (1994) *Trade Yearbook*, 48. Food and Agriculture Organization, United Nations.

IMF (1993a) *Georgia*, IMF Economic Reviews. International Monetary Fund, Washington DC, 91 pp.

IMF (1993b) *Kazakhstan*, IMF Economic Reviews. International Monetary Fund, Washington DC, 121 pp.

IMF (1993c) *Kyrgyz Republic*, IMF Economic Reviews. International Monetary Fund, Washington DC, 98 pp.

IMF (1994a) *Turkmenistan*, IMF Economic Reviews. International Monetary Fund, Washington DC, 109 pp.

IMF (1994b) *Uzbekistan*, IMF Economic Reviews. International Monetary Fund, Washington DC, 111 pp.

IMF (1995) *International Financial Statistics*, December. International Monetary Fund.

Köves, A. (1992) *Central and East European Economies in Transition*. Westview Press, Boulder, Colorado, 150 pp.

Marx, K. (1859) Preface to the Critique of Political Economy. In: Marx, K. and Engels, F. (eds) *Selected Works*. Progress Publishers, Moscow, 1969, pp. 502–516.

OECD (1995a) *Agricultural Policies, Markets and Trade in the Central and Eastern European Countries, Selected New Independent States, Mongolia and China: Monitoring and Outlook 1995*. Centre for Co-operation with the

Economies in Transition, Organization for Economic Co-operation and Development, Paris, 232 pp.

OECD (1995b) *Review of Agricultural Policies; Czech Republic.* Centre for Co-operation with the Economies in Transition, Organization for Economic Co-operation and Development, Paris, 298 pp.

Oxenstierna, S. (1993) Labour Issues in the USSR. *European Economy,* 49.

Population Reference Bureau (1996) *1996 World Population Fact Sheet.* Population Reference Bureau, Washington DC.

UN Economic Commission for Europe (1993) *Economic Survey of Europe in 1992–1993.* United Nations, New York, 291 pp.

UN Economic Commission for Europe (1994) *Economic Survey of Europe in 1993–1994.* United Nations, New York, 216 pp.

UN Economic Commission for Europe (1995) *Economic Survey of Europe in 1994–1995.* United Nations, New York, 254 pp.

World Bank (1993a) *Azerbaijan; from Crisis to Sustained Growth.* A World Bank Country Study. International Bank for Reconstruction and Development, Washington DC, 223 pp.

World Bank (1993b) *Latvia; the Transition to a Market Economy.* A World Bank Country Study. International Bank for Reconstruction and Development, Washington DC, 305 pp.

World Bank (1993c) *Uzbekistan; an Agenda for Economic Reform.* A World Bank Country Study. International Bank for Reconstruction and Development, Washington DC, 318 pp.

World Bank (1994a) *Moldova; Moving to a Market Economy.* A World Bank Country Study. International Bank for Reconstruction and Development, Washington DC, 104 pp.

World Bank (1994b) *Tajikistan; the Transition to a Market Economy.* A World Bank Country Study. International Bank for Reconstruction and Development, Washington DC, 240 pp.

World Bank (1994c) *Turkmenistan.* A World Bank Country Study. International Bank for Reconstruction and Development, Washington DC, 244 pp.

World Bank (1995a) *Armenia; the Challenge for Reform in the Agricultural Sector.* A World Bank Country Study. International Bank for Reconstruction and Development, Washington DC, 198 pp.

World Bank (1995b) *World Tables.* The Johns Hopkins University Press, Baltimore, Maryland, 750 pp.

World Bank (1996) *From Plan to Market.* World Development Report 1996. Oxford University Press, Oxford, 241 pp.

# 2 Description of Command Economy Agrifood Systems

To begin to understand the effect on the agrifood systems of the Central and Eastern European Countries (CEECs) and the New Independent States (NISs) of the forces of change set in motion as a result of the political events of the late 1980s and early 1990s, it is first necessary to examine how these sectors functioned within command economies. While much has been written about both the differences which existed among the agrifood sectors of communist countries and their subsequent varied approaches to liberalization, we would like to begin by stressing those elements which were common to all of their agrifood systems. These common elements of command economy agrifood systems have led to a number of fundamental problems which have had to be addressed by those charged with shepherding the process of transition. In many cases, the differences which were observable among the various command economies prior to the overthrow of their communist regimes simply represented attempts to solve the same set of basic problems within the structure of a command economy. While considerable differences did exist in the organization of primary agricultural production,[1] when one looks at the entire agrifood system from the provision of agricultural inputs through to the final consumer, the differences are actually small. All command economies applied the same fundamental organizational principles and those principles have led to a common set of problems which constrain the process of transition.

The legacy left by the command system will influence the functioning of the agrifood system in CEECs and the NISs for the foreseeable future. This is because, while it is possible to identify the problems created by the command economies' legacy, it is difficult to provide policy solutions which can foster the almost instantaneous creation of

19

the economic institutions which have slowly evolved to support the functioning of modern market economies elsewhere. While the same legacy applies to other sectors of the former command economies, the delays and mistakes which will inevitably arise from the restructuring of these sectors of an economy are less disruptive and possibly less life threatening than in the case of food (and energy). Short-run disruptions which prevent reasonably priced food supplies reaching even a segment of the population, can cause considerable hardship. As a result, there is simply less room for experimentation in the agrifood sector. For example, if the normal supply channels for shirts are disrupted, it is possible to make do until the normal supply channel begins to function again or until alternative channels arise. This is not the case with food. As a result, in some countries there has been considerable reticence by political authorities to push forward with reforms in the agrifood system. Further, when faced with supply disruptions or rapidly rising prices, consumers may reflect fondly on the security provided by the command system, however minimal, and be willing to support those who wish its return. Hence, there is a real need to understand the workings of the command system so that it can be compared to the realities of the operation of agrifood systems in transition.

## 2.1. The Food Chain in Command Economies

The metaphor of the food chain is biological in origin. It is used to convey the idea of dependence and hierarchy through the linking of species through diets. However, this metaphor has often been adopted by agricultural economists in market economies and has been adapted to describe the progression of inputs, intermediate outputs and final outputs through the various vertical stages of production and distribution. The use of the metaphor is expected to induce the reader into thinking of the agrifood industry as a system rather than as discrete tasks which can be considered in isolation. To the extent that an understanding of interdependence is important, it is a useful metaphor. The image of a chain, with its linear links forged in steel, conveys both the idea of logical progression and of interdependence. Of course, a broken link in a chain is one of our most powerful images of an interdependent system's failure.

In a smoothly functioning market economy, inputs, intermediate outputs and final outputs move through the various stages of production and distribution but, unlike a chain, the linkages are fluid and, in many cases, change frequently. This is the essence of markets. Competition leads to opportunities for finding new outlets for production and for identifying new sources of inputs. Prices provide the signals whereby alternative opportunities can be compared. The interaction of prices also determines when new enterprises should arise and existing enterprises should exit. It is a fluid system where, in most cases, potential alternatives exist. As a result, the image of the broken chain has little relevance in a market economy because no individual link is crucial. The market system both fosters the existence of alternatives and provides institutions whereby the alternatives can be easily identified.

The metaphor of a chain is much more appropriate, however, for agrifood systems as they existed in command economies. The combination of inputs in production, as well as some of the movement of intermediate outputs, was accomplished within the structure of command firms.[2] The transfers between command firms were directed by the planning ministry. Command firms were tied (or linked) to their officially designated suppliers and customers. They were not allowed to seek out alternative suppliers when the required quantity of inputs was not forthcoming or when the quality of the inputs offered was deficient.[3] Command firms were not allowed to seek out alternative customers when their designated customers could not use the quantity of output mandated in the plan or when they produced quantities in excess of those stipulated in the plan (Gaisford *et al.*, 1995). The managers of command firms were focused exclusively on their designated suppliers of inputs and designated customers. As there was no need to seek out alternatives, no infrastructure developed to facilitate the identification of alternative suppliers and customers or to assess the opportunities which they might provide. Following communist ideology, large-scale operations were stressed with the result that for each commodity a single distribution hub supplied a geographic area. As retail prices were fixed, there was no need for consumers to shop comparatively, although they were often forced to visit a number of retail outlets due to local shortages.

Hence, the agrifood system in a command economy is much closer in concept to a chain than is the case in market economies. The command agrifood system can be envisioned as a series of large-scale, vertically linked command firms running from the providers of agricultural inputs through primary production and processing, with distribution ending in geographic monopolies. Of course, at certain points a number of input chains fed into a central production activity and multiple retail outlets were connected to regional distribution centres, but the correct image is one of parallel chains which did not interact horizontally. Consequently, there was no need for infrastructure which could facilitate interaction between the chains. Hence, consumers' food security depended upon the entire chain functioning as a unit. The image of a broken chain is still a powerful metaphor in the post-command era. The spectre of food shortages haunts both consumers and those charged with shepherding the transition process.

## 2.2. Vertical Coordination in the Agrifood Sector

Vertical coordination[4]

> includes all the ways of harmonizing the successive vertical stages of production and marketing. The market price-system, vertical integration, contracting, cooperation singly or in combination are some of the alternative means of coordination.
>
> (Mighell and Jones, 1963, p. 1)

There is always some kind of vertical coordination if any production takes place. In market economies, vertical coordination can be viewed as a continuum. At one extreme are spot markets where goods are exchanged between multiple buyers and sellers in the current time period with price as the sole determinant of the transaction. Examples of spot markets are auction markets, stock markets and most consumer good purchases (e.g. purchases of food in a supermarket). At the other end of the spectrum is full vertical integration where products move between the various stages of production–processing–distribution as a result of within-firm managerial orders rather than being directed by relative prices. Of course, between the two extremes lie a myriad of alternative ways of coordinating economic activity from strategic alliances to formal written contracts, to quasi and tapered vertical integration.[5]

Viewed in this context, the command economy agrifood sector represents a vertically integrated system from agroinput suppliers through to retail outlets. The coordination of production and distribution activities within command firms was the responsibility of the managing director. The planning authority determined each command firm's suppliers and customers.[6] It also determined the quantities of output which were to be transferred between command firms – production targets. These production targets and the quantities to be transferred were determined through the method of material balances which formed the basis of central planning. The planners gave the managerial orders which provided the vertical coordination between command firms. In theory, prices played only a passive accounting role in the vertical coordination process.

In the agrifood sector of command economies, the rigid material balances approach was, at best, of limited importance (Kwiecinski and Quaisser, 1993). The major reason why the material balances approach could not be strictly applied in agriculture stems from the variability of output inherent in biological systems, particularly when weather is a major determining variable. The inherent variability of agricultural production makes formal planning exceedingly complex; sufficiently complex that the planning ministries were never able to overcome the problem. Since the quantity of output could not be controlled, beyond the establishment of pre-planting acreage targets and the allocation of inputs such as fertilizer, the planning bureaucracy's role was largely one of coordinating the movement of available supplies through the various command firms in the vertical chain.

Of course, the variation in agricultural output created problems for the planning process in other sectors. In the Stalinist period, when self-sufficiency was the official policy, harvest failures meant great hardship for consumers. To some extent, the livestock industry was used as a buffer with increased slaughtering and diversion of grain from livestock feed to human consumption in times of poor harvest. The use of the livestock industry as a food security safety net, however, inhibited the long-term development of that industry. In the post-Stalinist era, harvest shortfalls were supplemented by grain purchases in the world market. The purchase of foreign grain, of course, required the use of foreign currency reserves. As a result, hard currency purchases of technology and other inputs had to be delayed, meaning that other areas of the plan went unfulfilled.

## 2.3. Price Incentives for Farm Production

In command economies, compulsory deliveries from the farm to the state were the primary method of procuring food supplies for urban consumers. In the Soviet Union, the early Bolshevik regime turned to compulsory requisitions from peasants in order to feed the Red Army and urban centres. Committees of poor peasants were unleashed upon the countryside with the power to confiscate whatever food they could uncover (Wegren, 1992). While this procurement by confiscation policy was sufficiently successful to overcome the food supply difficulties created by the Russian civil war, farmers responded negatively to the removal of the property rights to their crops and reduced their output. As a result, the area under cultivation declined by 20% relative to its pre-Revolution levels (Wegren, 1992). In the wake of the disruptions to food production which resulted from the confiscation policy, price always had an incentive role in Soviet agricultural procurement and, hence, was never relegated to the strict accounting role envisioned in the material balances approach to economic planning. While the agricultural procurement policies changed – sometimes taxes in kind, sometimes monetary taxes, sometimes delivery quotas combined with prices, sometimes mandated deliveries at low prices plus incentive prices for over quota deliveries – prices retained an incentive role.

As the number of exogenous factors which can affect agricultural output are greater than is the case with most industrial processes, it was far more difficult to apportion blame when agricultural output failed to meet targets. Further, the decentralized and dispersed nature of agricultural production meant that shirking was much more costly to deter.[7] As a result, it became apparent to those charged with the procurement of agricultural commodities that, party doctrine aside, incentives had to be provided to farms if additional output was desired. The motive for the establishment of state farms (*sovkhoz* – farms organized like industrial factories with workers paid a wage) can, in part, be seen as an experiment which attempted to break the tie between incentives and agricultural productivity that was clearly evident in the case of collective farms (*kolkhoz*). The entire early history of agricultural policy in the Soviet Union, and subsequently in the CEECs, represented a precarious balancing act between setting the returns in agriculture high enough to provide a positive incentive for farms to expand output, while keeping them sufficiently depressed so that a surplus to finance industrialization could be extracted and consumer prices kept low. When prices were set too low, output declined and urban diets deteriorated. Improved incentives meant increased output but rising prices or increased food consumption subsidies. In the latter part of the Soviet era, consumer prices were kept low and farm subsidies became the norm. In the 1980s, direct agricultural subsidies increased from 8 to 17% of national income (Wegren, 1992).

While there was considerable experimentation with different procurement mechanisms, and mixes of procurement mechanisms, the objective was to induce farms to supply the Ministry of the Food Industry (which was responsible for the distribution of food) with the farm products it required. There were three main sources of agricultural products. The first was the output of collective and state farms specified in the planning targets. Prices had to be sufficiently high to induce farmers to produce up to the target levels and not to divert this output into black or grey market channels. The second source of supply was output from state and collective farms which exceeded the targeted quantities. Again, prices had to be sufficiently high to induce farms to exceed their mandated quantity of output and to attract this over-target production into the formal state distribution system. Third, prices had to provide an incentive for those who exercised their option to cultivate personal plots to supply a portion of their output to the official distribution system rather than consuming it themselves or selling it in local markets.

The official name for the latter was personal auxiliary agriculture (Hedlund, 1989). From the early 1930s onward, farm workers were allowed to use small plots of land for private production.[8] These plots were originally conceived as a means to supplement the food consumption of collective farm members and farm workers. They subsequently became an extremely important component of the Soviet agricultural system and a visible (and embarrassing) reminder of the motivating role of profits. Numerous policy changes were initiated which attempted to suppress the use of plots, only to see them reinstated when the extent of the reductions in output that followed became apparent. Private plots became particularly important in livestock and vegetable production. Hedlund (1989) reports that 60% of potato production (41% of marketings), 29% of vegetable production (14% of marketings) and 28% of meat production (13% of marketings) originated in private plots. This output arose from 1.3% of all agricultural land and represented only 2.7% of the total area sown. These figures underestimate the commitment of resources to this type of production, however, as personal livestock enjoyed some common grazing rights on collective farms and benefited from considerable unofficially diverted fodder from state and collective farms. The productivity of the private plots is striking and exceeded that of state and collective farms by a wide margin.

Prices were the primary incentive used to induce the operators of private plots to deliver their products to the official state distribution system. Local markets were allowed, primarily out of the realization that restricting sales to the official distribution system with its poor storage (particularly refrigeration) and transportation infrastructure, would lead to spoilage and waste. These alternative markets, however, remained primarily local due to a lack of officially sanctioned transport. As a result, local markets never became an alternative price information system because of the absence of inter-market arbitrage.

Of course, there were deviations from this stylized vertical coordination mechanism among the command economy countries. As suggested above, in some countries – particularly Poland – private farming survived to a significant degree. Price incentives were the major procurement mechanism for this fourth possible source of foodstuffs.[9]

## 2.4. The Evolution of the Agrifood System in Command Economies

If one wishes to characterize the command economy agrifood system, the metaphor of parallel chains seems appropriate. For both the inputs industry and the processing–distribution system, movements of products were coordinated by managerial orders originating either within command firms or planning ministries. Prices served only an accounting role. The only exceptions were at the point of transfer of inputs to farms, where prices of inputs were used as part of the general incentives package for on-farm production, and also the prices farms received for their outputs.

The rural credit system was used primarily to transfer income to those farms unable to make a profit within the state established price regime for inputs and outputs. The banking system also provided a mechanism for the planning agency to transfer investment funds to farms and other command firms in the agrifood chain. Hence, credit was a major means of subsidization in the command era. This credit was not expected to be repaid but the existence of the accumulated debts provided the planning ministry with additional leverage over the management of farms and other command firms in the agrifood sector. The liberal extension of credit, however, has subsequently left much of the sector with considerable debt.

While the basic structure – tied suppliers and customers – remained in place from its inception in the Soviet Union with the completion of collectivization in the 1930s and its introduction into the CEECs with the establishment of communist regimes, the organizational boundaries and management structures were changed frequently. The basic approach of coordination by managerial orders, combined with the limited use of prices as farm level incentives, remain unchanged throughout all the organizational changes. Further, within this general model there were also considerable opportunities for organizational experimentation among the various command economy countries.

As with all aspects of economic life in command economies, the Communist Party, with its political agenda, controlled the sector through an agricultural department at each level of the administrative hierarchy. With the Party holding the effective power, administration of the agrifood sector was usually divided between two broad ministries. One ministry (sometimes a set of ministries) was responsible for the supply of agricultural inputs and farm level production. A second ministry (or grouping of ministries) administered farm level procurement and downstream processing, transport, storage and sales to consumers. The former was normally designated the Ministry of Agriculture, the latter the Ministry of Procurements.

Separate agricultural inputs ministries were at times split off, only to be rolled back into the Ministry of Agriculture upon a subsequent reorganization. Ministries were created to handle certain crops. Consumer cooperatives run by the state handled some procurement. Whatever the sectoral organization, however, there was no competition. Farms had no choice as to who they sold to and food moved through the chain as directed by the various agencies and planning departments.

Van Atta describes one of the many reorganizations in the USSR.

> In 1985, the USSR *Gosagroprom* was created to unify all . . . agricultural
> agencies. But even then the division between production and
> procurement remained; the grain procurement agencies were not made a
> part of the 'superministry'. So coordination of the entire agricultural
> production chain, from 'upstream' input production to 'downstream'
> marketing and ultimately retailing, still fell not to the Minister of
> Agriculture, but to a deputy or first deputy chairman of the Council of
> Ministers responsible for the entire sector.
>
> (van Atta, 1994, p. 163)

One of the last experiments with reorganization was the creation of
agricultural-industrial-complexes (AICs) or *agrokombinates*. These were
horizontally, and often vertically, integrated management units within
a particular geographic area. In some cases, AICs controlled the entire
vertical chain from input provision through to retail. They were often
the sole distributor for entire sections of large cities. For example,
approximately 400 state and collective farms came under the umbrella
of an AIC near Moscow. In addition to its farming operations, this AIC
had a milk processing plant, a meat plant, maintenance facilities and
retail outlets in Moscow. It employed 55,000 people. The dairy operation alone had 16,500 cows and was the sole supplier of milk for one of
Moscow's districts. The management of an AIC was vested in a managing director with authority extending down through a hierarchy.
Organizationally, an AIC resembles a large vertically integrated firm in
a modern market economy.

Even when AICs were not vertically integrated, they represent very
large economic entities at the farm level. To the extent that they had
been implemented prior to liberalization, AICs underline the large-scale
operations that were stressed in command economies. Even when not
administratively bundled within an AIC, both input supply command
firms and the processing/distribution links in the chain tended to be
large scale. They represent regional monopolies for consumers. Not only
was the administrative structure large scale, the physical facilities also
tended to be large scale. Hence, at the onset of liberalization, the system
of command firms in place could be characterized as geographic monopolies reinforced by past investments in large-scale physical infrastructure. This legacy from the command system has had significant implications for the process of transition to a market economy.

## 2.5. Conclusions

Understanding the effects of the legacy of command firms as geo-
graphical monopolies when reforms entailing privatization and price

liberalization were implemented, provides the central theme of this book. Privatization has meant breaking the links in the chain. Prices are expected to replace much of the bureaucratic system of coordination over the length of the chain. The effectiveness of markets and private property as a means of coordinating economic activity, and as the spur for improvements to efficiency, relies on a number of concomitant institutions being in place or being easily created. The legacy of a command economy's agrifood system is that few of those institutions are in place nor can they be easily created. Without taking account of the importance of those institutions, proceeding with liberalization may greatly lengthen the transition process and extend and/or deepen the disruptive effects of the process of transition itself. It may be that when the period of transition is over, the agrifood system that emerges could be considerably different than that envisioned when the process of liberalization was set in motion.

It should be remembered that, for all its inefficiencies and shortcomings, the command economy agrifood system provided the overwhelming majority of citizens with a considerable degree of food security. Food security (or its absence) is a powerful political weapon. If the process of transition is mishandled, the level of food security provided to the populace may decline. While communism itself is discredited, there are still strong political elements in all the former communist states who advocate reimposition of elements of the command economy. Declining food security can provide them with a potentially powerful argument. As a result, governments in the CEECs and the NISs have been cautious in their approaches to liberalizing the agrifood system. Many prices remain controlled, privatization remains incomplete and regulation has replaced bureaucratic direction. Given the importance of food security, the success of the process of transition in the agrifood sector may be crucial to the entire liberalization process. The central question is, what is required to turn the very rigid food chain inherited from the command system into the much more fluid and flexible system which constitutes the agrifood sector of modern market economies?

# Notes

1. For example, unlike most command economy farming systems which stressed large production units, in Poland approximately 80% of farmland remained in small, privately-owned farm units.
2. This distinction is made so that the arbitrary bundling of production and distribution activities under a separate (or quasi-independent) management unit in command economies is not confused with the explanation of a firm put forward by Coase (1937). According to Coase, the limit to the activities undertaken by a firm is determined by the relative efficiencies of using within-firm managerial orders or external markets to

direct economic activities. When markets are more efficient, they will be the mechanism used to direct economic activity. When the same task can be accomplished more efficiently through directions given by the firm's managers, then the activity will take place within a firm.

3. Of course, command firms faced with fulfilling planned output targets when the required inputs were not forthcoming resorted to the use of expediters (the closest English term is probably scroungers) and bribery to obtain scarce items (van Atta, 1994).

4. Some economists refer to vertical coordination as the governance structure of contractual relations. See for example Williamson (1979).

5. Quasi vertical integration refers to a relationship between a buyer and a seller that involves some form of long-term contractual obligation where both parties invest resources in the relationship. It differs from full vertical integration because the arrangement ceases at the end of an agreed period of time and the firms remain independent of one another. A joint venture is one example of quasi-integration. Tapered vertical integration occurs when a firm obtains a proportion of its inputs through backward integration to the supplier. For example, a beef processing firm integrated backwards into beef production could obtain a proportion of its beef supplies from its own farms with the remainder procured from auction markets or directly from beef producers.

6. Except in the case of final consumers.

7. This is not to suggest that there are not exogenous factors which can affect a factory's industrial output or that shirking is costless to deter on the factory floor. Command factories, when they attempted to ignore the role of incentives, also suffered from low productivity. In agriculture, however, the costs of acquiring the information to apportion blame and to reduce shirking are higher.

8. Those dwelling in urban areas were also allowed to cultivate private plots.

9. At times, delivery quotas were also specified for private farmers.

# References

Coase, R.H. (1937) The nature of the firm. *Economica* n.s. 4, 386–405.

Gaisford, J.D., Hobbs, J.E. and Kerr, W.A. (1995) If the food doesn't come – vertical coordination problems in the CIS food system: some perils of privatization. *Agribusiness: An International Journal* 11(2), 179–186.

Hedlund, S. (1989) *Private Agriculture in the Soviet Union*. Routledge, London, 208 pp.

Kwiecinski, A. and Quaisser, W. (1993) Agricultural prices and subsidies in the transformation process of the Polish economy. *Economic Systems* 17(2), 125–154.

Mighell, R.L. and Jones, L.A. (1963) *Vertical Coordination in Agriculture*. USDA ERS-19, Washington DC.

van Atta, D. (1994) Agrarian reform in post-Soviet Russia. *Post-Soviet Affairs* 10(2), 159–190.

Wegren, S.K. (1992) Two steps forward, one step back: the politics of an emerging new rural social policy in Russia. *The Soviet and Post-Soviet Review* 19, 1–51.

Williamson, O.E. (1979) Transaction cost economics: the governance of contractual relations. *Journal of Law and Economics* 22, 233–262.

# 3 The Problem of Prices

The abandonment of central planning as the allocative decision making institution for economic activity is the fundamental change underlying the process of economic liberalization in the former command economies. The desire to move away from central planning is shared by all of the Central and Eastern European Countries (CEECs) and the New Independent States (NISs). Even where former communists remain in (or have been returned to) power, a reinstitution of formal central planning is not part of the political agenda.[1] The great debate among the governments, and for that matter among economists, has centred on the rate at which central planning should be abandoned and the degree to which government should withdraw from its role in directing the economy. Virtually the entire spectrum of transformation possibilities has been advocated, from the big bang or shock treatment approach whereby all prices should be freed and state property moved into private hands as rapidly as possible, to timid gradualism that appears to move at a snail's pace both on price liberalization and privatization. Each country has followed a different approach, with the state withdrawing from different sectors at different rates. The path of transition has not been a smooth one with liberalization often followed by the reimposition of some government controls.

While the process of economic liberalization has been embraced by all of the former command economies, less attention has been paid to the question: where is the process of transition leading? or, a transition to what? If central planning is not, in future, to be the mechanism by which economic allocation is to be decided, what will replace it? The simple answer to this question is that the allocation of resources and consumer goods should be decided by market forces. No consensus exists, however, regarding the desired degree of state intervention in the economy at the end of the transition process any more than there is a consensus in the modern market economies, which are all mixed economies with greater or lesser degrees of state intervention. Regardless of the eventual degree of state intervention envisioned, markets are expected to have a central role in the process of economic allocation in the former command economies.

When markets are the allocation mechanism, prices are expected to perform a signalling role for the movement of resources among economic activities. In the stylized market economy, maximizing individuals are expected to continually attempt to improve their lot by acting in their own self-interest. They do this by following price signals from the market. To be able to do this, however, they need to have a system of well defined property rights, including property rights to their labour. In the CEECs and the NISs, the first step towards a system of private ownership has been the disposal of the state's assets. Privatization alone, however, does not lead to an economy based on property rights. With help and advice from the modern market economies, legal systems are being put in place to: (i) vest property rights in legal economic persons (including corporations, cooperatives, etc. as well as individuals); and (ii) to protect those property rights.

While most countries have enacted property rights legislation, enforcement of those property rights by the courts and by police is by and large ineffective. To the degree that individuals feel unsure about their property rights, their responses to price signals will be inhibited. This is particularly true in the case of investment, where fear of official confiscation by the state or unofficial confiscation by bureaucrats or other individuals will prevent potential investors from acting fully on the price signals received from the market. At best, investment activity will lack balance as the funds available will be channelled into activities where property rights are secure or at least where confiscation is difficult. At worst, expenditures will be channelled into direct consumption activities. While the form which privatization actually takes may not be important to long run allocative efficiency,[2] the poor enforcement of property rights will reduce the effectiveness of markets as allocative mechanisms. However, many developing countries also have poorly defined or enforced property rights, yet markets and prices still act as the primary allocation mechanism in these economies.

## 3.1. A Role for Prices

The central question to be answered, subject to the constraints imposed by poorly defined and/or enforced property rights is, to what extent can prices act as a guide to resource allocation in the agrifood sectors of the CEECs and the NISs? The role of prices in the agrifood sector, however, is part of the wider problem of prices in the former command economies. To examine the problem of prices, it is first necessary to examine the use of prices in a command economy.

As suggested in Chapter 2, the major use of prices in command economies was to facilitate accounting. This is because the allocation of resources was accomplished by managerial orders either within the planning ministry or within command firms themselves. In some sectors, primary agriculture in particular, prices did play a limited role in providing incentives for production and procurement activities. Planning, based on the material balances approach, was the command economy's allocation mechanism and one in which prices had no allocative function. The inability of central planning to rationally choose among allocation alternatives was central to the command economies' failure to provide for sustained improvements to living standards.

It was Ludwig von Mises (1981)[3] who first articulated the calculation problem of command economies. The essence of his argument is that once government intervenes in the economy to set prices by fiat, it is no longer possible to use prices as a means of allocation. Further, it is not possible to make allocative decisions based on an objective efficiency criteria. This is because the set of relative prices promulgated by the state does not reflect the relative value (based on opportunity cost) placed on goods and services by their users whether they be producers of goods or final consumers.[4] Planners in command economies realized that their official prices could not be used for the purposes of allocating resources. The material balances approach, whereby technical input coefficients were used to determine the ratio of resources in production could theoretically be used as a static allocation rule. It is not useful for the dynamic decision-making processes which characterize a modern economy. Questions relating to when an industry should decline or expand as resource scarcity changes or consumer preferences evolve cannot be answered. The viability of new products and processes cannot be assessed objectively.

The problem of determining relative value, as identified by von Mises in 1922, was never solved by communist intellectuals or planners. As a result, the official prices (including wages) which existed at the beginning of the liberalization process were artificial and did not reflect the relative values of goods and services. They simply reflected the arbitrary decisions of those in the planning ministries. Hence, official prices did not provide even rough approximations of value upon which to base the liberalization process.

Governments in the former command economies were all faced with a dilemma. The existing set of prices conferred benefits to some members of society. For example, low food prices provided a degree of food security for those on fixed incomes such as pensioners. If they did not free prices, however, correct signals would not be conveyed upon which to base resource allocation decisions.

Freeing prices without the institutions necessary to support markets was, unfortunately, not likely to produce the set of prices which reflected relative values as was envisioned by those who put their faith in the process of allocation through market forces. The rapid rates of inflation which followed the freeing of prices in some former command economies is likely to have further reduced the ability of individuals to discern relative value from the available prices.[5] Where markets are not developed, as in the former command economies, the ability to use prices as a decision criteria is severely limited.

The problem of prices in an economy in transition is extremely complex and requires an understanding of the conditions that allow prices to act as a guide for economic allocation in market systems. Certainly, the problems discussed below all exist in market economies but they represent small amounts of friction in the allocative system. In the former command economies, where markets are not fully developed, they are likely to be endemic. Prices are not perfect guides to resource allocation in market economies but the existence of well developed markets and the possibility of government intervention in the case of market failure allow prices to fulfil their signalling role reasonably well.

## 3.2. The General Equilibrium Problem

Market economies have enjoyed a long evolution. The observed relative prices can probably be assumed to be close to equilibrium. Most price shocks, say those which arise from a failed harvest in a major crop or from the disruptions arising from an industrial labour dispute, are small relative to the entire economy. This is important because market economies are general equilibrium systems where all markets are interconnected. A price shock in one sector or industry will affect all other prices. In reality, of course, the effect on relative prices is likely to be widely diffused except in markets which are heavily dependent upon one another. A shock which alters the price of fertilizer is not likely to be discernable in the market for cutlery. It is still true, however, that altering any price will have an effect on all other prices and it will take time for adjustments to interactively work their way through to a new equilibrium. Of course, a new equilibrium will probably never be reached because new price shocks will arise before the completion of the process.[6]

Due to its interdependent nature, even a well-developed market economy may not be sufficiently resilient to easily accommodate large and/ or multiple price shocks. One has only to remember the difficulties experienced by developed market economies in adjusting to the OPEC oil price shocks of the 1970s. The increase in the price of this important input to production created the need for a major adjustment to relative values with price changes in one sector forcing adjustments to be made in other sectors which eventually fed back to the original markets. Due to the lags in adjustment, relative prices did not provide an accurate guide to long-run relative values. Remember, this was in a system which was probably close to equilibrium at the beginning of the first oil price shock.

Now consider the position of command economies at the beginning of the liberalization process. After almost 70 years of prices being established by fiat in planning ministries (40 years in the CEECs), the existing prices bore no relationship to relative value as understood in market economies. Freeing prices simultaneously means that each market must find its own equilibrium in the short run. Once that equilibrium is established, however, it becomes relevant price information in the next decision period in related markets causing further adjustments. At the same time, the price adjustments in all other newly freed related markets become relevant price information for pricing decisions in the original market which again leads to price changes. As some of the short-run price changes in the former command economies were much larger, in relative terms, than the OPEC oil price shocks, e.g. the prices of some food staples, energy and housing, it is unlikely that prices reflected long-term relative values in the early period of transition. Economists have very few insights into the rate at which economies experiencing large multiple price shocks return to equilibrium, i.e. where prices will approximate long-run relative value.

The pricing problems in the former command economies are exacerbated by the creation of new markets where none were allowed before, such as markets for land. The consumer/firm choice set is expanded with the creation of new markets and a reallocation of available budgets must be made. New products are always arising in market economies, but seldom do they require the large-scale budgetary reallocations suggested by the opening up of a land market. Markets for private automobiles and imported appliances and electronics also represent significant expansions of consumers' choice sets in the former command economies.

Further, while the governments of the former command economies have differed considerably in the degree to which they have been willing to free prices, none were willing to free all prices at once. This meant that part of the economy began the general equilibrium adjustment process but as some relative prices remain fixed at distorted levels, the equilibrium to which the liberalized portion of the economy converges will not accurately reflect long-run relative value. When the prices of new sets of commodities are subsequently freed, they provide new shocks to the system which set off additional rounds of adjustment. While this may sound as if the simultaneous freeing of all prices is being advocated, in fact, economists cannot provide a definitive answer to the question of whether the simultaneous freeing of all prices will lead to a convergence near to a general equilibrium at a faster rate than a more gradualist approach.

One possible means by which the long adjustment to a general equilibrium may be shortened is to open the economy to foreign competition. This will be particularly relevant for small economies with good links to modern market economies and where non-tradeable goods do not represent a significant proportion of the economy. The apparent success of the Czech Republic's transition may, in part, be attributed to its ability to establish external market linkages. For a small open economy relative value can, for the most part, be determined in the international market.

In an open economy, domestic prices which deviate from world prices will either fall as a result of competition from imports or be bid upwards by foreign customers. With clear foreign market signals for outputs, tradeable inputs and capital goods, it is easier to identify the relative value of non-tradeable inputs such as labour or land.[7] Of course, unfettered currency convertibility, floating exchange rates and liberalized capital markets will facilitate the movement to a new equilibrium.[8] For large, geographically dispersed economies which have only poor infrastructures for interacting with modern market economies, the effect of foreign markets on the process of transition will be limited.

## 3.3. The Monopoly Problem

In market economies, the role of prices as a guide to resource allocation is based on the premise that they represent, for the goods in question, a convergence of the value of the opportunities foregone in the goods' production and the value placed on the goods by consumers. In other words, price is set where the marginal (resource) cost of producing a good equals the marginal valuation which consumers put on that good. This is the familiar intersection of supply and demand curves. Disequilibrium in a market sets economic forces in motion to reallocate resources. For example, when excess demand exists, rising prices provide an incentive for increased resource allocations to the goods supplied to that market. Falling prices suggest a reduction in the resources committed to the production of the goods supplied to markets exhibiting excess supply. This basic tenet of market economies bears repeating in the case of liberalizing command economies.

In market economies it is recognized that monopolies lead to inefficient levels of output because the price (and, hence, valuation of consumers) exceeds the marginal cost of producing the good or service. The observed monopoly price does provide a signal for more resources to be transferred to the production of the good but barriers to entry prevent that transfer from taking place. Prices are not allowed to play their role as a guide to resource allocation.

In modern market economies, monopoly inefficiencies are often tolerated because the effect of their perceived distortions is small (or the gains from regulating such monopolies do not justify the resource costs associated with the regulatory process). Where the inefficiencies or price distortions created by a monopoly (or potential monopoly) are considered to be unacceptable, government policy has generally been: (i) to prevent monopolies from arising – anti-merger provisions in anti-trust or competitions legislation; (ii) to break up existing monopolies; or (iii) to regulate the output and price of monopolies. The intent, in each case, is to keep or move the observed price nearer to the price which would arise in a competitive market. It is recognized, however, that a price established through a regulatory process will only be an approximation to the theoretically desired price. This is because acquiring the information to determine the regulated price is not costless. The problem of setting a regulated price is especially difficult when the firm whose price is being regulated has an incentive to provide misinformation.

In the liberalizing former command economies, monopolies may be far more widespread than in market economies. There are two reasons for this: (i) the former command system stressed large-scale production/distribution facilities; and (ii) the ability of firms to identify and conclude transactions with alternative suppliers/customers is limited by the lack of institutions to support the process of broadening markets, i.e. the costs associated with firms broadening markets are very high. As discussed in the previous chapter, the system of food provision in command economies can be viewed as a series of parallel chains where command firms were tied to both suppliers and customers. Each chain represented a geographic monopoly. A belief in the inherent efficiency of large-scale operations led to investments in large-scale physical facilities. Privatization of chains (usually by selling off the former administrative entities we identified as command firms) and the liberalization of prices has led to the creation of: (i) geographic monopolies with unregulated prices; and (ii) bilateral monopolies at the now so-called market interfaces between privatized former command firms. These bilateral monopoly interfaces extend over the length of the production/distribution chain (Gaisford *et al.*, 1995). The problems created by chains of bilateral monopolies is discussed at greater length in Chapter 5.

The geographic monopolies enjoyed by now privatized former command firms can probably be maintained because the size of the entities created by the investment patterns of the former regimes gives existing enterprises the ability to erect barriers to entry. Further, it is costly for firms to seek out alternative suppliers/customers when they find themselves faced with monopoly/monopsony prices. As market information systems and infrastructure have not been fully developed (or do not exist) the costs faced by those seeking to identify and evaluate alternative suppliers/customers are very high. These transactions costs are discussed in greater detail in Chapter 4. As a result, the forces of competition are blunted. Monopolization in the agrifood systems of the former command economies is endemic. According to Gady and Peyton, in the CEECs and the NISs:

> Many of the large state monopolies still exist. Consequently, there is little competition to bring about the operating efficiencies that need to occur. Furthermore, many of the existing monopolies have been able to raise prices to punitive levels because of lack of competition, making life difficult for businesses having to buy or sell to these monopolies.
>
> (Gady and Peyton, 1992, p. 1180)

The creation of monopolies through the process of privatization[9] and their ability to maintain their monopoly position, means that prices are not able to carry out their resource allocating role to the same degree as is the case in market economies. Endemic monopolies will distort the perception of relative value which any observed set of prices creates.

Governments may feel obligated to intervene in some monopolized markets. In the absence of effective anti-trust legislation and/or enforcement, this intervention is likely to take the form of price fixing. Intervention may be expected in markets for products considered to be necessities such as food and energy. For example, in January 1993 the Russian government reimposed price controls on a range of food products including bread, meat and milk. According to *The Economist* (January 9, 1993) this intervention was motivated by a belief among policy makers:

> that the main source of inflation is not the growth in the money supply but structural factors, like the ability of monopoly companies to jack up prices as if consumers did not matter. If this were true, then administrative controls on monopoly profits (which are great in the food industry) might possibly succeed in reducing inflation.
>
> (Anon., 1993, p. 43)

Authorities in the former command economies are unlikely to be well versed in the theory of monopoly regulation nor are they likely to have experience with the implementation of regulatory policy. As a result, their ability to approximate accurately the theoretically desirable regulated price may be less than is the case in market economies and prices will remain considerably distorted.


## 3.4. The Disequilibrium Problem

The ability to rely on prices as an efficient mechanism for allocating resources in market economies is dependent upon prices never being far from equilibrium. When prices in market economies exhibit considerable disequilibrium, alternating periods of over and under-production relative to a market equilibrium may result. This phenomenon is well known in the agricultural sectors of market economies with, for example, persistent (but not regular) pork and beef cycles being observed. In most modern market economies, there is considerable intervention in the grain sector, for example, which is aimed at stabilizing market prices.

Agricultural markets are subject to a range of external shocks such as those arising from inclement weather. The price movements initiated by these external shocks become part of the price information which farmers may use in their decisions in subsequent production periods. While the process by which farmers form their price expectations is not well understood and probably very complex, it appears that nearby (in time) observed prices are an important component of their expected price. As a result, a high disequilibrium price in production period, $t_0$, leads to increased levels of output in $t_1$. This overshooting of production, and possibly investments in productive capacity, leads to excess supply at the expected price and a subsequent collapse in price. Of course, the decline in price feeds back into price expectations for the following production period. Excessive reductions in output and cancelled investments may be the result. Unwarranted entry and exit by firms may also be encouraged by disequilibrium prices. It has long been recognized that such market instability can be wasteful.

As suggested above, the set of prices observed in former command economies are not likely to reflect long-run equilibrium conditions. As prices in command economies did not represent equilibrium values, freeing those prices may induce supply responses which destabilize the agrifood sector. Agricultural stabilization policies in modern market economies usually require a considerable commitment of fiscal resources from government. This may represent a budgetary expenditure which exceeds that which governments in former command economies are willing to commit. As a result, if prices exhibit considerable instability, they will not be able to efficiently fulfil their resource allocation role.

## 3.5. The Problem of Quality Differentiation

Prices also fulfil a role in providing information on relative quality. Prices aid consumers and industrial purchasers in differentiating non-homogeneous goods. In market economies when incomplete or asymmetric information[10] is present, it is well known that gains may be available to one party in a transaction if they behave opportunistically. The threat of opportunistic behaviour can inhibit investment and reduces the efficiency of markets.

Well developed markets can remove some of the difficulties which arise when an individual attempts to assess quality. Competition leads sellers or buyers to put mechanisms in place so that relative quality can be discerned through prices. Often, these mechanisms have high fixed costs associated with them and, hence, require a large number of market participants before they can be justified. For example, laboratory testing facilities may require a certain volume of throughput before their costs decline to the point where they are exceeded by the benefits gained from testing. When markets have few participants and information is expensive, price may not provide a signal that differentiates quality. As a result, more costly methods of differentiating quality may be required. For example, if fruit or vegetable products are not graded and priced accordingly, each customer is forced to individually sort the produce prior to buying. As the same lot of produce may be sorted a number of times, ascertaining quality is more costly and the efficiency of the marketplace declines. As quality played almost no part in the command system, governments, former command firms and consumers have little experience with the mechanisms used in market economies to facilitate the use of price as a means to differentiate relative quality. Hence, a long period of learning may be required and prices will have only a limited role in providing information on relative product quality in the near future.

## 3.6. Conclusions

It should be obvious that prices, even if fully liberalized, will only be able to undertake the resource allocation role they play in modern market economies to a very limited degree in economies in transition. The discussion in this chapter also suggests that the transition to an economy which is efficiently directed by market prices will be long, fraught with difficulties and plagued with false starts and partial returns to state control of prices. Privatization and the freeing of prices are not sufficient conditions for the establishment of a market economy. Market institutions will also be required. The development of these new institutions will require far more resources, both private and public, than the simple task of freeing prices and the relatively tractable problem of privatizing the assets held by the state. While there is a grudging realization that market institutions are necessary, there are, as yet, few concrete proposals as to how institutions that have evolved over a very long period of time in modern market economies are to be fostered in the short run.

This is not to suggest that prices should have remained under state control. Planners in command economies in the post-communist era would have faced the same problems as their predecessors if they had attempted to fix all prices and government ownership of the means of production was retained. Freeing prices and privatization are, however, only the first steps towards a market economy. The problem of prices is likely to mean a process of transition which is much longer than that envisioned when liberalization was initiated.

What does the general problem of prices faced by liberalizing former command economies mean for the agribusiness sector? Little can be said with any certainty. The economic transformation taking place in the CEECs and the NISs has no precedents. The 'planner's conscious hand' has never before been removed. The expectation, both in the former command economies and by those interested in transition observing from modern market economies, is that Adam Smith's 'invisible hand' should be the replacement. Exactly how and when a new equilibrium will be approached is unknown. Some necessary conditions are suggested: (i) well specified and enforceable property rights;[11] (ii) privatization;[12] (iii) the freeing of prices; and (iv) the coming into being of the institutions to support a market economy.

Whether these represent sufficient conditions for a transition to a market economy, no one can say for sure. While conditions (ii) and (iii) are being pursued more or less vigorously in all former command economies, only limited progress has been made on condition (i), particularly with respect to enforcement, and little is known about condition (iv). While there is considerable discussion of the need for market institutions, less advice is proffered regarding how they should come into being.

In modern market economies, the institutions which support the market based system have evolved over a very long period and are primarily non-governmental. In some former command economies, institutions have arisen which, at least in their formal appearance, have counterparts in modern market economies. One example is stock markets. Whether these institutions function as they do in modern market economies, however, is less clear. New market institutions have also arisen that have, for the most part, been abandoned in modern market economies. Physical marketplaces where food is exchanged directly between farmers and consumers have sprung up or have been expanded all over the CEECs and the NISs. Physical food markets no longer play a significant part in modern market economies, presumably because the costs associated with assembling both goods and consumers as well as those associated with haggling are too high.[13] The full range of market institutions in the agrifood sector seems to be evolving only slowly and there are few ideas offered as to how they might be fostered by public policy. Further, the institutions which are often provided by governments as public goods in modern market economies are largely absent in the former command economies.

What does it mean when the 'planner's conscious hand' has been lifted but the 'invisible hand' does not yet exist? Fundamentally, it means that the prices which do exist will often give false signals. In the short run, they will give false signals about what to produce, in the intermediate run they will give false signals about where to invest. If producers and investors are unaware that the existing prices give false signals, and do not learn, the market clearing price is not likely to represent a stable equilibrium and there will be considerable wasted investment. Little is understood about the actions of producers and investors when they are aware that prices give false signals. If they are risk averse, they are likely to under-produce and under-invest when they cannot believe in the price signals they receive.

The agrifood sector will not be exempt from false signalling by observed prices. As a result, investment in new farm and agribusiness facilities is likely to be curtailed. The effect on primary agricultural production in the short run is less clear. Farmers in modern market economies often appear to make production decisions based on prices which do not represent market equilibriums. This is the case even when the existence of market cycles are well established and well understood. Hence, agricultural markets in the former command economies may be subject to considerable volatility. While there is currently considerable volatility in these markets, it is not yet possible to discern to what degree this can be attributed to decisions based on false price signals. This is because the agrifood sector has experienced a number of major exogenous shocks since the process of liberalization began – disruptions in input supplies, restrictions on credit, fuel shortages, labour shortages, lack of spare parts, product diversion from official marketing channels to black and grey markets, etc.

The performance of the agrifood sector is likely to be closely watched by politicians. Food security is an important political issue. If the price mechanism does not perform sufficiently well in its resource allocation role to provide a politically acceptable degree of food security, then some form of allocation based on command will be reinstated. Of course, the CEECs and the NISs have not yet been willing to fully entrust food security to the price mechanism. This chapter suggests that it is unlikely that prices will be able to efficiently fulfil the resource allocation role that they play in modern market economies for some time into the future. Hence, the agrifood sector may be faced with varying degrees of command style intervention during the period of transition. In some countries, particularly those which are able to integrate with modern market economies, there will be only a limited degree of intervention. In others, it will remain significant. The greater the degree of intervention in the agrifood sector, the longer will be the transition period – not just in the sector but in the entire economy.

The problem of prices will be a theme that runs through this entire volume. Its effect on specific industries and the particular problems it creates will be discussed in the chapters that follow this section.

# Notes

1. Of course there are political parties or groups within most countries who advocate a return to the old ways.
2. The method by which privatization is conducted has both short and long-run distributional implications and it is the distributional aspects of privatization which has been the subject of fierce debate across the CEECs and the NISs. How privatization is undertaken may affect short-run allocative efficiency as the unbundling of the existing state assets may not be efficient. For example, distributing agricultural land among all the members of collective farms may lead to sub-optimal farm units and inefficient use of the collective farm's existing infrastructure – dairy barns, tractors and machinery only suitable for large-scale operations, fixed irrigation systems, etc.
3. von Mises originally published his work in German in 1922 under the title *Die Gemeinwirtschaft: Untersuchungen über den Sozialismus*, Gustav Fisher, Jena. The first English translation (with a few additions) was published in 1936 as *Socialism: An Economic and Sociological Analysis*, Jonathan Cape, London (translator J. Kahane).
4. von Mises eloquently spells out the problem for what he called socialist communities as well as the central role of prices in a market economy:

> Let us try to imagine the position of a socialist community. There will be hundreds and thousands of establishments in which work is going on. A minority of these will produce goods ready for use. The majority will produce capital goods and semi-manufacturers. All these establishments will be closely

connected. Each commodity produced will pass through a whole series of such establishments before it is ready for consumption. Yet in the incessant press of all these processes the economic administration will have no real sense of direction. It will have no means of ascertaining whether a given piece of work is really necessary, whether labour or material are not being wasted in completing it. How would it discover which of two processes was the most satisfactory? At best, it could compare the quantity of ultimate products. But only rarely could it compare the expenditure incurred in their production. It would know exactly – or it would imagine it knew – what it wanted to produce. It ought therefore to set about obtaining the desired results with the smallest possible expenditure. But to do this it would have to be able to make calculations. And such calculations must be calculations of value. They could not be merely 'technical,' they could not be calculations of the objective use-value of goods and services; this is so obvious that it needs no further demonstration.

Under a system based upon private ownership in the means of production, the scale of values is the outcome of the actions of every independent member of society. Everyone plays a two-fold part in its establishment first as a consumer, secondly as a producer. As consumer, he establishes the valuation of goods ready for consumption. As producer, he guides production-goods into those uses in which they yield the highest product. In this way all goods of higher orders are graded in the way appropriate to them under the existing conditions of production and the demands of society. The interplay of these two processes ensures that the economic principle is observed in both consumption and production. And, in this way, arises the exactly graded system of prices which enables everyone to frame his demand on economic lines. Under socialism, all this must necessarily be lacking.

(von Mises, 1981, pp. 103–104)

5. The effect of inflation on economies with underdeveloped markets is not unambiguous. It can be argued that an inflationary burst followed by monetary stability could actually aid in the process of establishing prices which reflect relative value. An inflationary burst could be instrumental in altering the illusion of relative value which was conveyed over long periods of time by the controlled prices of the command era. Expansion of the money supply combined with the freeing of previously controlled prices will cause all prices to move, breaking down previous perceptions of relative value. Hence, inflation acts as grease to the wheel. However, once the former perceptions of relative value have been altered, a period of monetary stability is required so that new relative values can be established.
6. We are assuming here that the system converges to a stable equilibrium.
7. This assumes that immigration/emigration is, to some extent, restricted.
8. von Mises envisioned this external role for market economies in the establishment of relative value in centrally planned economies.

[If Socialism] exists only, one might say, in socialistic oases in what, for the rest, is a system based on free exchange and the use of money... For the existence of a surrounding system of free pricing supports such concerns in their business affairs to such an extent that in them the essential peculiarity of economic activity under Socialism does not come to light. In State and Municipal undertakings it is still possible to carry out technical improvements, because it is possible to observe the effects of similar improvements in similar private undertakings at home and abroad. In such concerns it is still possible to ascertain the advantages of reorganization because they are surrounded by a society which is still based upon private ownership in the means of production and the use of money. It is still possible for them to keep books and make calculations for which similar concerns in a purely Socialist environment would be entirely out of the question.

(von Mises, 1981, p. 102)

9. In many cases prices were liberalized along the production and distribution chain without privatization. In other words, allocation between command firms by planners has been replaced by price bargaining between state enterprises. To the extent that the managers of state controlled enterprises can act independently to set prices, monopoly behaviour has been observed.

10. Incomplete information arises when both parties are unable to fully discern a product's quality. Asymmetric information arises when one party to a transaction has more information than the other.

11. The observations by Steven Cheung (1982) on transition in China may shed some additional light on the role of property rights.

12. The need to privatize the means of production is still the subject of some economic debate (see for example Pejovich, 1994). However, events have largely overtaken this academic debate and all states are implementing some form of privatization policy.

13. Of course, physical markets for some food products have enjoyed somewhat of a revival in modern market economies. However, these tend to be low volume markets where products are differentiated by freshness, production method (e.g. organic), etc., or the trade is in specialized boutique-type food products. One suspects that there may be a considerable consumption good aspect to the experience of shopping at these markets in modern market economies.

# References

Anon. (1993) Russia slides away from reform. *The Economist* 326, 7793 (January 9).

Cheung, S.N.S. (1982) *Will China Go Capitalist?* Hobart Paper No. 94, The Institute of Economic Affairs, London.

Gady, R.L. and Peyton, R.H. (1992) A food processor's perspective on trade and investment opportunities in Eastern Europe and the former Soviet Union. *American Journal of Agricultural Economics* 74, 1175–1183.

Gaisford, J.D., Hobbs, J.E. and Kerr, W.A. (1995) If the food doesn't come – vertical coordination problems in the CIS food system: some perils of privatization. *Agribusiness: An International Journal* 11, 179–186.

Pejovich, S. (1994) A property right's analysis of alternative methods of organizing production. *Communist Economies and Economic Transformation* 6, 219–230.

von Mises, L. (1981) *Socialism*. Liberty Classics, Indianapolis, 569 pp.

# 4 Doing Business – The High Cost of Transactions

An important constraint to the successful and smooth transformation of agrifood systems in the Central and Eastern Europe Countries (CEECs) and the New Independent States (NISs) is the high cost of transactions associated with doing business in these economies. These costs stem from the lack of effective or functioning market institutions to facilitate business transactions. These institutions were alluded to at the end of Chapter 2. In section III, the process of economic transformation at different stages of the agrifood chain is discussed at length. One of the recurring themes is the need to develop effective market institutions. These include the financial, legal, communication and marketing institutions which underpin agribusiness transactions in modern market economies. They are institutions which business people in developed economies can, for the most part, take for granted but without which the costs of doing business would often become prohibitive. Why are these institutions important? What is the nature of these transaction costs? How is the organization of economic activity affected by these transaction costs? This chapter provides an overview of transaction cost economics (TCE) and seeks to answer these questions. Later chapters draw on the theoretical concepts explained in this chapter. The high costs of doing business provide one explanation for the existence of monopolies in the agrifood systems of the CEECs and the NISs. The problem of monopolies is discussed in Chapter 5.

# 4.1. The Development of Transaction Cost Economics – An Historical Perspective

### 4.1.1. Neoclassical economics

The core of the current ruling paradigm for economic analysis of markets, industries and the firm is provided by neoclassical economics. Central to the neoclassical theory is the concept of a single product firm, operating in a perfectly competitive industry with a large number of competitor firms all producing the same product under the same cost conditions and all facing the same market demand curve.[1] The standard neoclassical transaction involves the exchange of a homogeneous product – it is assumed that there are no quality variations among outputs of one industry and consequently no costs involved in measuring the value of a product. Where quality variations do exist, they are treated as separate products serving distinct markets. Economic agents are assumed to possess full information, hence, there is no uncertainty regarding prices, product characteristics or the behaviour of competitors and trading partners. The neoclassical transaction occurs in the current time period between multiple buyers and sellers, thereby ruling out the possibility that one firm could exercise market power over others since many alternative buyers and sellers exist. Thus, neoclassical economic analysis concentrates on equilibrium market outcomes. There is no consideration of how transactions occur, instead, transactions are treated as though they occur in a frictionless economic environment, somewhat analogous to the physicist's perfect vacuum. Somewhat ironically, the neoclassical theory of the firm has little to say about the firm; it does not provide a rationale for the existence of firms, an explanation for the growth of firms nor an analysis of the internal organization of firms. The firm is, instead, treated as a black box – a featureless production function which turns inputs into outputs; as such it is a component of the neoclassical explanation of the workings of a competitive market economy, but one which is little understood. It was these perceived shortcomings of neoclassical theory which led some economists to search for a more realistic explanation of the existence of firms, eventually culminating in a theory of transaction costs.

### 4.1.2. The nature of the firm

In 1937 Ronald Coase published his seminal paper *The Nature of the Firm*. This paper was to form the basis of what became known as transaction cost economics. Coase argued that in order to understand what a firm does, one must first understand why a firm exists and, therefore, what forces govern the organization of economic activity. Unlike standard neoclassical economics, the Coasian approach recognized that there are costs to using the market mechanism. These include the costs of discovering what prices should be, the costs of negotiating individual contracts for each exchange transaction and the costs of accurately specifying the details of a transaction in a long-term contract (Coase, 1937). When the CEECs and the NISs liberalized their economic systems, the costs of using the market system suddenly became important in shaping the type of firms which emerged within industries and in shaping the relationships between these firms.

The costs of using the market can be avoided if a firm becomes vertically integrated and assumes the burden of coordinating economic activity internally through within-firm managerial direction. However, this means that a firm must assume the alternative costs of administering vertical flows of products and organizing factors of production. Provided that a firm can carry out these activities at a lower cost than the market-pricing mechanism, then one would expect the organization of economic activities to be carried out by a vertically integrated firm, *ceteris paribus*, i.e. everything else held constant. Borrowing the Marshallian concept of substitution at the margin, Coase argued that:

> ... a firm will tend to expand until the costs of organising an extra transaction within the firm become equal to the costs of carrying out the same transaction by means of exchange on the open market or the costs of organising in another firm.
>
> (Coase, 1937, p. 395)

Hence, he provided a rationale for the existence of the firm which was based on the costs of carrying out a transaction. These insights, however, did not have a major impact on economic thought until more than thirty-five years later.

### 4.1.3. The development of mainstream theories of industrial organization

In the ensuing period between Coase's original insights and the development of a theory of transaction costs in the 1970s, mainstream economists expanded the neoclassical model of economic activity, developing new theories of industrial organization. Stigler (1968) and Bain (1968) analysed the pricing and output policies of oligopolistic firms, leading to the development of the Structure–Conduct–Performance (SCP) approach to industrial analysis. This approach has been criticized on the grounds that it assumes the boundaries of the firm (where inter-firm transactions replace within-firm managerial coordination) to be given. Structure–Conduct–Performance theory explains the performance of firms (in terms of profit, growth in sales, the adoption of new techno- logy, etc.) on the basis of their existing structure; it does not question why firms are of a particular size (i.e. why certain activities are carried out within the firm rather than in the market) and does not study the dynamic process by which a firm itself grows or shrinks.

Other post-war developments in economic thought, however, influenced the development of a theory of transaction cost economics, for example, the work of Stigler (1961) on the economics of information and Lancaster's (1966) characteristics approach to assessing the relative attributes of different goods. A concept which was to become central to the transaction cost theory was originally developed in the human behaviour literature by Simon (1961). Bounded rationality refers to human behaviour that is intendedly rational but only limitedly so, i.e. individuals do not have perfect knowledge and, hence, can only make the best decision they can with the information which is available. The relevance of this concept to TCE is explained in section 4.2.

The concept of transaction costs was used again by Coase in his 1960 seminal paper *The Problem of Social Cost* (Coase, 1960). Although he does not actually use the term transaction cost,[2] the Coase Theorem, as it was later to be named, states that if transaction costs are zero, the initial assignments of property rights between a polluter and a victim of pollution will not affect the efficiency with which resources are alloc- ated. This paper served to increase interest in the costs of transacting through the market and was part of a growing literature on property rights (Alchian, 1965; Demsetz, 1967). Hence, although there was, at best, limited interest shown in transaction costs in the 1950s and 1960s, some of the core concepts upon which the modern theory of transaction costs rests were being developed during this period.

### 4.1.4. New institutional economics

In the 1970s, interest increased in applying a transaction cost focus to analysing micro-economic problems. Oliver Williamson (1975, 1979) pioneered much of the development of a theory of transaction cost eco- nomics. Gradually, a body of theories based on the concept of transac- tion costs emerged. Often referred to as new institutional economics, these include the transaction cost economics of Oliver Williamson, the property rights school (Alchian and Demsetz, 1972), agency theory (Jensen and Meckling, 1976); the economics of the multinational enterprise (Teece, 1986; Casson, 1990) and a transaction cost approach to economic history (North, 1987). Although focusing on separate economic problems, these approaches all have their roots in the original ideas of Coase (1937) and use the concept of transaction costs to explain the organization of economic behaviour. As Cheung explains:

> If we call a certain branch of economics 'new institutional economics,' then we are obligated to tell people what is 'new.' Institutional economics has had a long tradition of concern for the real world. What is new is that we are now able to explain why the observed institutional arrangements are as they are ... First ... the new practitioners are well versed in modern economic theory and they use it routinely. Second, the new practitioners are far more interested in the constraints of the real world. More specifically, the presence of transaction (institutional) costs is now taken as a constraint of paramount importance.
>
> (Cheung, 1992, p. 62)

Many of the transition problems faced by CEECs and the NISs discussed in later chapters stem from the lack of effective market institutions. Understanding why these problems arise will assist in the development of solutions.

## 4.2. Transaction Cost Economics – Key Concepts

Although the term transaction cost economics is often used to refer to the work of Williamson (1975, 1979, 1986, 1991), the core concepts of TCE are fundamental to all branches of new institutional economics. Two of the key concepts are: (i) bounded rationality and; (ii) opportunism. As was explained earlier, bounded rationality, the idea that people are intendedly rational but only limitedly so, was first introduced by Simon (1961). Bounded rationality means that although people may intend to make a rational decision, their capacity to accurately evaluate all possible decision alternatives is physically limited. A useful analogy is that of a chess player who, although able to view the position of all playing pieces on the chess board, cannot feasibly evaluate all potential outcomes of a move, given an opponent's counter-moves, and given their own counter-counter moves, and so on (Douma and Schreuder, 1992). Bounded rationality poses a problem only in situations of complexity or uncertainty which impede the ability of people to make a fully rational decision. Clearly, in the environment of rapid changes and structural dislocation which surrounds economies in transition, complexity and uncertainty are rife. Within the old planned economy, bureaucratic decisions replaced market transactions. There was little of the complexity and uncertainty associated with locating potential suppliers or buyers, in assessing the suitability of their products, etc. Neither firms nor consumers had choice; with choice comes the complexity of

decision-making under situations of uncertainty. The bounded rationality of firms and individuals was far less important in the planned economy, although the bounded rationality of the planners was a crucial
determinant of the poor performance of command economies. Supply
gluts and disruptions were partly the result of the inability of bureaucrats to account for every contingency when drawing up their economic
plans.

Opportunism, or self-interest seeking with guile (Williamson, 1979,
p. 234), recognizes that sometimes economic agents will seek to exploit
a situation to their own advantage. This does not imply that all individuals act opportunistically all of the time, rather, it recognizes that
the risk of opportunism is often present. This risk is greater when there
exists a small numbers bargaining problem (Williamson, 1979). This
problem may be present if privatization of agricultural inputs has
resulted in the industry being dominated by a small number of firms or
by a monopoly. For example, the fewer the alternative equipment suppliers available to a processing company, the more likely it is that an
existing supplier will act opportunistically to alter the terms of the business relationship to their own advantage, such as by demanding a
higher price than that previously agreed. Williamson stresses the
importance of bounded rationality and opportunism to the study of
transaction cost economics:

> But for the *simultaneous* existence of both bounded rationality and
> opportunism, all economic contracting problems are trivial and the study
> of economic institutions is unimportant.
>
> (Williamson, 1986, p. 140)

Another key concept of TCE is asset specificity. Asset specificity
arises when one partner to an exchange (firm A) has invested resources
specific to that exchange which have little or no value in an alternative
use, for example, the installation of specialized machinery in a meat
processing plant which serves the requirements of a single customer or
the development and promotion of a product unique to one export
market. Firm A then faces the risk that its trading partner (firm B) will
act opportunistically by trying to appropriate some of the rent from this
investment. Knowing that firm A has made a specialized investment
and is therefore locked into the exchange, firm B could renege on the
previous agreement by offering firm A a lower price for the product.
Provided that this lower price covers A's variable costs of production
and makes a contribution towards the fixed cost of the investment, A
will have little choice but to accept B's discounted price offer. The
economic rent captured by firm B is known as specialized appropriable quasi-rent. The opportunistic behaviour displayed by firm B is
often termed post-contractual opportunistic behaviour (PCOB) or

opportunistic recontracting. In countries with well developed commercial legal systems, firm A would have recourse to contract law to enforce the contract and could sue for damages if the terms of the contract were not met. However, in the CEECs and the NISs, contract law is either non-existent or is poorly enforced. This leaves firms particularly vulnerable if they have made asset-specific investments.

The mere existence of contract law, however, may not entirely remove the threat of opportunistic behaviour. A firm's decision to utilize contract law in the event of a problem will depend on whether the benefits to be gained from forcing the opportunistic firm (firm B) to uphold its obligations outweigh the costs of legal action, allowing for the uncertainty of the judicial process.

To further illustrate the asset specificity problem, suppose a hypothetical firm, firm A, located in a small town in the Ukraine, owns a controlled atmosphere meat packaging system and contracts to provide firm B (a retailer) with packaging services for a rate of $5500 per week.[3] Suppose also that firm B is the only retailer in the area willing to sell meat in controlled atmosphere packages. The transportation infrastructure is such that alternative markets are currently beyond the reach of firm A. Assume that the fixed cost (purchase price) of the machine, depreciated over a number of years, is $3000 per week and the variable costs of operating the machine are $2500 per week.[4] Firm A will only make this investment if it can negotiate a price with the retailer which is no less than $5500 per week – where it is breaking even on its investment. The specialized appropriable quasi-rent on the packaging machine is $3000, which is the revenue required to cover total cost ($5500) minus operating costs ($2500). Once firm A has made the investment and installed the machine, firm B could act opportunistically to capture the quasi-rent of $3000 by cutting its offer price for the packaging service to just over $2500. This offer would still cover Firm A's variable costs in running the packaging system. Any additional rent above $2500 would contribute something towards the fixed costs of the machine. Since the investment was specialized to firm B (the only local retailer willing to sell meat in controlled atmosphere packages), Firm A would have little choice but to accept the discounted offer of firm B. Firm B is practising PCOB in appropriating the $3000 quasi-rent from firm A.

Assume now that there is a second retailer, firm C, who will pay a maximum of $3500 rent for the use of the packaging machine. The specialized appropriable quasi-rent which firm B can now capture, given the existence of firm C, is $2000 which is the full rent ($5500) minus the value of the machine in its next best alternative use ($3500). Firm B could still practice PCOB by lowering its price to just over $3500, thereby capturing most of the $2000 quasi-rent.

Also central to TCE is the relaxation of the full or perfect information assumptions of neoclassical theory. Drawing on the economics of
information literature (Stigler, 1961; Akerlof, 1970; Arrow, 1973), TCE
recognizes that many economic exchanges are characterized by incomplete, imperfect or asymmetrical information.[5] Information asymmetries
can lead to opportunistic behaviour in two ways. The first, *ex ante*
opportunism, occurs where information is hidden prior to a transaction.
This is known as adverse selection and was first defined by Akerlof in
his 1970 seminal paper on the market for 'lemons'. Akerlof suggested
that in a situation of asymmetric information a seller may possess
information about defects in a product (e.g. a faulty second-hand car or
lemon) that is not available to the potential buyer. As a result, the seller
can act opportunistically by failing to reveal these defects to the buyer
prior to the transaction. Buyers of second-hand cars always face the risk
that the seller is acting opportunistically, trying to sell them a lemon.
They cannot tell the difference between a good car and a lemon, therefore, both cars must sell at the same market price. At this price owners
of good cars are less likely to sell them on the second-hand market,
whereas selling lemons may still be a viable option. Lemons tend to
self-select the used car market and there is a higher probability of purchasing a lemon in this market. Akerlof used this reasoning to explain
why the price of new cars depreciates so rapidly once they have been
sold. Hidden information can lead to adverse selection and problems of
opportunistic behaviour. In some countries, contract law can provide a
degree of protection to the buyer. In the CEECs and the NISs, adverse
selection may be an acute problem in markets in which there is little
information about the product or in which the characteristics of a product cannot be easily detected by visual inspection.

Moral hazard can also exist when there are informational asymmetries. This is *ex post* opportunism which arises because of the hidden
actions of economic agents. Individual agents have the incentive to act
opportunistically to increase their economic welfare because their
actions are not directly observable by other parties. For example, suppose an insurance company offers fire insurance which covers the
replacement of articles damaged by contact with cigarettes. Since the
policyholders' actions cannot be observed by the insurance company,
they may take less care in the prevention of fires or they may even act
opportunistically by deliberately damaging items which they wish to
replace and claiming the accidental damage against their insurance policies. The insurance company cannot differentiate between the claims
made by people whose possessions genuinely suffered accidental
damage and those claims arising as a result of deliberate damage. This
type of insurance was available for some time in the Netherlands but

was eventually abandoned because of moral hazard problems (Douma and Schreuder, 1992).

To summarize, the existence of bounded rationality, opportunism, asset specificity and the lack of full information leads to transaction costs.

## 4.3. Transaction Costs

Transaction costs are the costs of carrying out a transfer of goods between technically separable phases of production or distribution when we no longer assume information costs to be zero. They arise wherever there is any form of economic organization, be it within a vertically integrated firm, in a market or in the previous command economies of the CEECs and the USSR (in which market transactions were largely absent).[6] Cheung offers the following definition of transaction costs:

> a spectrum of institutional costs including those of information, of negotiation, of drawing up and enforcing contracts, of delineating and policing property rights, of monitoring performance, and of changing institutional arrangements.
>
> (Cheung, 1987, p. 56)

Unlike the frictionless economic system implied by neoclassical theory, TCE recognizes that transactions do not occur without friction and labels the costs which arise from the interaction between and within firms as transaction costs.

We can divide transaction costs into three main classifications: (i) information costs; (ii) negotiation costs and; (iii) monitoring and enforcement costs. Information costs arise when we drop the neoclassical assumption of perfect information and recognize that economic agents face costs in the search for information about products, prices, inputs and buyers or sellers. Market institutions which reduce these information costs have evolved in modern market economies. If one needs to locate alternative suppliers or potential buyers, there are numerous sources of information to which one can turn – the yellow pages, industry associations, local chambers of commerce, etc. Price information is available in many forms – agricultural commodity prices are published regularly in the farming press, in local newspapers, etc. A variety of industry associations collate and publish information on prices and market trends. In contrast, in the CEECs and the NISs, economic agents may face considerable information costs in simply locating and assessing suitable suppliers, in discovering reliable price information, in finding products of a suitable quality, etc. Immediately following liberalization, Moscow did not have a telephone directory. Private firms now compile lists of telephone numbers to sell to other firms. Clearly, the information costs involved in locating other firms, much less in establishing their suitability as trading partners, are far higher than in modern market economies which have well-developed information infrastructures.

Negotiation costs arise from the physical act of the transaction, whether it be negotiating and writing contracts (costs in terms of managerial expertise, perhaps the hiring of lawyers, etc.), or paying for the services of an intermediary to the transaction (such as an auctioneer or a broker). Although these costs still exist in modern market economies, negotiation costs are reduced by a variety of market institutions. For example, financial transactions are facilitated by efficient banking services. Many transactions in former command economies are now carried out on a cash-on-delivery basis because the financial instruments common in developed market economies are not yet available or are unreliable. Private businesses have developed more rapidly in some of the CEECs than in others. For example, Poland has experienced growth in the small business sector. The relative success of these small businesses is possible because transactions can still be conducted on a personal level – people know one another, they can more easily establish the reliability of trading partners, more easily agree upon and enforce the terms of a transaction, etc. As the size of businesses expand, transactions are carried out with a larger number of economic agents and negotiation costs rise. Increasingly, the business must rely on market institutions for information and for the regulation of transactions.

Monitoring or enforcement costs arise *ex post* to the transaction. These costs arise from monitoring the quality of goods from a supplier or monitoring the behaviour of a supplier or buyer to ensure that all pre-agreed terms of the transaction are met. Also included are the costs of legally enforcing a broken contract, should the need arise. High monitoring and enforcement costs may be another reason for the use of cash-based transactions in the CEECs and the NISs, i.e. the difficulty of securing payment when the legal and financial institutions common to market economies do not exist or do not function effectively. If the terms of a contractual agreement are not adhered to by the other party, economic agents in market economies can often turn (or threaten to turn) to litigation. Economic agents in many economies in transition cannot yet rely on legal institutions to enforce contractual agreements.

The importance of informational asymmetries should be apparent; in a world of perfect costless information, the allocation of resources to the enforcement of contracts would be unnecessary. The potential of economic agents to renege on a contract would be known in a perfect information world, as would the outcome of any legal processes. All possible outcomes of cheating could be calculated in advance and the results of all transactions would be known with certainty (Eggertssen, 1990). In reality, information is not costless, informational asymmetries exist and transaction costs arise as a result of the movement of goods through the various stages of production, processing and distribution.

## 4.4. The Relationship Between Transaction Costs and Vertical Coordination

### 4.4.1. Characteristics of transactions

The significance of TCE lies in its insights into the effect of transaction costs on the organization of economic activity, or in other words, on vertical coordination. According to TCE, one of the determinants of the choice of vertical coordination is the nature and level of transaction costs, wherein, a change in the transaction costs arising from the exchange of a product may lead to a change in the means used to accomplish the vertical coordination of that product. The concept of vertical coordination was introduced in Chapter 2, where it was explained that vertical coordination varies from spot markets on the one hand to full vertical integration on the other and includes all intermediary methods of coordinating transactions, e.g. contracts, strategic alliances, quasi and tapered vertical integration. Williamson (1979) identifies the core characteristics of transactions as: (i) the degree of uncertainty surrounding the transaction; (ii) the degree of asset specificity, and (iii) the frequency with which transactions occur. He links these characteristics to vertical coordination outcomes.

A low level of uncertainty lends itself to spot market transactions. In situations in which aspects of the transaction (such as quality characteristics) are highly uncertain, however, a more formal type of vertical coordination where one party can have more control over the outcome of the transaction may result, e.g. a contract or vertical integration.

Goods that are non-specific in nature, or produced with non-specific assets, have many alternative uses and would tend to be sold in a spot market. As asset specificity increases, we move along the spectrum of vertical coordination towards vertical integration. Whether the result is a long-term contract or full vertical integration may depend on whether one party, or both make an asset specific investment.[7]

For transactions carried out frequently, both buyer and seller will probably value repeat business and will not wish to tarnish their reputations by acting opportunistically. Frequent transactions also provide buyers and sellers with information about one another. For these reasons, transactions repeated frequently, *ceteris paribus*, tend to be carried out in the spot market (this would also mean lower negotiation costs than, say, a contract). As transactions become more infrequent, however, the incentive to act opportunistically and exploit any informational asymmetries that may be present increases and we move further along the continuum of vertical coordination towards the extreme of vertical integration. The ultimate effect of uncertainty, asset specificity and frequency on governance structures (vertical coordination) depends crucially on the particular combination of these characteristics. Table 4.1 illustrates how the asset specificity and uncertainty characteristics of a transaction influence vertical coordination.

Where asset specificity is high for only one party to a transaction, that party is likely to integrate vertically with the other party, whether or not uncertainty is also present. If asset specificity is high for both parties, vertical integration may result in situations of uncertainty, whereas a long-term contract may be sufficient if uncertainty is low. When asset specificity is low for both parties and there is little uncertainty, the spot market is likely to be the governance structure that prevails. A transaction which involves low asset specificity on both sides but a high level of uncertainty may be carried out under a contract if it is an occasional transaction or through vertical integration if it is highly frequent. The level of asset specificity is the crucial factor here. As indicated earlier, if asset specificity and uncertainty were both low, highly frequent transactions would tend to be carried out in the spot market.

Similarly, we can combine the frequency aspect of a transaction with the extent to which an asset specific investment has been made and draw conclusions about the expected governance structure. Table 4.2 illustrates this and provides examples of the type of transactions with these two characteristics.

The purchase of standard raw materials for a production process occurs frequently and is a non-specific asset. We would expect this transaction to occur in the spot market, as we would occasional purchases of standard equipment. On the other hand, the transfer of an idiosyncratic intermediate product across successive production stages, such as the supply of a unique type of iron-ore to a steel smelting plant, is highly asset specific (assuming no other steel plants in the vicinity) and is carried out frequently, suggesting that full vertical integration of the two stages would be the appropriate form for vertical coordination.

The remaining entries in Table 4.2 represent other types of contractual relationships between separate firms. For example, customized

**Table 4.1.** The relationship between characteristics of transactions and vertical coordination: case 1 – asset specificity and uncertainty.

| Uncertainty | Asset specificity | | |
|---|---|---|---|
| | Low for both parties | High for both parties | High for one party, low for other party |
| High | (Depends on frequency) | Vertical integration | Vertical integration |
| Low | Spot market | Long-term contract | Vertical integration |

Source: Douma and Schreuder (1992), p. 142.

**Table 4.2.** The relationship between characteristics of transactions and vertical coordination: case 2 – asset specificity and frequency.

| Transaction frequency | Asset specificity of transaction | | |
|---|---|---|---|
| | Non-specific | Mixed | Idiosyncratic |
| Occasional | Purchasing standard equipment (spot market) | Purchasing customized equipment (contract) | Constructing a factory (contract) |
| Frequent | Purchasing standard material (spot market) | Purchasing customized material (strategic alliance, quasi or tapered vertical integration) | Site-specific transfer of intermediate product across successive stages (full vertical integration) |

Source: adapted from Williamson (1986) p. 112, p. 117.

equipment purchased on an occasional basis means that the transaction to some extent involves a specialized investment. There is an interest in sustaining the business relationship to recover any set-up costs that may already have been incurred by one of the parties. In these circumstances, we might expect to see a long-term contract between the two parties to the transaction. This case may also involve the use of arbitration to settle any contractual disputes rather than relying on litigation which, in certain business relationships, might damage the prospects for future transactions (Williamson, 1979).

Purchasing customized material would occur on a frequent basis and would also involve a degree of asset specificity. In this situation we might expect a strategic alliance, or some form of quasi-integration such as a joint venture or tapered integration. The final outcome may depend on the level of uncertainty present. This would reduce the incentive for

either party to act opportunistically but would still leave the relationship flexible enough to adapt to changes in the market environment
given that the investment is not totally specialized.

Finally, the construction of a factory represents an idiosyncratic
infrequent transaction. Despite the fact that a highly specific investment
has been made, the once-only nature of the transaction precludes vertical integration as the method of vertical coordination. As with the
occasional purchase of customized equipment, a contract would be
more likely. Thus, it can be seen that different types of transaction costs
will lead to different vertical coordination arrangements.

### 4.4.2. Transaction costs and valuing product quality

Chapter 3 discussed quality issues in relation to pricing. This section
deals with the implications of quality issues for transaction costs and,
by implication, for vertical coordination in an industry. In addition to
being affected by the type of transaction, the vertical coordination outcome is affected by changes in transaction costs. For example, the addition of a quality characteristic to a product which cannot easily be visually assessed by buyers will raise the information costs of that
transaction because buyers cannot detect the characteristic or measure
its value. This might be illustrated by the addition of a high animal
welfare standard or disease-free characteristic to beef which cannot be
detected visually by cattle buyers. Other methods of demonstrating the
presence of this characteristic must be used, such as evidence that the
animal was reared subject to the guidelines of an accredited animal welfare or disease prevention scheme.

An insightful discussion of the information costs involved in measuring the value of a good is provided by Barzel (1982). He argues that
the costs of valuing the non-visible attributes of a product are particularly high because the valuation is subject to measurement errors and
that there is a difference between the price of a good and the value of
that good to an individual buyer. One of the parties to the transaction –
either the buyer or the seller – will face sorting costs in establishing the
value of the good. As a result of informational asymmetries, suppose
the seller has more information about the true value of the product than
the buyer. For example, in determining the value of oranges at the retail
level, the seller may possess information about how they have been
produced, transported and stored, whereas the buyer (the consumer in
this case) is unlikely to have this information. The consumer incurs
sorting costs in picking over the oranges in an effort to ascertain through
visual inspection of the oranges, which ones are good and which ones
are bad.

Furthermore, oranges are often sold on the basis of weight, yet one of the characteristics of interest to the consumer may be the juiciness of the orange; weight is only a proxy measure of the true value of the orange to the consumer. In this way, errors of measurement arise and a divergence exists between the price of a product and its valuation by a consumer. The amount of a product purchased will increase as the demand for that good increases and as the price of the good falls. When measurement costs are introduced into the analysis of a transaction, we are also saying that the amount purchased will increase as the cost of sorting falls and as the variability in the quality of the good increases. This is because there is no additional penalty in inspecting a poor quality item but there is an additional benefit in finding a high quality item; a product whose quality is variable at least offers buyers the potential of discovering high quality items (Barzel, 1982).

Each buyer must go through the same sorting process when making a purchase decision and incur the same sorting costs. When buyers incur the costs of value measurement, a particular product may be sorted a number of times. If the task of measuring the value of a product is difficult, meaning that measurement costs for the buyers are high, the net price (the posted price net of the costs of measurement) which they are prepared to pay for the commodity will be reduced. When the seller incurs the sorting costs, however, products will be measured just once. In circumstances where it is difficult to ascertain the true value of a product through visual inspection, it is in the seller's interest to incur the sorting costs and provide the buyer with a guarantee as to the true quality of the product. This is one of the roles of a grading system for agricultural products. For example, eggs are often graded according to size or beef carcasses graded on the basis of conformation and fat content. Grades can be private, or as in the examples above, established by a public body, i.e. a national (or supra-national, e.g. European Union) government. Grades reduce buyers' sorting costs.

In the newly liberalized agricultural markets of the CEECs and the NISs, buyers now incur sorting costs. Under the command economy system, neither firms nor consumers could exercise freedom of choice. Firms had designated suppliers and designated buyers. Retail outlets were limited in number and scope. There was little, if any, opportunity for sorting goods to obtain a higher quality and prices did not reflect quality or the true worth of the good. If apples were available in a store in any particular week, consumers purchased apples because they might not have been available the following week. Meat was often simply apportioned to queuing consumers in the order in which the cuts came off the carcass. With the move towards open market economies and the use of price as an indicator of value, sorting costs and the problem of measuring the true value of a good become important.

There are a number of steps that an individual seller can take to reduce the buyer's incentive to measure the value of a product, and hence, to reduce a buyer's information (transaction) costs. For example, a seller can offer a warranty with the product whereby goods that prove to be defective are later replaced. This is applicable in situations where the true value of a product becomes immediately apparent when it is used by the buyer. It may be extremely costly for a seller to check every aspect of every product for defects; a warranty allows the buyer to do this but protects the buyer from loss. An example is the warranty issued with new cars or electrical items.

An alternative method of reducing a buyer's sorting costs is through the use of brand names. The intention is to persuade a buyer to rely on a seller's reputation as an indication of the quality of a product. Branding works on the premise that the loss to a seller arising from the sale of a substandard product would be substantial since there is a risk that the entire brand name will be tarnished and the seller's investment in establishing the brand name lost.

> A canner known to change the quality of peas (e.g. size, tenderness, sweetness) from one season to another will induce buyers to conduct a fresh, costly test every season. If, on the other hand, the canner is known to maintain tight quality control, much less testing is required. The canner's reputation, or brand name, serves here to guarantee that the product is, and will remain, uniformly good.
>
>                                                    (Barzel, 1982, p. 36)

The use of warranties and brand names to create trust between the consumer and the seller means that the buyer will be willing to pay a higher average price for uninspected products because less of their time and resources were spent on sorting prior to the sale; as a result the buyer has incurred lower transaction costs.

A quality assurance scheme[8] works in a similar way. Sellers incur costs in maintaining uniform quality standards and in establishing the reputation of the quality assurance mark. This investment induces buyers to trust the quality assurance mark as a guide to the true value of the product. A quality assurance scheme under which quality is highly variable will quickly lose its value and fail to reduce buyers' sorting costs.

Firms also incur sorting costs in measuring the quality of inputs to their production processes. When the value of some inputs can be established relatively easily through visual inspection, information costs will be low. However, if inputs possess characteristics that are not easy to establish through visual inspection, for example, where the quality of an input is affected by the production processes used further back in the production chain, costs of measurement will be high. High measurement costs may occur at a number of stages along the production chain if several firms are involved in production; firms further down-stream incur information costs in measuring the quality of inputs that may have already been measured by previous firms upstream in the chain. An example could be a food product produced to new food safety standards. This food safety characteristic is impossible to verify visually, hence, downstream firms incur costs in monitoring the production practices of upstream firms and/or subjecting the output of these firms to various laboratory tests to establish the food safety claims of the seller. If the production process extends over a number of firms, it is likely that the same input tests will be carried out by consecutive firms. Duplicate testing represents an increase in transaction costs.

Alternatively, the process could be carried out within a single vertically integrated firm; this would reduce the need for duplicate testing of inputs. When the marginal costs of measurement from successive inter-firm transactions exceed the marginal costs of within-firm measurement, vertical integration should emerge as the dominant form of vertical coordination in that market. Thus, for commodities such as fresh vegetables which involve little processing before sale, ownership is likely to change frequently because of the low costs of measuring the value of the product. Commodities whose form has undergone a large degree of processing, for example bread, may change hands less often, tending to be processed within firms which are vertically integrated to some degree.

To summarize, transaction costs arise when there are costs to measuring the value of a good. Often, these are information costs incurred *ex ante* to the transaction. In the case of intermediate goods exchanged on a frequent basis, monitoring costs arise because of the need to measure the value of successive inputs to a production process. Economic agents will take steps to alleviate these measurement costs, such as through the use of warranties, brand names and quality assurance guarantees to reduce the sorting costs of consumers, or vertical integration to reduce the monitoring costs of firms. Barzel observes that:

> The problems and costs of measurement pervade and significantly affect all economic transactions. Errors of measurement are too costly to eliminate entirely. The value of equally priced items will differ, then, and people will spend resources to acquire the difference. Such resource expenditure is wasteful, and it is hypothesized that exchange parties will form such contracts and engage in such activities that reduce this kind of resource use.
>
> (Barzel, 1982, p. 48)

## 4.5. A Critical View of Transaction Cost Economics

One of the problems with TCE is that its theoretical development was not accompanied by the successful measurement of transaction costs.

This is perhaps not surprising, since, unlike physical production costs, transaction costs (the costs of economic organization) are often not easy to separate from other managerial costs[9] or readily measurable. The complex nature of economic institutions means that the costs of their operation are not easy to quantify and the data which one might use to measure transaction costs are not usually collected by governments or by the standard accountancy practices of firms. Although one can recognize that there are indeed costs involved in valuing a good or in monitoring the actions of a buyer or seller, it is difficult to measure these costs in financial terms. Until, and indeed if, the data necessary for this type of estimation is ever recorded by firms, an accounting approach to empiricizing the transaction cost approach is impractical and, therefore, not particularly useful.

Economists have turned instead to other ways of empirically analysing transaction costs. Empirical investigations of TCE, its application to agribusiness and those studies relevant for the transformation of the agrifood sectors of the CEECs and the NISs are discussed in the Appendix to this chapter. A further hurdle to the wider acceptance and use of TCE is that it has lacked clear definitions of both the concepts and terminology central to the theory. The theory developed as a gradual evolution from Coase's original 1937 paper, with contributions by many different economists. Coase did not set out in 1937 to present a theory of transaction cost economics, so that the concept of transaction costs was initially extremely vague and it has been left to other economists to provide more specific definitions and develop a coherent theory. As Posner observes:

> Coase's influence in economics has been diminished by the fact that his articles do not speak the language of modern economics, which is mathematics, and by the fact that he has not attempted to develop a *theory* of transaction costs.
>
> (Posner, 1993, p. 207)

Even a casual glance at the current literature reveals that a commonly accepted definition of transaction costs has yet to emerge. Definitions of transaction costs are all fairly similar and it remains for the individual researcher to be clear as to the definition being used. This can be extremely confusing for those new to the transaction cost literature and has led to the complaint that almost anything can be explained by appropriately defining problems as transaction costs (Fischer, 1977). This problem is probably caused more by shortcomings in the presentation of the transaction cost approach to a sceptical economics profession than by failings of the theory itself. The work of Oliver Williamson in developing a theory of transaction cost economics has helped to alleviate some of these criticisms. Much of Williamson's work involved a careful specification of the distinguishing features of a transaction and the nature of governance structures – a term he introduced to the literature (Williamson, 1979).

The introduction of new concepts often leads to confusion. For instance, a foundation of Williamson's transaction cost theory is the assumption that individuals behave opportunistically, which has been criticized as an unnecessarily negative view of human behaviour. This criticism is mistaken, since it is only necessary for some individuals to act opportunistically some of the time for the risk of opportunism to be real. It may have been because TCE represents a departure from the familiar neoclassical theory of the firm, thereby necessitating the definition of a myriad of new concepts, that it has been misunderstood in the past. This problem should diminish as economists become more familiar with TCE and as advances in the application of the theory demonstrate its usefulness.

The important contribution which TCE makes is in the identification of different types of transaction costs and in an explanation of their impact on vertical coordination. In modern market economies, market institutions have evolved to minimize transaction costs. In the newly emerging markets in the CEECs and the NISs, a failure to understand and take account of high transaction costs when providing advice or initiating policy reforms can set in motion changes whose outcome will be inaccurately forecast. At best, the process of transition will be considerably lengthened and, at worst, it will be thwarted.

## 4.6. Conclusions

The long and detailed review of TCE in this chapter has been provided for two reasons:

**1.** Questions of market efficiency and the industrial organization of the agrifood system lie at the very heart of the transition process. Every aspect of transaction costs discussed above has direct application to the problems being experienced in transforming agribusinesses in the CEECs and the NISs.
**2.** Agricultural economists and others interested in reform of the agribusiness sector in the former command economies tend to be less familiar with the theory of transaction costs than other areas of economic analysis. Hence, it is important to lay out formally its main theoretical tenets, to familiarize readers with the constraints which have limited the usefulness of TCE and to outline its major criticisms.

Only in this way can the later chapters, which draw heavily on TCE, be put fully into context.

Many of the problems facing former command economies as they allow market forces to begin to direct economic activity stem from the existence of high transaction costs. It is important for firms and governments operating within the CEECs and the NISs and for those providing advice to these countries to identify high transaction costs and to understand how and why they arise. Transaction costs create friction in business exchanges. Smoothly functioning markets, in which the only important features of a transaction are that demand equals supply giving an equilibrium price and quantity, exist only in the stylized world of the introductory microeconomics textbook. In reality, there are costs to using the market mechanism. These costs influence how a transaction is carried out. The unfettered forces of supply and demand are still important in establishing market clearing prices, however, economists, business people and governments should also understand the importance of transaction costs in shaping the organization of economic activity. Policies and market institutions are required which reduce high transaction costs and/or create the necessary economic incentives to guard against opportunistic behaviour. If these policies or institutions are not put in place, the costs of transacting through the marketplace may become prohibitive. If, at the margin, the costs of using the price mechanism exceed the costs of directing that economic activity within the firm, then *ceteris paribus*, vertical integration will result. One danger is that massive privatization and liberalization without the necessary institutions, laws and policies which enable markets to function effectively will simply result in the creation of monopolies due, in part, to high transaction costs. The problem of monopolies is discussed in the next chapter.

# *Appendix to Chapter 4*

## 4A.1. Empirical Applications of Transaction Cost Economics – Data Limitations

A range of different methodological approaches have been used to indirectly measure the effects of transactions costs. Some of these evaluate the effect of transaction costs on vertical coordination (primarily vertical integration) across several industries (for example, Levy, 1985; Lieberman, 1991). Others are industry-specific, for example, Globerman and Schwindt (1986) and Leffler and Rucker (1991) analyse the organization of transactions in the forest products and forestry industries. Another industry-specific application of TCE is Hallwood's (1990) analysis of the organization of the offshore oil production industry in the UK.

Evaluating the determinants of vertical coordination on a general, non-industry specific level is extremely difficult due to data limitations. The available published data sources often require the researcher to construct indirect and potentially confusing proxy measures of transaction costs. For this reason, most empirical work regarding transaction costs has been carried out on an individual industry level, or even more specifically, on a case study basis. This approach has been criticized on the grounds that the results are not necessarily representative of the wider economic environment (Levy, 1985; Matthews, 1986). However, for a credible empirical alternative to the measurement of transaction costs to be developed, it would require that the necessary transaction cost information be collected on a routine basis. Currently it is not. As Coase has lamented:

... just as important at the present stage would be the gathering in a
systematic way of new data on the organization of industry so that we can
be better aware of what it is that we must explain.

(Coase, 1972, p. 69)

Attempts to apply the transaction cost paradigm empirically range
from general applications to those that are specific to a particular industry. Data limitations are overcome either by the use of proxy variables
drawn from published data sources or by the collection of primary data
through the use of surveys. The latter appear to yield the most fruitful
results. The following section will describe some of the applications of
TCE to agricultural issues.

## 4A.2. Agricultural Applications of Transaction Cost Economics

The transaction cost framework has been applied to only a limited
degree to agricultural questions. These applications can be divided into
three main groups: (i) studies of land ownership, share-cropping and
wage contract issues; (ii) applications to agriculture in developing
market economies; and (iii) applications to vertical coordination in agricultural marketing channels. These questions all arise in the CEECs and
the NISs. Examples of each of these are given below.

### 4A.2.1. Applications to land ownership issues

Land ownership and the structure of agricultural land holdings is a contentious yet vitally important issue in the CEECs and the NISs. Private
land was often appropriated into large state or collective farms. The
decisions over how to divide land among private citizens after liberalization – whether it should belong to the families of former owners, to
existing farm labourers, etc. and the creation of land markets all have
transaction cost implications. These issues are discussed in more detail
in Chapter 7.

There have been a few applications of TCE to agricultural land-
ownership issues. Economic historians have turned to transaction costs
to explain the survival of the apparently inefficient open-field system of
agriculture in Northern Europe throughout the Middle Ages. Under this
system, peasant farmers cultivated their land in small unfenced strips
within the village; a practice which many economic historians have criticized as wasteful and inefficient (McCloskey, 1972). However, it has
been suggested that the open-field system of scattering a peasant's land

into dispersed strips acted to reduce the risk of crop damage due to adverse weather conditions or to reduce the risk of cultivating poor quality land (McCloskey, 1986). The peasants exchanged a lower average income for a reduced variance in income. Other less expensive methods of insurance were unavailable due to high transaction costs which meant that, for example, storing grain and trading between villages to offset the effects of unpredictable local harvests was not a viable option because of high interest rates, poor storage technology and only rudimentary transportation infrastructure. These are common problems in some CEECs and the NISs.

Another explanation of the survival of the open-field system which also relies on transaction cost reasoning is provided by Dahlman (1980). Villagers benefited from economies of scale in grazing their animals on the common land of the open-field system. Dahlman argues that it was necessary for the peasants' land to be scattered in individual strips to avoid the danger of one farmer expropriating quasi-rents from other farmers by refusing grazing rights to a large piece of land. Particularly, if one farmer owned a strategic piece of land that allowed access to other fields, the farmer could behave opportunistically by refusing to join the common grazing unless paid a high rent. Worries about similar strategic use of assets might explain some of the reluctance to abandon large scale farming units in favour of individual farms in the CEECs and the NISs.

The disappearance of the open-field system in the late 18th and early 19th centuries suggests that the system eventually lost its transaction cost advantage. Technological change and expanding markets placed a strain on the decision-making process of collectively organized villages resulting in prohibitive increases in the cost of reaching an agreement among poorly educated peasant farmers. Eventually the open-field village was replaced by a smaller, more flexible unit – the modern capitalist farm (Eggertssen, 1990).

Transaction cost theory has been applied to share-cropping as an agricultural land ownership issue. Under a share-cropping system, a landowner leases land to the tenant in return for a share of the crops. Exogenous factors such as climatic variation affect yield. This risk is borne by the land owner if control of the means of production is retained and a wage contract used, whereas under a fixed rent contract, the tenant bears the risk of yield variation. Share-cropping contracts share this risk. If tenants are highly risk averse they will avoid fixed rent contracts; this may explain the existence of share-cropping arrangements in many developing countries since the consequences of a bad harvest can be particularly devastating for very poor farmers. Extensive work on the issue of share-cropping has been carried out by Cheung (1969). He suggests that the choice of contractual arrangement is made so as to maximize the gains from spreading risk subject to the constraint of transaction costs.

Chew (1993) has also used a transaction cost framework to investigate the economics of share-cropping arrangements. Much of the work on share-cropping makes use of the principal-agent theory. When the actions of the agent (tenant) cannot easily be observed by the principal (landowner), i.e. monitoring costs are high, share-cropping arrangements result. Chew (1993) observes that share-cropping often occurs when the marginal product of monitoring is low compared with owner cultivation, for example, in areas of low soil fertility where the value of output is low and, therefore, the potential returns to monitoring are low. This may explain why share-cropping tends to decline and wage contracts increase following technological innovations that raise output value, making closer monitoring economically viable. Many applications of TCE to the share-cropping issue tend to be conceptual and/or anecdotal in nature with little attempt, perhaps due to data limitations, to carry out an empirical analysis.

In a somewhat related approach, empirical studies have been made of the structure of wage tenancy contracts in US agriculture (Alston and Higgs, 1982). These studies have achieved some success in predicting marginal changes in the contractual mix both over time and between regions. Emphasis has been placed on contractual arrangements between the farm unit and input suppliers. Modelling these contracts is very complex due to the diverse range of contractual arrangements used, even on an individual farm, for the purchase of inputs and disposal of outputs. For this reason, research has tended to focus on the use of land as an input to the production process. This has been accomplished through studies of the relationship between the value of land and the frequency of wage contracts or fixed rent contracts relative to share-cropping contracts (Eggertssen, 1990). As property rights and markets for agricultural land have been slow to develop in the CEECs and the NISs, one might expect a range of contractual relationships to evolve.

### 4A.2.2. Applications to agricultural development

An institutional approach to problems of economic development has expanded the use of TCE. The fact that institutional costs have often been ignored by agricultural development economists is noted by Thomson and Smith (1991). They argue that the inclusion of transaction costs in an analysis allows for a more comprehensive appraisal of the role of the state (as one of the institutions operating in a market system) in agricultural development. They advocate an approach which identifies the levels and types of costs associated with providing goods and services through alternative vertical coordination mechanisms, including through the market or by the government. This might involve obtaining estimates of transaction costs or, if that were not feasible, measuring the impact of a reduction in transaction costs. They observe that:

> The whole issue of transaction cost measurement is quite undeveloped, in spite of its potential importance. Economists have tended to comment on the issue in passing, rather than undertake serious analysis on the incidence and magnitude of transaction costs.
>
> (Thomson and Smith, 1991, p. 13)

They discuss the impact of transaction costs on the liberalization of agricultural markets in the context of the reforms of the cereal marketing policy in Mali in the 1980s. An analysis of both the transaction costs and the transformation costs[10] which arose under alternative levels of state intervention in the marketing of cereals suggests that the reforms mainly resulted in switching the burden of these costs between groups rather than a reduction in the overall costs. Several authors have stressed the importance of attention to transaction cost inefficiencies in any analysis of liberalizing command economies (see for example, Cheung, 1992; Hobbs *et al.*, 1993; Williamson, 1994).

### 4A.2.3. Applications to agricultural marketing channels

There has been much discussion of vertical coordination in agricultural marketing, with references to TCE occurring to a greater or lesser degree. Some economists have focused primarily on vertical integration in agriculture. Kilmer (1986) develops a methodology to project the extent of backward and forward vertical integration at various levels of the agrifood chain. This work acknowledges the transaction cost approach but does not use it explicitly. Rather than focusing on vertical integration as a single method of organizing transactions, however, many other writers base their analyses on the wider concept of vertical coordination.

One seminal paper by agricultural economists regarding vertical coordination in agribusiness is provided by the insightful and oft-quoted study by Mighell and Jones (1963). Although too early to benefit from the formal development of TCE, the authors approach the study of vertical coordination in agriculture by discussing the concept of the firm and differentiating between transactions carried out by managerial direction and those carried out in the marketplace. Economic efficiency, risk, uncertainty and financing agricultural activities are all discussed in relation to vertical coordination. Subsequent applications of TCE to agricultural marketing made use of many of these insights.

That the development of transaction cost economics has provided a theoretical base for examining vertical coordination in agriculture has been acknowledged by a number of agricultural economists. Barry *et al.* (1992) discuss these theoretical developments within the context of financing different methods of vertical coordination in agriculture. Sporleder (1992) discusses a range of vertical coordination alternatives including contracts, cooperatives and vertical integration and compares the orthodox neoclassical view of the firm with other views which focus on transaction costs or on strategic alliances as a particular form of vertical coordination.

Barkema (1993) describes how vertical coordination within the US food system is evolving, with greater use of contracting and vertical integration. Using the pork industry as a case study, he suggests a number of explanations for these growing trends including increased concentration at the producer level, new production technologies and the desire to reduce risk. Central to his descriptive analysis are many of the concepts of TCE. Sonka (1992) highlights the importance of information technology in reducing information and monitoring costs by improving the flow of information between different stages of the agri-food chain.

Masten (1991) provides a useful synthesis of TCE, suggesting its applicability to the analysis of vertical coordination in agricultural markets. In particular, he discusses how the breach of contracts can be endemic when asset specificity is present. In addition to physical, human and site specificity, in which capital is designed or located for a specific user with little salvage value, he identifies temporal specificity. Temporal specificity arises when the timing of transactions is important so that the negative consequences of any delays are exacerbated. Highly perishable agricultural commodities are subject to temporal specificity. When perishable goods are ready for market, they must be moved rapidly through the production–distribution system regardless of the offer price. This leaves the seller vulnerable to the risk that the buyer will behave opportunistically by attempting to capture the quasi-rents created by the temporal specificity of the product. Masten provides agricultural examples of the problems of temporal specificity, for example the existence of producer cooperatives and forward contracts between growers and processors of fruit and vegetables. The grower has a highly perishable crop and if the processor were to fail to purchase the crop it may be lost altogether or a hurried sale on the spot market arranged, probably at a discount and involving deterioration of the crop. On the other hand, perishability prevents processors holding stocks of the raw product to guard against the risk of the grower failing to supply the processor. Since both parties are at risk from these types of delay, forward contracts and agricultural bargaining cooperatives are widespread in fruit and vegetable transactions (Knoeber, 1983; Masten, 1991). In the CEECs and the USSR, losses from spoilage of perishable agricultural commodities were very large during the command era and continue to be large. Market institutions have not yet developed which can significantly reduce these losses.

Lang (1980) presents a number of case study examples of how collective bargaining affects vertical coordination through its influence on the negotiation between producers and processors of agricultural commodities. For example, sugar beets often deteriorated because of shrinkage whilst lying in a processor's receiving yard waiting to be processed. Despite the fact that the deterioration occurred after the product was delivered, growers were charged for any shrinkage losses, leaving processors no incentive to reduce shrinkage. As a result of this problem, a growers association negotiated a maximum producer penalty for shrinkage so that processors bore the cost of any shrinkage losses above five per cent. The processors' reaction was to improve storage facilities for the sugar beets. Lang argues that the existence of information asymmetries in marketing perishable agricultural commodities means that joint costs are minimized when they are borne by the party possessing the most information. Collective bargaining alters the contractual relationship between grower and processor leading to a reallocation of resources.

Although providing sound logical reasoning and anecdotal evidence of the importance of transaction costs in determining vertical coordination, the studies discussed above do not attempt an empirical validation of their hypotheses. As with applications of TCE to other sectors, agricultural applications have been hampered by data limitations.

An empirical analysis of the effect of transaction costs on vertical coordination in US food industries is provided by Frank and Henderson (1992). Working at a fairly disaggregated level across several food industries, they developed a vertical coordination index to measure the extent of integration in an industry. The index has two parts, firstly it includes an input–output transactions matrix to calculate interdependencies between firms in an industry. Secondly, it includes a measure of the degree of administrative control over transactions. This is measured by the percentages of farm commodities procured through spot markets, contracts or through vertical integration. The resulting vertical coordination index ranges from zero (spot markets) to one (vertical integration). The index is regressed against proxy measurements of transaction costs divided into four categories: uncertainty, industry concentration, asset specificity and the costs of administering vertical coordination:

**1.** As uncertainty rises we expect integration to increase because it is no longer possible to fully specify all contingencies in a contract. The uncertainty surrounding food manufacturers' input supply is measured by the percentage change in farm output supply between 1981 and 1982.

**2.** Increasing concentration leads to a small numbers bargaining problem. A four firm concentration ratio measures concentration in food manufacturing industries.

**3.** As asset specificity rises, an increase in non-market vertical coordination is expected. Variables used to measure this include an advertising to sales ratio and a research and development expenditure to sales ratio for food industries.

**4.** As the costs of administering vertical coordination rise, a reduction in vertical integration is expected. The authors identify a number of characteristics of firms which affect the internalization of transactions (e.g. capital intensity) and use proxy measures of these characteristics (e.g. a capital to sales ratio) to measure the costs of administering vertical coordination.

Although not all of the variables used were significant, the authors conclude that transaction costs are a primary determinant of vertical coordination. The influential transaction costs are identified as those related to uncertainty, input supplier concentration, asset specificity and scale economies. Due to the lack of adequate data, Frank and Henderson (1992) resorted to the use of industry characteristics as indirect proxy measurements of transaction costs.

Hobbs (1995, 1996a) used TCE to analyse the UK beef marketing chain. Empirical estimates of the importance of a number of transaction costs in farmers' selection of marketing channels[11] were obtained. The beef purchase decision of major supermarket retailers and the selection of a beef supply channel by UK beef processing firms were also analysed. Data for these three empirical analyses of transaction costs were collected through surveys of farmers, processors and retailers.[12] There are many pitfalls to the use of primary survey data to measure transaction costs. The subtle nature of these costs means that great care must be taken in the design of questions. However, until firms, industry associations and governments routinely collect the type of information necessary for the measurement of transaction costs, cooperation between researchers, firms and governments in the provision of survey data is required. This may be the only way of obtaining empirical predictions of the effects of transaction costs on vertical coordination.

# Notes

1. Of course, as discussed later, neoclassical theory has been successfully extended to cover monopolies and, with less success, to other intermediate forms of industrial organization such as monopolistic competition and oligopoly.
2. The term transaction cost was not coined until a later date and is usually attributed to Arrow (1970).
3. This illustration is based on an example given by Klein *et al.* (1978).
4. Assume the salvage value of the packaging machine is zero.
5. Information incompleteness and uncertainty refer to the situation where all economic agents face the same, but incomplete, levels of information. Therefore, they all face the same uncertainty. Information asymmetry arises when there is public information available to all agents but also private information which is only available to selected agents, so that economic agents no longer possess the same levels of information.
6. Note, including within-firm transfers across technically separable phases of production broadens the definition of a transaction beyond the common English use of the term. In common English usage a transaction takes place between an organizationally separate seller and buyer.
7. If only one party (A) has made an asset specific investment, it is probably party A that will vertically integrate with the other party (B) because of the risk of B opportunistically recontracting. Where both parties have made an asset specific investment, either vertical integration or a long term contract may result. The outcome will depend on other features of the transaction, such as the level of uncertainty.
8. Under a quality assurance scheme, firms adhere to widely recognized and often independently established production, processing and distribution standards.
9. The problem is analogous to the difficulty accountants have in assigning costs of jointly used assets to individual enterprises in a multiple enterprise firm. For example, a tractor may be used by a farm's beef enterprise to move feed, by the farm's cereal enterprise for ploughing and harvesting and by a farm's hay enterprise for baling. If the beef enterprise needs to be evaluated then a portion of the tractor's total costs must be assigned to the beef enterprise. This process is always somewhat arbitrary and prone to measurement errors.
10. Transformation costs are defined by the authors as the physical costs of marketing such as transportation, storage and processing.
11. Live-ring auction, electronic auction, direct sale to processor or sale through a cooperative group marketing scheme.
12. Further details of the research can be found in Hobbs (1995, 1996a) or summarized in Hobbs (1996b).

# References

Akerlof, G.A. (1970) The market for 'lemons': qualitative uncertainty and the market mechanism. *Quarterly Journal of Economics* 84, 488–500.

Alchian, A.A. (1965) Some economics of property rights. *Il Politico* 30, 816–829.

Alchian, A.A. and Demsetz, H. (1972) Production, information costs, and economic organization. *American Economic Review* 62, 777–795.

Alston, L.J. and Higgs, R. (1982) Contractual mix in southern agriculture since the civil war: facts, hypotheses, and tests. *Journal of Economic History* 42, 327–353.

Arrow, K.J. (1970) The organization of economic activity: issues pertinent to the choice of market versus nonmarket allocation. In: Haveman, R.H. and

Marglois, J. (eds) *Public Expenditure and Policy Analysis*. Markham Publishing Company, Chicago, pp. 59–71.

Arrow, K.J. (1973) *Information and Economic Behavior*. Federation of Swedish Industries, Stockholm, 284 pp.

Bain, J.S. (1968) *Industrial Organization*. John Wiley and Sons, Inc., New York, 678 pp.

Barkema, A.D. (1993) New roles and alliances in the US food system. Presented at the Spring Meeting of the Federal Reserve System Committee on Agriculture and Rural Development, Kansas City, Missouri, 20 May.

Barry, P.J., Sonka, S.T. and Lajili, K. (1992) Vertical coordination, financial structure, and the changing theory of the firm. *American Journal of Agricultural Economics* 74, 1219–1225.

Barzel, Y. (1982) Measurement cost and the organization of markets. *Journal of Law and Economics* 25, 27–48.

Casson, M. (1990) *Enterprise and Competitiveness: A Systems View of International Business*, Clarendon Press, Oxford, 229 pp.

Cheung, S.N.S. (1969) Transaction costs, risk aversion, and the choice of contractual arrangements. *Journal of Law and Economics* 12 , 23–42.

Cheung, S.N.S (1987) Economic organization and transaction costs. In: Eatwell, J., Milgate, M. and Newman, P. (eds) *The New Palgrave – a Dictionary of Economics*. Macmillan Press, London; Stockholm Press, New York, (2) pp. 55–57.

Cheung, S.N.S. (1992) On the new institutional economics. In: Werin, L. and Wijkander, K. (eds) *Contract Economics*. Blackwell, Oxford, pp. 48–71.

Chew, T.A. (1993) The transactional framework of sharecropping: further implications. *Canadian Journal of Agricultural Economics* 41, 209–221.

Coase, R.H. (1937) The nature of the firm. *Economica* n.s. 4, 386–405.

Coase, R.H. (1960) The problem of social cost. *Journal of Law and Economics* 3, 1–44.

Coase, R.H. (1972) Industrial organization: a proposal for research. In: Fuchs, V.R. (ed.) *Policy Issues and Research Opportunities in Industrial Organization*. National Bureau of Economic Research, New York, pp. 59–73.

Dahlman, C.J. (1980) *The Open-Field System and Beyond: a Property Rights Analysis of an Economic Institution*. Cambridge University Press, Cambridge, 234 pp.

Demsetz, H. (1967) Toward a theory of property rights. *American Economic Review* 57, 347–359.

Douma, S. and Schreuder, H. (1992) *Economic Approaches to Organizations*. Prentice Hall International (UK) Ltd, Hemel Hempstead, Hertfordshire, UK, 185 pp.

Eggertssen, T. (1990) *Economic Behaviour and Institutions*. Cambridge Surveys of Economic Literature, Cambridge University Press, Cambridge, UK, 385 pp.

Fischer, S. (1977) Long-term contracting, sticky prices, and monetary policy: a comment. *Journal of Monetary Economics* 3, 317–323.

Frank, S.D. and Henderson, D. (1992) Transaction costs as determinants of vertical coordination in US food industries. *American Journal of Agricultural Economics* 74, 941–950.

Globerman, S. and Schwindt, R. (1986) The organization of vertically related transactions in the Canadian forest products industries. *Journal of Economic Behavior and Organization* 7, 199–212.

Hallwood, C.P. (1990) *Transaction Costs and Trade Between Multinational Corporations: a Study of Offshore Oil Production*. Unwin Hyman, Boston, 202 pp.

Hobbs, J.E. (1995) A transaction cost analysis of finished beef marketing in the United Kingdom. PhD thesis, University of Aberdeen, Scotland, 386 pp.

Hobbs, J.E. (1996a) Transaction costs and slaughter cattle procurement: processors' selection of supply channels. *Agribusiness: an International Journal* 12, 509–523.

Hobbs, J.E. (1996b) A transaction cost approach to supply chain management. *Supply Chain Management* 1(2), 15–27.

Hobbs, J.E., Kerr, W.A. and Gaisford, J.D. (1993) Transforming command economy distribution systems. *Scottish Agricultural Economics Review* 7, 135–140.

Jensen, M.C. and Meckling, W.H. (1976) Theory of the firm: managerial behavior, agency costs and ownership structure. *Journal of Financial Economics* 3, 305–360.

Kilmer, R.L. (1986) Vertical integration in agricultural and food marketing. *American Journal of Agricultural Economics* 68, 1155–1160.

Klein, B., Crawford, R.G. and Alchian, A.A. (1978) Vertical integration, appropriable rents and the competitive contracting process. *Journal of Law and Economics* 21, 297–326.

Knoeber, C. (1983) An alternative mechanism to assure contractual reliability. *Journal of Legal Studies* 12, 333–343.

Lancaster, K.J. (1966) A new approach to consumer theory. *Journal of Political Economy* 74, 132–157.

Lang, M.G. (1980) Marketing alternatives and resource allocation: case studies of collective bargaining. *American Journal of Agricultural Economics* 62, 760–765.

Leffler, K.B. and Rucker, R.R. (1991) Transactions costs and the efficient organization of production: a study of timber-harvesting contracts. *Journal of Political Economy* 99, 1060–1087.

Levy, D.T. (1985) The transactions cost approach to vertical integration: an empirical examination. *Review of Economics and Statistics* 67, 438–445.

Lieberman, M. (1991) Determinants of vertical integration: an empirical test. *Journal of Industrial Economics* 39, 451–466.

Masten, S.E. (1991) Transaction-cost economics and the organization of agricultural transactions. Presented at the NC-194 World Food Systems Project Symposium entitled: Examining the Economic Theory Base for Vertical Coordination, Chicago, Illinois, 17–18 October.

Matthews, R.C.O. (1986) The economics of institutions and the sources of growth. *The Economic Journal* 96, 903–918.

McCloskey, D.N. (1972) The enclosure of open fields: preface to a study of its impact on the efficiency of English agriculture in the eighteenth century. *Journal of Economic History* 32, 15–35.

McCloskey, D.N. (1986) *The Open Fields of England: Rent, Risk and the Rate of Interest, 1300–1815*. University of Iowa, Department of Economics and

History. As cited in Eggertssen, T. (1990) *Economic Behavior and Institutions*. Cambridge Surveys of Economic Literature, Cambridge University Press, Cambridge, UK, 385 pp.

Mighell, R.L. and Jones, L.A. (1963) *Vertical Coordination in Agriculture*. USDA ERS-19, Washington DC, 90 pp.

North, D.C. (1987) Institutions, transaction costs and economic growth. *Economic Inquiry* 25, 419–428.

Posner, R.A. (1993) Nobel Laureate: Ronald Coase and methodology. *Journal of Economic Perspectives* 7(4), 195–210.

Simon, H. (1961) *Administrative Behavior*. 2nd edn, Macmillan, New York, 364 pp.

Sonka, S.T. (1992) *New Industries in Agriculture: Concepts, Issues and Opportunities*. Mimeo, Department of Agricultural Economics, University of Illinois at Urbana-Champaign, Illinois.

Sporleder, T.L. (1992) Managerial economics of vertically coordinated agricultural firms. *American Journal of Agricultural Economics* 74, 1226–1231.

Stigler, G.J. (1961) The economics of information. *Journal of Political Economy* 69, 213–215.

Stigler, G.J. (1968) *The Organization of Industry*. Richard Dr. Irwin, Inc., Homewood, Illinois, 328 pp.

Teece, D.J. (1986) Transaction cost economics and the multinational enterprise: an assessment. *Journal of Economic Behavior and Organization* 7, 21–45.

Thomson, A. and Smith, L.D. (1991) *Liberalization of Agricultural Markets: an Institutional Approach*. University of Glasgow, Centre for Development Studies, Occasional Paper No. 8, 35 pp.

Williamson, O.E. (1975) *Markets and Hierarchies: Analysis and Anti-Trust Implications: a Study in the Economics of Organization*. Free Press, New York, 286 pp.

Williamson, O.E. (1979) Transaction cost economics: the governance of contractual relations. *Journal of Law and Economics* 22, 233–262.

Williamson, O.E. (1986) *Economic Organization – Firms, Markets and Policy Control*. Harvester Wheatsheaf, Hemel Hempstead, Hertfordshire, UK, 310 pp.

Williamson, O.E. (1991) The logic of economic organization. In: Williamson, O.E. and Winter, S.G. (eds) *The Nature of the Firm: Origins, Evolution and Development*. Oxford University Press, New York, pp. 90–116.

Williamson, O.E. (1994) Research needs and opportunities in transaction cost economics. *Journal of the Economics of Business* 1, 45–46.

# 5 Privatization and Monopolization of the Agrifood Chain

Market-oriented reform and privatization has created firms that are typically large by the standards in comparable industries in modern market economies. As a result, policy makers and observers have expressed concern over monopoly issues. Monopolies, of course, restrict production below socially efficient levels to increase profits by raising the prices of their outputs relative to the cost of their inputs.

> The price liberalization frees processors with market power to act like
> monopolists, and many respond by raising prices to consumers *and*
> *pressuring producer prices*. The price increase that accompanies
> liberalization is thus in part due to the removal of subsidies and in part
> due to the exercise of market power.
>
> (Brooks *et al.*, 1991, p. 155; italics added)

Thus, part of the initial burst of inflation at the onset of price liberalization arose from the exercise of monopoly power. Many commentators have recommended some standard solutions such as monopoly-busting and the formulation and implementation of anti-trust law. Most advisors, however, have wisely counselled against the use of price regulation which is a another standard solution to the problem of monopoly in modern market economies. Renewed price regulation, even when motivated on the basis of efficiency rather than equity, provokes well-founded fears of a general reversal in the process of liberalization.

In this chapter, we argue that the standard textbook analysis of monopoly is deficient as a description of the problems that now beset the agrifood systems, and the economies at large, in the Central and Eastern European Countries (CEECs) and the New Independent States

(NISs). As a consequence, the simple monopoly-busting and anti-trust solutions based on the standard analysis of a one-sided monopoly also require reconsideration. In the wake of the command system, there are frequently few buyers as well as few sellers. Rather than standard monopoly or monopsony situations, it is normal to find more complex bilateral monopolies and bilateral oligopolies where there are strategic interactions among a small number of firms. In a bilateral monopoly, there is a single seller and a single buyer, while in a bilateral oligopoly, there are a small number of both buyers and sellers. For example, a bilateral monopoly might consist of a single fertilizer supplier that sells to a single former state farm in a geographically segmented market. In such bilateral monopoly situations, the profit-maximizing supply actions of the seller depend on the demand actions being pursued by the buyer and vice versa. Thus, bilateral monopolies – and, by extension, bilateral oligopolies – are so-called game situations that involve strategic interaction amongst the players. Buyers strive to reduce prices at the same time that sellers try to increase them. The italicized part of the passage from Brooks *et al.* (1991) which was quoted in the previous paragraph makes it clear that processors strive to reduce the prices of their inputs from agricultural producers as well as increase the prices of their own products.

While bilateral monopolies, such as union-management situations, have conventionally been viewed through the lens of cooperative game theory in mainstream economics, a non-cooperative game-theoretic approach is much more appropriate in the CEECs and the NISs. In a cooperative bilateral-monopoly game, the players may be able to split the maximum available surplus in accordance with their bargaining positions (e.g. threat points). While *ex post* bargaining will not generate an efficient volume of exchange when information is asymmetric (e.g. Myerson and Satterthwaite, 1983) or when specific investments are involved, efficiency can typically be restored through an appropriate, possibly complex, *ex ante* contract (see Tirole, 1988).

The prognosis of efficient contracts, however, is overly optimistic especially for the CEECs and the NISs. In the absence of an omniscient judiciary to enforce such contracts, as outlined in Chapter 4, there are likely to be many examples of opportunism where firms strategically renege on contracts (see Williamson, 1975, and Klein *et al.*, 1978). In the liberalizing command economies, where mechanisms to enforce contracts are few, it would be particularly rash to expect cooperative behaviour and efficient levels of transactions between firms. In this chapter we will develop a model of a non-cooperative bilateral monopoly game where efficient volumes of exchange do not arise even when problems of asymmetric information and specific investments are absent.

## 5.1. The Advent and Persistence of Bilateral Monopolies

The command system that operated in the Soviet Union and Eastern and Central Europe was, of course, intrinsically inefficient. As outlined in Chapter 2, the economic philosophy underlying the organization of production and distribution favoured large command firms with a minimum of linkages to command firms at other stages of the vertical chain in order to simplify planning. The resources absorbed by the central planning bureaucracy and the difficulties resulting from the internal governance of artificially large firms were coupled with the frequent and often severe misallocation of resources across sectors of the economy that arose primarily as a result of the pricing problems discussed in Chapter 3. The elimination of the central planning bureaucracy in favour of market forces, whether gradually or rapidly, is a central tenet of the reform process. While the prospects for efficiency gains over the long term are significant, it would be both politically and economically naive to think that programmes of economic reform can deliver rapid gains in efficiency.

At the outset of the process of liberalization and privatization, each ex-command firm has tended to have established business connections with a very small number of buyers and suppliers. Further, creating new business linkages has been extremely difficult because of the high transaction costs discussed in Chapter 4. Many of the means of keeping transaction costs to a minimum that are taken for granted in modern market economies are missing in the CEECs and the NISs simply because they were unnecessary under the command system. In particular, the financial, legal, communication, and even transportation infrastructure tends to be very rudimentary.

Since there are few existing business connections between firms and there are serious impediments to the development of new linkages, thin markets on which a small number of sellers and buyers transact business are an endemic feature of the early years of the transition in the CEECs and, particularly, the NISs. This problem is particularly acute in rural areas where the command system favoured geographic monopolies. Thus, the initial stages of liberalization and privatization must take place in an absence of the competitive forces that generate efficiency. In order to understand the problems posed by newly created markets on which a small number of sellers and buyers transact business, we construct a simple diagrammatic model of a bilateral monopoly where a single seller or monopolist confronts a single buyer or monopsonist. We choose the extreme bilateral monopoly case in order to highlight the problems that arise from thin markets in the clearest possible manner. In the Appendix to this chapter, we provide a more rigorous mathematical version of the bilateral monopoly model and also extend the analysis to the more general case of a bilateral oligopoly where there are a small number of buyers and sellers.

## 5.2. Measuring Market Power

To provide a concrete frame of reference for our analysis of thin markets, let us consider a geographic bilateral monopoly in Russia or the Ukraine where the seller is a fertilizer supplier and the buyer is a former state farm. The fertilizer supplier may well sell to farms in other geographic areas, but we will assume that information and transport costs prevent arbitrage across geographic markets and, hence, allow different fertilizer prices to prevail in different locations. In the New Independent States of the former Soviet Union in particular, large farms remain predominant even where official reorganization has taken place (see Csaki 1995).[1] Nonetheless, in some geographic markets there might now be a limited number of small private family farms. For analytic clarity, we will abstract from the minor complications that would be posed by a competitive fringe on the demand side of the fertilizer market. We also assume that both firms maximize profits. In subsequent chapters, we argue that economic reform may lead to a separation between ownership and control where management acts to maximize employment or the wage bill or some more complex hybrid goal. We also temporarily assume that buyer and seller behave competitively on all other markets.

Suppose that the buyer and the seller of fertilizer transact quantity $x$ at price $p$ on the particular geographic market in question. The seller's profit is $px - C(x)$, where $C(x)$ is the seller's cost function associated with the production and delivery of fertilizer for the particular market, and the buyer's profit is $R(x) - px$, where $R(x)$ is the buyer's revenue function associated with the use of fertilizer. Notice that since the seller is likely to be a multi-product firm (e.g. if it sells fertilizer in other locations), $C(x)$ represents the seller's input costs net of revenue from other (locationally-distinct) products. Similarly, in the likely event that the state farm uses many variable inputs, $R(x)$ denotes the revenue net of all factor costs other than fertilizer. We will let $C'(x)$ denote the seller's marginal cost function associated with producing fertilizer and $R'(x)$ denote the buyer's marginal revenue function associated with using fertilizer.[2] We will assume that the marginal cost curve of the seller intersects the marginal revenue curve of the buyer at a unique positive quantity, $x^*$. For the purposes of our discussion, we will also assume that the marginal cost curve of the seller is always positively sloped or horizontal and that the marginal revenue of the buyer is always negatively sloped or horizontal.

In Fig. 5.1, the marginal cost curve of the seller is $C'$ and the marginal revenue curve of the buyer is $R'$. The efficient volume of transactions is $x^*$ where the marginal cost curve of the seller intersects the marginal revenue curve of the buyer. That is to say, when the quantity of fertilizer that is traded is equal to $x^*$, the total surplus is at a maximum. The total surplus consists of the sum of buyer's (consumer) plus seller's (producer) surplus and it is equal to the area above the marginal cost curve but below the marginal revenue curve. A competitive seller chooses its output such that its marginal cost is equal to the price, while a competitive buyer sets its price such that its marginal revenue would be equal to price. Thus, if the buyer and seller both behaved competitively, they would exchange the efficient quantity, $x^*$, at the competitive

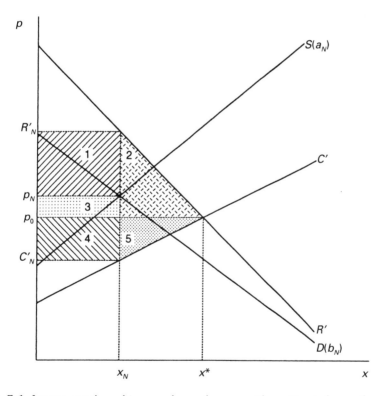

**Fig. 5.1.** Inverse supply and inverse demand curves with positive indexes of market power.

price of $p_0$. Of course, since the buyer and seller both possess market power in situations with thin markets such as in the former command economies, this competitive result is not likely to be forthcoming.

Notice that in Fig. 5.1 the magnitude of the slope of the seller's marginal cost curve happens to be less than that of the buyer's marginal revenue curve. Consequently, the seller's producer surplus would be smaller than the buyer's consumer surplus if the efficient quantity of fertilizer, $x^*$, was transacted. This suggests that the seller is less dependent than the buyer on the geographic market that is under analysis. For example, it may be relatively easy for the seller to substitute into other geographic markets that it serves, but relatively difficult for the former state or collective farm to substitute other inputs for fertilizer. Under these circumstances, the seller will have a stronger position than the buyer in the bilateral monopoly.

Lerner indexes provide a simple, well-known measure of the degree of market power exercised by the seller and the buyer.

$$a \equiv \frac{p - C'(x)}{p} \tag{1}$$

$$b \equiv \frac{R'(x) - p}{R'(x)} \tag{2}$$

Here, $a$ is the Lerner index of monopoly power for the seller and $b$ is the Lerner index of monopsony power for the buyer. If the seller exhibits no monopoly power, it will set price equal to marginal cost and the Lerner index of monopoly power will be equal to zero (i.e. as $p = C'(x)$, then $p - C'(x) = 0$). On the other hand, the highest possible value of the Lerner index is one. For example, $a$ would be equal to one if the price were positive and the marginal cost were equal to zero. Thus, the higher the value of the Lerner index of monopoly power on the interval from zero to one, the more market power that is being exercised by the seller. Analogously, the higher the value of the Lerner index of monopsony power on the interval from zero to one, the more market power that is being exercised by the buyer.

Since the actions of the seller affect the buyer and vice versa, the bilateral monopoly situation can be considered to be a game and it can be analysed using the game-theory approach of economists who study industrial organization. Broadly speaking, a game is any situation where the choices made by any one economic agent or player affect the welfare of the other(s). Thus, in a game the players interact strategically in the sense that they must anticipate what their rival(s) will choose to do when they determine their own best actions.[3] To begin with, we will analyse a bilateral monopoly game where the seller and buyer behave non-cooperatively. This means that the buyer and seller do not collude to jointly maximize the benefits arising from a transaction. Later we will explore the prospects for cooperation.

It is important to assign the firms or players decision-making instruments that are symmetric and independent. The instruments are symmetric if both firms have the same type of choice variable, and the instruments are independent if each firm has its own choice variable.[4] Consider the implications of asymmetric instruments. For example, suppose that the seller were modelled as price setter and the buyer were modelled as a quantity setter. In such a case, the buyer would be a competitive price taker choosing its best quantity at any given price set by the seller. Thus, the seller would be a single price setter and act as a pure monopolist (see Spengler, 1950; Tirole, 1988). All of the market power would be arbitrarily assigned to the seller simply by assumption. On the other hand, if we assumed that the buyer was price setter and the seller was a quantity setter, the seller would be competitive and the buyer would be a pure monopsonist. Clearly, this too is arbitrary. The fact that all of the market power ends up on one or other side of the market is an artefact of the assumption that the buyer and seller have different types of choice variables. Yet, the seller and buyer cannot both simultaneously choose the price at which they transact nor can they both choose the quantity.[5] Since the price and/or quantity are unsatisfactory choice variables for the analysis of the bilateral monopoly game, we must adopt alternative, but still intuitively appealing, choice variables and allow the price and quantity to be determined endogenously. To this end, we will consider a bilateral monopoly game where the buyer and seller choose how much market power to wield.[6]

## 5.3. Supply and Demand Revisited

It is intuitively appealing to have indexes of market power as instruments or choice variables for the firm and it also allows us to adapt the usual supply and demand analysis that is applied to competitive markets. In order to facilitate this supply and demand, we temporarily treat the Lerner indexes of the two firms as parameters (givens) rather than choice variables. Figure 5.1 is drawn such that the Lerner indexes of the seller and buyer are positive and equal to $a_N$ and $b_N$ respectively. These indexes of market power are effectively (proportional) pricing wedges that cause differences between the price and the firm's internal marginal valuation.[7]

Given that the seller's index of market power is set at the positive value, $a_N$, the price that the seller is willing to offer when bargaining with a buyer will exceed its marginal cost for any quantity of fertilizer that is transacted. Thus, the seller's effective (inverse) supply curve, which relates its offer price to the quantity of fertilizer that is transacted, will lie above its marginal cost curve. Figure 5.1 shows that there is a proportional, i.e. increasing with the quantity of fertilizer transacted, vertical displacement of the seller's inverse supply curve, $S(a_N)$, above its marginal cost curve, $C'$, that is caused by its index of monopoly power, $a_N$. In other words, the exercise of market power implies that the seller extracts a price that is higher than its marginal cost at any quantity. Further, an increase in the seller's index of monopoly power above $a_N$ would shift its supply curve upward.

When the buyer's index of market power is set at the positive level, $b_N$, its offer price will be less than its marginal revenue for each quantity of fertilizer, and its inverse demand curve will lie below the marginal revenue curve. In Fig. 5.1, there is a proportional vertical displacement of the buyer's inverse demand curve, $D(b_N)$, below its marginal revenue curve, $R'$, that is caused by the index of monopsony power, $b_N$. Thus, the ability to exercise market power allows the buyer to set an offer price that is lower than its marginal revenue at any quantity. Moreover, an increase in the buyer's index of monopsony power would shift the demand curve downward.

In equilibrium, the offer prices of the seller and buyer must be equal, otherwise no transaction will take place. In Fig. 5.1, the intersection of the seller's supply curve, $S(a_N)$, and the buyer's demand curve, $D(b_N)$, determines the wedge-contingent equilibrium quantity of fertilizer, $x_N$, and the wedge-contingent equilibrium price, $p_N$. It is important to emphasize that both $x_N$ and $p_N$ depend on the given values of the two wedges or indexes of market power, $a_N$ and $b_N$. For different values of either firm's index, a different wedge-contingent equilibrium would arise. For example, by increasing its index of monopoly power and shifting up the supply curve, the seller would be able to raise the price by reducing the quantity. In effect, this is the usual situation where a monopolist is able to raise the price by restricting the quantity. Similarly, by increasing its index of monopsony power and shifting the demand curve downward, the buyer would be able to lower the price by restricting the quantity.

## 5.4. The Profit-maximizing Degree of Market Power

Of course, the indexes of market power are not parameters but instruments or choice variables that firms attempt to optimize. In particular, each firm chooses to maximize its profits conditional on any given value of the other firm's index of market power. Thus, each firm attempts to increase its index of market power to the point where its total profit is at a peak or the marginal addition to its profit is equal to zero.

Each firm's profit maximization problem can also be viewed as an indirect choice of the quantity of fertilizer. The profit-maximizing value of the seller's index of monopoly power is chosen so as to equate its own internal marginal cost of producing fertilizer with the external marginal revenue that it faces as a result of the buyer's demand. Similarly, the profit-maximizing value of the buyer's index of monopsony power is chosen so as to equate its own internal marginal revenue from using fertilizer with the external marginal cost that it faces. Thus, the seller behaves much like a standard monopolist while the buyer behaves much like a standard monopsonist. Nevertheless, it is very important to emphasize that the external marginal revenue that the seller faces is not the same thing as the buyer's internal marginal revenue even though the two are related. Likewise, the external marginal cost that the buyer faces is not the same thing as the seller's internal marginal cost.

Figure 5.2 illustrates the seller's profit-maximizing choice. Given that the buyer happens to have an index of market power that is equal to $b_N$, the seller faces the demand curve, $D(b_N)$, from the buyer. Thus, $MR_S(b_N)$ is the external marginal revenue curve facing the seller given the demands of the buyer. The marginal revenue curve facing the seller must lie below the demand curve because, as the seller increases the quantity of fertilizer that it sells, it bids down the price on all infra-marginal units sold. Since the buyer's demand curve, $D(b_N)$, is itself determined by the appropriate downward displacement from its marginal revenue function, $R'$, the marginal revenue curve faced by the seller, $MR_S(b_N)$, is related to the buyer's internal marginal revenue curve, but only in an indirect way. The two curves are not the same.

In Fig. 5.2, the seller's optimum quantity of fertilizer is $x_N$ which serves to equate the external marginal revenue that it faces from the buyer with its own internal marginal cost. Since the associated values of the price and the seller's internal marginal cost are $p_N$ and $C'_N$, the profit-maximizing value of the seller's index of market power that gives rise to $x_N$ is $a_N = [p_N - C'_N]/p_N$. This pricing wedge displaces the seller's supply curve upward from its marginal cost curve $C'$ to $S(a_N)$. Thus, the seller's supply curve intersects the demand curve that it faces from the buyer at the seller's profit-maximizing output. Since the quantity of fertilizer and thence the values of the price and marginal cost, $p_N$ and $C'_N$, depend on the indexes of market power of both firms, the optimum value of the seller's index implicitly depends on the buyer's index. In Fig. 5.4, the reaction curve of the seller, AA, shows the seller's best

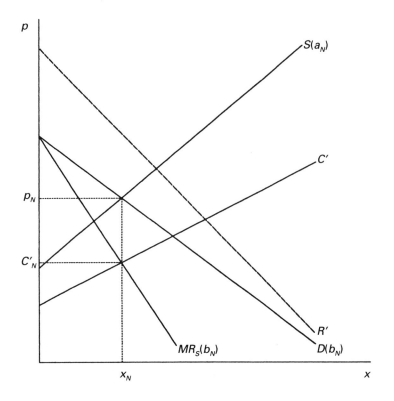

**Fig. 5.2.** The monopolistic behaviour of the supplier.

choice of its index of market power for each possible value of the buyer's index.[8] We discuss this reaction curve more fully below but, first, we consider the profit-maximization decision of the buyer.

In Fig. 5.3, the buyer's own internal marginal revenue curve is $R'$, the supply curve that it faces from the seller is $S(a_N)$, and the external marginal input cost curve that it faces is $MC_B(a_N)$. Given that the seller's index of market power is $a_N$, the buyer's optimum quantity of fertilizer is $x_N$ where the $R'$ curve intersects the $MC_B(a_N)$ curve. Since the price is $p_N$ and the buyer's marginal revenue is $R'_N$ the buyer's optimum wedge is $b_N = (R'_N - p_N)/R'_N$. When the buyer's wedge is set optimally at $b_N$, its demand curve $D(b_N)$ intersects the supply curve that it faces from the seller at the profit-maximizing output. Since the quantity of fertilizer and thence the values of the price and marginal cost, $p_N$ and $R'_N$, depend on the indexes of market power of both firms, the optimum value of the buyer's index implicitly depends on the seller's index. In Fig. 5.4, the reaction curve of the buyer, BB, shows the best choice of its index of monopsony power for each possible value of the seller's index.

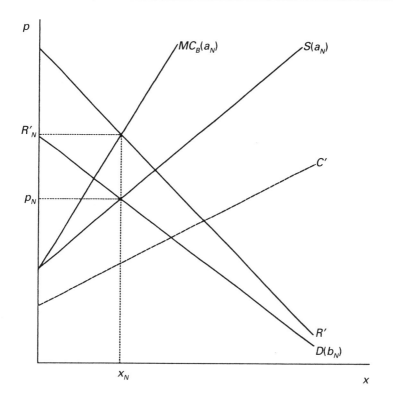

**Fig. 5.3.** The monopsonistic behaviour of the buyer.

## 5.5. Strategic Behaviour

Remembering that $a$ is the Lerner index of monopoly power for the seller and $b$ is the Lerner index of monopoly power for the buyer, in Fig. 5.4, point 0 is the competitive position where neither firm exercises market power (i.e. $a = 0$ and $b = 0$). The efficient volume of trade between the seller and buyer, $x^*$, arises at point 0. At the other extreme, the TT curve is a trade (transaction) termination curve. At all points on the TT curve or above and to the right of it, the pricing wedges of the two firms are so high as to entirely curtail exchange between the seller and the buyer. This would represent the complete break of the fertilizer link in the chain of production and distribution. If either firm sets its index of market power above $t$, all exchange between the seller and the buyer would be cut off even if its opponent exercised no market power.

Provided that trade is possible because the buyer's index of market power is set less than $t$, it would never be optimal for the seller to set

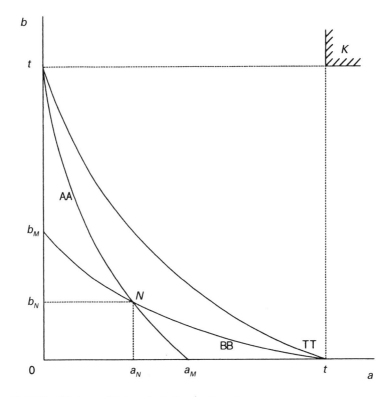

**Fig. 5.4.** The Nash equilibrium in indexes of market power.

its wedge so high that trade is entirely cut off. Thus, in Fig. 5.4 the seller's reaction curve, AA, must remain to the left of the trade termination curve, TT. Further, the seller's reaction function does not cross to the left of the vertical axis because it is never optimal for the seller to exercise a negative degree of market power. An increase in the buyer's index of monopsony power will typically reduce the marginal as well as total profit that the seller obtains from any particular value of its own index. Consequently, the increase in the buyer's index will typically lead to a decrease in the seller's optimum index of monopoly power. Thus, the seller's reaction curve typically has a negative slope.

The bliss point or pure monopoly point for the seller in Fig. 5.4 is where it sets its index of market power equal to $a_M$ while the buyer does not exercise market power (i.e. $b = 0$). The seller's profits decline continuously as it is moved upward along its reaction curve and away from this pure monopoly point. The quantity traded also declines continuously as the seller is moved upward along its reaction function towards the trade termination curve. If the buyer switched to a higher index of market power, this would adversely affect the seller by lowering both the price and the quantity. Even if the profit-maximizing response of the seller is to reduce its index of market power (i.e. even if the seller's reaction function is negative sloped), which causes a further reduction in price, it will only partially offset the reduction in quantity.

In Fig. 5.4, the pure monopsony point for the buyer is where its index of market power is equal to $b_M$ and that of the seller is equal to zero. The buyer's reaction curve, BB, extends to the right from the monopsony point and it remains below the TT curve. Both the buyer's profit and the quantity exchanged decline as the buyer is moved to the right along its reaction function. Further, the buyer's reaction function, like that of the seller, will typically have a negative slope. If the seller increases its index, the marginal profit of the buyer usually declines and, therefore, the buyer's optimum index of market power usually falls.

## 5.6. The Nash Equilibria in Indexes of Market Power

When neither player regrets its action given the action of its opponent, one has what is known as a Nash Equilibrium (see Gravelle and Rees, 1992). Fig. 5.4 shows directly that there is such a Nash equilibrium at point $N$ where the reaction curves of the seller and buyer intersect. In this Nash equilibrium, the degree of market power exercised by the seller is $a_N$ while that of the buyer is $b_N$. In other words, $a_N$ happens to be the profit-maximizing degree of market power for seller when the buyer's index of market power is set at $b_N$. At the same time, $b_N$ happens to be the optimum degree of market power for the buyer given that $a_N$ is chosen by the seller. Figures 5.1, 5.2 and 5.3 happen to have been drawn to be consistent with the Nash equilibrium at $N$ in Fig. 5.4.

At a typical internal Nash equilibrium such as $N$ in Fig. 5.4 each of the two firms sets a strictly positive index of market power. Given that the buyer's marginal revenue curve is negatively sloped (rather than horizontal), the demand curve and the marginal revenue curve facing the seller will be negatively sloped as shown in Fig. 5.2. Consequently, the profit-maximizing index of market power for the seller will always be positive (strictly positive) because the price will exceed its marginal cost. Analogously, when the seller's marginal cost curve is positively sloped, the supply curve and the marginal cost curve facing the buyer will be positively sloped as shown in Fig. 5.3. Thus, the profit-maximizing index of market power for the buyer will be strictly positive because its marginal revenue will exceed the price.

Boundary Nash equilibria, where only one firm exercises market power, are also possible. In such a case, the equilibrium would lie on one of the axes. For example, in the situation where its marginal cost curve is flat, the seller will be able to act as a pure monopolist. In this case the supply curve and marginal cost curve facing the buyer would coincide and be horizontal. Thus, the buyer would be a price taker. Since the price would be equal to the buyer's marginal revenue, its reaction curve would lie on the horizontal axis. Consequently, the equilibrium index of market power for the buyer would equal to zero and the seller would act as a pure monopolist. Analogously, in the situation where its marginal revenue curve is flat, the buyer will be able to act as a pure monopsonist.

Since at least one firm will exercise market power in a Nash equilibrium, the non-cooperative Nash equilibrium is inefficient. In Fig. 5.1, the efficiency loss is equal to the sum of areas 2 and 5. Moreover, in the case of a typical internal equilibrium such as point $N$ in Fig. 5.4, quantity exchanged is restricted to a greater extent than under either a pure monopoly or a pure monopsony. The quantity exchanged declines moving up the seller's reaction curve from the pure monopoly point at $a_M$ to the Nash equilibrium at $N$. Likewise, the quantity declines moving to the right from the pure monopsony point at $b_M$ to the equilibrium at $N$. In the case of a boundary equilibrium, either the pure monopoly or the pure monopsony quantity will be exchanged. Thus, a non-cooperative bilateral monopoly is always at least as inefficient as either a pure monopoly or a pure monopsony and it is typically less efficient.[9]

At least one of the firms must be worse off (i.e. less profitable) at the Nash equilibrium of the non-cooperative bilateral-monopoly game (i.e. at point $N$ in Fig. 5.4) than it would at the competitive position (i.e. at point 0) because of the efficiency loss. In Fig. 5.1, the net gain in the seller's 'producer' surplus is equal to area 3 minus area 5, while the net loss in the buyer's 'consumer' surplus is equal to the sum of areas 2 and 3. Area 5 is the seller's share of the distortionary loss, area 2 is the buyer's share of the distortionary loss and area 3 represents a terms of trade effect that is transferred from the buyer to the seller because of the higher price that prevails in the Nash equilibrium (i.e. $p_N$ rather than $p_0$).[10] In Fig. 5.1, the price that arises in the Nash equilibrium is higher than the competitive price because the seller is less dependent on the given fertilizer market.

It should also be observed that there are other Nash equilibria that are less benign than the Nash equilibrium at point $N$ in Fig. 5.4. If both firms simultaneously were to set their indexes of market power high enough to independently curtail exchange (i.e. if $a \geq t$ and $b \geq t$), then neither firm would regret its own action. If, for example, the buyer cuts off exchange, there is nothing that the seller can do. Thus, any degree of market power for the seller, including values equal to or above *t*, is equally beneficial. Thus, all points in area *K* in Fig. 5.4 are Nash equilibria in which trade in fertilizer is entirely cut off. Of course, it is reasonable to expect the firms to focus on the equilibrium at point *N* provided that they each earn higher profits at *N* than they would if no trade occurs. Situations can arise, however, where one or other of the firms would prefer to curtail trade because it is unable to cover its variable costs at point *N*. In such situations, the only Nash equilibria will involve the cessation of trade. Thus, though both firms might still be profitable if they cooperated, no exchange between the firms will take place.

A complete break in a single link in the agrifood chain could, of course, have far-reaching consequences. Suppose that the fertilizer link is broken because the supplier rather than the state farm is driven to cease operations. In the absence of trade in fertilizer on the particular geographic market in question, the subsequent downstream production and distribution of many food commodities might be adversely affected to some degree. Moreover, other geographic markets serviced by the fertilizer supplier will be affected in a similar manner. In the alternative case where the state farm rather than the fertilizer producer ceases operation, many products moving to a single downstream market could be affected and a serious, albeit highly localized, food crisis could arise.

To this point we have focused on an isolated bilateral monopoly and in the Appendix to this chapter, we adapt the analysis to handle a single bilateral oligopoly situation. In reality, however, chains and webs of bilateral monopolies and bilateral oligopolies are pervasive in the transitional economies of the CEECs and the NISs. Spengler (1950) has shown that chains of product-market monopolies can lead to a multiple-margin phenomenon where each firm sets a positive price-cost margin (i.e. a positive index of monopoly power) on its output market. Chains of bilateral monopolies add to the severity of this multiple-margin problem because each firm would typically set a positive index of market power on its input markets as well as its output markets. Indeed, one of the most prevalent problems in the former command economies is the constriction of the quantity transacted at each successive stage of production and distribution.

To summarize, the important conclusion of these theoretical explorations is that non-cooperative bilateral monopolies will, at best, transact quantities equal to monopoly (monopsony) quantities. They will likely transact even smaller quantities. When bilateral monopolies are found along the links in the agrifood supply chains of the CEECs and the NISs, the reduction in output can be significant.

## 5.7. Solutions: Regulation, Integration, Competition or Cooperation?

At least in theory, price regulation is one possible means of restoring efficiency in the face of non-cooperative bilateral monopoly. If the regulated price was set at $p_0$ on this particular geographic market, then both firms would be reduced to the status of price takers (i.e. their optimum indexes of market power would be equal to zero) and the efficient volume of fertilizer, $x^*$, would be transacted.

There are, however, a number of very compelling arguments against the reintroduction of widespread price regulation in the CEECs and the NISs. First, price regulation could well reduce rather than enhance efficiency. Suppose that in the event that the regulator happened to choose a disequilibrium price, the short side of the market (i.e. the minimum of the quantity demanded and the quantity supplied) would determine the amount of fertilizer that would be exchanged. For simplicity, also assume that black market activities would not occur. If the regulated price was mistakenly set above $R'_N$ or below $C'_N$ in Fig. 5.1, the efficiency loss would be larger than under bilateral monopoly.

The second argument against re-regulation relates to information requirements. Each and every geographic market on which bilateral monopoly issues arise would need to be regulated separately. Of course, sellers would have an incentive to overstate their own marginal costs and buyers would have an incentive to understate their own marginal revenues in order to influence the setting of the regulated price to their own advantage. Further, political expediency would suggest a return to pricing in accord with equity rather than efficiency criteria. Thus, re-regulation would be a significant step backward towards the command system with its inherent information and efficiency problems.

The third major problem with re-regulation of prices is that it would discourage rather than encourage the integration of markets and competition. Since each market would be regulated separately, bilateral monopoly situations would remain an on-going feature of the economic landscape. Not only would a command-type structure be re-imposed but the need for such a structure would tend to be made permanent.

Vertical integration is another possible means of resolving the problems associated with bilateral monopoly that will seldom be workable in the CEECs and the NISs. In many key areas including food production and distribution, ex-command firms already tend to be far too large when compared with what is standard elsewhere in the world. Thus, there would be significant increases in the costs of organizing vertical coordination within these very large firms (e.g. see Coase, 1937). Too many markets would be internalized. Moreover, it may be beneficial to avoid widespread recourse to greater vertical integration in order to avoid the equity problems that could arise with increased concentrations of wealth and power.

Increased competition and across-market cooperation between the buyer and seller are two more promising means of alleviating the problem of inefficiently low volumes of transactions that arises in a non-cooperative bilateral monopoly situation. In the CEECs and the NISs, the long-term prospects for increased competition at many points on agrifood chains are good. In order to promote the development of such competition, it is important for governments and off-shore aid donors to actively promote the development of the transportation, communication and financial infrastructure in order to reduce transaction costs and stimulate the development of new business linkages. Over the longer term, it should be possible to reduce the geographic fragmentation of agrifood chains that had been actively promoted under the command system in order to facilitate planning.

Active competition policy is another possible means of promoting competition. On the surface, monopoly busting has considerable attraction because of the fact that command firms in the food sector and the economy in general tended to be unduly large. In spite of the possibilities of improved internal governance in new smaller firms, competition policy needs to be implemented with considerable care. Where large ex-command firms can be divided into a number of competing new parallel firms, there may be increased efficiency both from enhanced competition and improved internal governance. For example, it would be worthwhile to actively promote the development of an independent trucking industry since this would also help expose producers on parallel food chains to greater competition.

In practice, there seems to have been little impetus for the voluntary dissolution of large enterprises into smaller entities that compete with one another. This is understandable since the new competitive firms would face monopolistic sellers for their inputs and monopsonistic buyers for their outputs. For example, where legislation has permitted, but not forced, the dissolution of state and collective farms, there have typically been few households that have exercised their prerogative to take a share of the land and strike out as independent private farms (see Brooks and Lerman, 1993; Csaki, 1995). Further, even countries such as Albania and Romania which have required the dissolution of state and collective farms have certainly not been immune to large declines in farm yields (see Zedgar, 1995).

If large ex-command firms were divided up into a vertical series of new firms, a sequence of new bilateral monopolies would be created. Unfortunately, there seems to be growing evidence that it is easiest to carve up large enterprises on a functional basis that creates this type of vertical segmentation. While vertical integration within the food sector holds little promise as a means of promoting efficiency, such vertical segmentation could seriously compound the existing inefficiencies associated with bilateral monopoly. Thus, a cavalier policy of monopoly busting is fraught with danger.

The inefficiencies that arise in a non-cooperative bilateral monopoly could potentially be eliminated by cooperation between the seller and the buyer such as the strategic alliances that are becoming common among agribusiness firms in modern market economies.[11] In a cooperative bilateral-monopoly game, the players may be able to split the maximum available surplus in accordance with their bargaining positions (see Gravelle and Rees, 1992, pp. 380–387).[12] Across-market cooperation is more likely in situations where contracts can be enforced at low cost and/or the possibility of repeat business is important, i.e. where the game is repeated at a low discount rate. Of course, both of these conditions are typically present in modern market economies.[13] In the CEECs and, especially, the NISs, however, economic relationships are in such a state of flux that the possibility of future cooperation is unlikely to give rise to self-enforcing efficient contracts at least in the short term. Further, there is very limited experience with contract law in these countries so that recourse to the courts as a means of enforcing an efficient contract, even where this is possible, is at present a lengthy, highly uncertain and costly process. Consequently, the development of the legal infrastructure for enforcing contracts at low costs between sellers and buyers should be a key priority for governments that have embarked on the process of economic reform. Further, it is an area where assistance from modern market economies could be particularly valuable. We will consider the development of legal institutions in more detail in Chapter 18.

## 5.8. Conclusions

Over the medium to long term, both enhanced competition and greater cooperation between sellers and buyers should gradually reduce the efficiency problems associated with bilateral monopolies in economies of the CEECs and the NISs. Moreover, steps can be taken now to help develop the transportation, communication, financial and legal infrastructure that will expedite future competition and across-market cooperation. Nevertheless, there is no short-term escape from the problems of non-cooperative bilateral monopoly. The nascent market systems in the CEECs and the NISs cannot be expected to generate major efficiency gains overnight. Inefficiently low volumes of transactions will remain typical over the coming years. Further, the collapse of particular markets remains a possibility and the equilibration process could be marked by significant temporary disruptions. In the food sector, even a localized short-term supply disruption could have catastrophic economic and political consequences.

# *Appendix to Chapter 5*

## 5A.1. A Formal Model of Bilateral Monopoly

Given the definitions of the Lerner indexes in equations (1) and (2) in the main text of the chapter, the inverse supply function of the seller and the inverse demand function of the buyer are as follows.

$$p_S = \frac{1}{1-a} C'(x) \quad \text{where:} \frac{\partial p_S}{\partial x} = \frac{1}{1-a} C''(x) \geq 0 \tag{3}$$

$$p_B = (1-b) R'(x) \quad \text{where:} \frac{\partial p_S}{\partial x} = (1-b) R''(x) \leq 0 \tag{4}$$

Here, $p_S$ is the offer price or supply price of the seller, and $p_B$ is the offer price or demand price of the seller. Although it is somewhat unconventional, it is simpler to work primarily in terms of elasticities of offer and acceptance prices rather than elasticities of supply and demand.

$$\psi \equiv \frac{x}{p_S} \frac{\partial p_S}{\partial x} = \frac{x C''(x)}{C'(x)} \equiv \Psi(x) \geq 0 \tag{5}$$

$$\phi \equiv \frac{x}{p_B} \frac{\partial p_B}{\partial x} = \frac{x R''(x)}{R'(x)} \equiv \Phi(x) \leq 0 \tag{6}$$

The elasticity of the seller's offer price is equal to zero if its marginal cost curve is flat but it is positive if its marginal cost curve is positively sloped. Analogously, the elasticity of the buyer's offer price is equal to zero if its marginal revenue curve is flat but it is negative if its marginal revenue curve is negatively sloped. If $\sigma \equiv (p_S/x)(\partial x/\partial p_S) \geq 0$ is the seller's elasticity of supply and $\delta \equiv (p_B/x)(\partial x/\partial p_B) \leq 0$ is the buyer's elasticity of demand, then $\psi = 1/\sigma$ and $\phi = 1/\delta$.

By equating the offer prices of the two firms given by the inverse supply and demand functions, we can ascertain the equilibrium volume of transactions conditional on the pricing wedges of the seller and the buyer.

$$x = X\left([1-a][1-b]\right) \text{ solves } \frac{1}{1-a} C'(x) = (1-b) R'(x) \tag{7}$$

$$\text{where}: X' = \frac{1}{(1-a)(1-b)\left[(C''/C')-(R''/R')\right]} > 0$$

If either $a$ or $b$ were to rise, $(1-a)(1-b)$ would fall and, therefore, $x$ would fall. It is also possible to establish the wedge-contingent equilibrium price by substituting the wedge-contingent equilibrium quantity into either the seller's inverse supply function or the buyer's inverse demand function.

$$p = P(a,b) = \begin{cases} \dfrac{1}{1-a} C'\left(X([1-a][1-b])\right) \\ (1-b)R'\left(X([1-a][1-b])\right) \end{cases} \tag{8}$$

The seller's and buyer's profits can be written as $f = p_B \cdot x - C(x)$ and $g = R(x) - p_S \cdot x$ in recognition that each firm is constrained by its opponent's offers. We can then use equations (3), (4) and (8) to write wedge-contingent profit functions for the two firms.

$$\begin{aligned} f &= F(a,b) \equiv (1-b) R'\left(X\left([1-a][1-b]\right)\right) \cdot X([1-a][1-b]) \\ &\quad -C\left(X([1-a][1-b])\right) \end{aligned} \tag{9}$$

$$g = G(a,b) \equiv R\left(X\left([1-a][1-b]\right)\right)$$

$$\quad - \frac{1}{1-a} C'\left(X([1-a][1-b])\right) \cdot X\left([1-a][1-b]\right) \tag{10}$$

The first-order conditions for profit maximization require that a firm's marginal profits are equal to zero.

$$\frac{\partial F(a,b)}{\partial a} = 0 \Leftrightarrow \left(p_B + x \frac{\partial p_B}{\partial x}\right) = C' \tag{11}$$

$$\frac{\partial G(a,b)}{\partial b} = 0 \Leftrightarrow R' = \left( p_S + x \frac{\partial p_S}{\partial x} \right) \tag{12}$$

Consequently, the marginal revenue facing the seller must be equal to its own marginal cost, and the buyer's own marginal revenue must be equal to the marginal cost that it faces. The second-order conditions for profit maximization stipulate that a firm's marginal profits are diminishing as its own index of monopoly power increases.

$$\frac{\partial^2 F(a,b)}{\partial a^2} < 0 \Leftrightarrow \left( 2\frac{\partial p_B}{\partial x} + x\frac{\partial^2 p_B}{\partial x^2} \right) < C'' \tag{13}$$

$$\frac{\partial^2 G(a,b)}{\partial b^2} < 0 \Leftrightarrow R'' < \left( 2\frac{\partial p_S}{\partial x} + x\frac{\partial^2 p_S}{\partial x^2} \right) \tag{14}$$

Thus, the slope of the marginal revenue function facing the seller must be less than the slope of its own marginal cost function, and the slope of the buyer's own marginal revenue function must be less than the slope of the marginal cost function that it faces.

The first-order conditions (i.e. equations 11 and 12) can be simplified using the definitions of the Lerner indexes of market power (i.e. equations 1 and 2) and the elasticities of offer prices (i.e. equations 5 and 6).

$$a = -\Phi\left( X([1-a][1-b]) \right) \qquad \text{or } a = -\frac{1}{\delta} \tag{15}$$

$$b = \frac{\Psi\left( X([1-a][1-b]) \right)}{1 + \Psi\left( X([1-a][1-b]) \right)} \qquad \text{or } b = \frac{1}{1+\sigma} \tag{16}$$

Equations 15 and 16 are implicitly the reaction functions of the seller and buyer. Since the elasticity of the seller's offer price takes on non-negative values, it is immediately clear from equation 16 that the buyer's profit-maximizing index of market power must lie on the closed interval from zero to one. Although the elasticity of the buyer's acceptance price can take on any non-positive value, the seller's optimum index of market power in equation 15 must also lie on the closed interval from zero to one. This is because the marginal revenue confronting the seller is non-negative if and only if the elasticity of the buyer's offer price lies on the closed interval between negative one and zero (i.e. the buyer's demand is elastic or $\delta \leq -1$).

The slopes of the reaction functions are as follows.

$$\frac{\partial b}{\partial a}\bigg|_{AA} = -\frac{(1-b)\left[\Phi' - \dfrac{1}{(1-b)X'}\right]}{(1-a)\Phi'} \tag{17}$$

$$\frac{\partial b}{\partial a}\bigg|_{BB} = -\frac{(1-b)\Psi'}{(1-a)\left[\Psi' + \dfrac{(1+\psi)^2}{(1-a)X'}\right]} \tag{18}$$

In equation 17 the term in square brackets has to be negative to satisfy the seller's second-order condition for profit maximization, while in equation 18 the term in square brackets has to be positive to satisfy the buyer's second-order condition. Thus, the seller's reaction function will be negatively sloped if (and only if) the elasticity of the buyer's offer price declines as the quantity exchanged is increased (i.e. if $\phi' < 0$). In the case where the buyer's demand curve is linear (i.e. its revenue function is quadratic), the seller's reaction curve must be negatively sloped over its full range as shown in Fig. 5.4. In the linear demand case, the elasticity of the buyer's offer price declines from zero to negative infinity as quantity is increased and, correspondingly, the elasticity of demand rises from negative infinity to zero. The buyer's reaction function will be negatively sloped if and only if the elasticity of the seller's offer price is increasing in the quantity exchanged (i.e. if $\psi' > 0$). For example, the buyer's reaction function must be negatively sloped over its full range as shown in Fig. 5.4 when the seller has a linear supply curve (i.e. the revenue function is quadratic) with a positive intercept. In such a case, the elasticity of the seller's offer price rises progressively and approaches one from below, and correspondingly, the elasticity of supply falls progressively and approaches one from above.

The reaction functions given by equations 15 and 16 can be solved simultaneously to determine mutually consistent or Nash equilibrium values of the indexes of market power. At a typical internal Nash equilibrium, each of the two firms sets a positive index of market power. Given that the buyer's marginal revenue curve is negatively sloped, the elasticity of its acceptance price will always be strictly negative and the profit-maximizing Lerner index for the seller will be strictly positive in accordance with equation 15. Analogously, when the seller's marginal cost curve is positively sloped, the elasticity of its offer price will be strictly positive and the buyer's optimum index of market power will be positive in accordance with equation 16.

Boundary Nash equilibria where only one firm exercises market power are also possible. In such a case, the equilibrium would lie on one of the axes in $(a,b)$ space. For example, in the situation where its own marginal cost curve is flat, the seller would be able to act as a pure monopolist. In this case: (i) the elasticity of the seller's offer price will always be equal to zero; (ii) the buyer's reaction curve will coincide with the horizontal axis; (iii) the buyer's Nash-equilibrium index of market power will be equal to zero; and (iv) the seller's index of market power will be set at its pure-monopoly value, $a_M$. Analogously, in the situation where its marginal revenue curve is flat, the buyer will be able to act as a pure monopsonist.

Efficiency requires that $x = x^*$. Thus, the efficiency requirement on the indexes of market power is:

$$(1-a)(1-b) = 1 \tag{19}$$

While the situation where both firms set their indexes of market power equal to zero is certainly efficient, there are other efficient points where one firm exercises a negative degree of market power to offset the positive market power of the other. Of course, non-cooperative profit-maximizing firms will typically not exercise a negative degree of market power.

Given that the intercepts of the seller's marginal cost curve and the buyer's marginal revenue curve are $C'(0)$ and $R'(0)$ respectively, trade would be cut off if either firm were to set its index of market power equal to $t \equiv [R'(0) - C'(0)]/R'(0)$ while the other firm set its index equal to zero. Consequently, the equation for the trade termination curve where $x = 0$ is:

$$(1-a)(1-b) = (1-t) \tag{20}$$

Thus, all trade will be curtailed whenever $1 - (1-a)(1-b) \geq t$ or $a + b - ab \geq t$.

If constant quantity contours were to be drawn in Fig. 5.4, they would have negative slopes.

$$\left. \frac{\partial b}{\partial a} \right|_{dx=0} = -\frac{(1-b)}{(1-a)} \tag{21}$$

Notice that the slope of the seller's (buyer's) reaction function given in equation 17 (equation 18), must be of a larger (smaller) magnitude than the slopes of the constant quantity contours. Thus, the quantity transacted as well as the seller's (buyer's) profit declines continuously as the seller (buyer) is moved upward (to the right) along its reaction function and away from the pure monopoly (monopsony) point. Consequently, the Nash equilibrium at point $N$ in Fig. 5.4 must involve a lower volume of transactions than either the pure monopoly or the pure monopsony point.

## 5A.2. A Formal Model of Bilateral Oligopoly

While thin markets are a pervasive legacy of the former command system, there may not be just one seller and one buyer. In this section of the Appendix, we generalize from the simple model of a pure bilateral monopoly to a more complex model of a bilateral oligopoly where there are a small number of buyers and sellers. We also show more formally that fostering competition helps remove and ultimately eliminate the efficiency problems associated with both bilateral monopoly and bilateral oligopoly.

Consider a bilateral oligopoly where there are $I$ sellers and $J$ buyers. For the '$i$'th seller: the output quantity is $y_i \geq 0$, the cost function is $C_i(y_i)$, and the marginal cost function is $C'_i(y_i)$. Similarly, for the '$j$'th buyer: the input quantity is $z_j \geq 0$, the revenue function is $R_j(z_j)$, and the marginal revenue function is $R'_j(z_j)$. For simplicity, we will assume that the marginal cost function of each seller is positively sloped (i.e. $C''_i(y_i) > 0$) and that the marginal revenue function of each buyer is negatively sloped (i.e. $R''_j(z_j) < 0$). We will continue to measure the degree of market power exercised by each player using Lerner indexes.

$$a_i \equiv \frac{p - C'_i(y_i)}{p} \quad i = 1, \ldots, I \tag{22}$$

$$b_j \equiv \frac{R'_j(z_j) - p}{R'_j(z_j)} \quad j = 1, \ldots, J \tag{23}$$

Here, $a_i$ is the Lerner index of monopoly power for the '$i$'th seller and $b_j$ is the Lerner index of monopsony power for the '$j$'th buyer.

The inverse supply function of each seller and the inverse demand function of each buyer can be obtained by rearranging the definitions of the indexes of market power.

$$p_i = \frac{1}{1 - a_i} C'_i(y_i) \quad i = 1, \ldots, I \tag{24}$$

$$p_j = (1 - b_j) R'_j(z_j) \quad j = 1, \ldots, J \tag{25}$$

Here, $p_i$ is the offer price of the '$i$'th seller, and $p_j$ is the offer price of the '$j$'th buyer. Once again, the Lerner indexes of market power are effectively proportional pricing wedges.

In order to aggregate across sellers and buyers, it is necessary to invert the inverse supply and demand functions. The regular as opposed to inverse supply function of each seller and the regular demand function of each buyer will be written in the following way.

$$y_i = S_i(p, a_i) \quad i = 1, \dots, I \tag{26}$$

$$z_j = D_j(p, b_j) \quad j = 1, \dots, J \tag{27}$$

Before proceeding further, it is useful to define various vectors that consist of the indexes of market power or pricing wedges.

$$\boldsymbol{a} \equiv (a_1, \dots, a_I), \quad \boldsymbol{a}_{-i} \equiv (a_1, \dots, a_{i-1}, a_{i+1}, \dots, a_I) \quad i = 1, \dots, I \tag{28}$$

$$\boldsymbol{b} \equiv (b_1, \dots, b_J), \quad \boldsymbol{b}_{-j} \equiv (b_1, \dots, b_{j-1}, b_{j+1}, \dots, b_J) \quad j = 1, \dots, J \tag{29}$$

We can now define the residual demand function facing the '$i$'th seller and the residual supply function facing the '$j$'th buyer.

$$D_{-i}(p, \boldsymbol{a}_{-i}, \boldsymbol{b}) \equiv \sum_{j=1}^{J} D_j(p, b_j) - \sum_{\substack{h=1 \\ h \neq i}}^{I} S_i(p, a_h) \, i = 1, \dots, I \tag{30}$$

$$S_{-j}(p, \boldsymbol{a}, \boldsymbol{b}_{-j}) \equiv \sum_{\substack{k=1 \\ k \neq j}}^{J} D_j(p, b_k) - \sum_{i=1}^{I} S_i(p, a_i) \, j = 1, \dots, J \tag{31}$$

The residual demand facing the '$i$'th seller consists of the demands of all the buyers minus the supplies of all of the other sellers, while the residual supply facing the '$j$'th buyer consists of the supplies of all the sellers minus the demands of all of the other buyers. Of course, the market constrains the output of each seller to be equal to the residual demand it faces, and the input of each buyer to be equal to the residual supply it faces.

$$y_i = D_{-i}(p, \boldsymbol{a}_{-i}, \boldsymbol{b}) \quad i = 1, \dots, I \tag{32}$$

$$z_j = S_{-j}(p, \boldsymbol{a}, \boldsymbol{b}_{-j}) \quad j = 1, \dots, J \tag{33}$$

In order to provide the most conventional form for the analysis, we will invert these market constraints in order to obtain the inverse residual-demand function facing each seller and inverse residual-supply function facing each buyer.

$$p_{-i} = P_{-i}(y_i, \boldsymbol{a}_{-i}, \boldsymbol{b}) \quad i = 1, \ldots, I \tag{34}$$

where: 
$$\frac{\partial P_{-i}(y_i, \boldsymbol{a}_{-i}, \boldsymbol{b})}{\partial y_i} = \left[ \sum_{j=1}^{J} \frac{1}{(1 - b_j) R_j''} - \sum_{\substack{h=1 \\ h \neq i}}^{I} \frac{1 - a_h}{C_h''} \right]^{-1}$$

$$p_{-j} = P_{-j}(z_j, \boldsymbol{a}, \boldsymbol{b}_{-j}) \quad j = 1, \ldots, J \tag{35}$$

where: 
$$\frac{\partial P_{-j}(z_j, \boldsymbol{a}, \boldsymbol{b}_{-j})}{\partial z_j} = \left[ \sum_{i=1}^{I} \frac{1 - a_i}{C_i''} - \sum_{\substack{k=1 \\ k \neq j}}^{J} \frac{1}{(1 - b_k) R_j''} \right]^{-1}$$

Here, $p_{-i}$ is the minimum offer price facing the '$i$'th buyer, and $p_{-j}$ is the maximum offer price facing the '$j$'th seller. The elasticity of the acceptance price facing each seller and the elasticity of the offer price facing each buyer can also be defined.

$$\phi_{-i} \equiv \frac{y_i}{P_{-i}(y_i, \boldsymbol{a}_{-i}, \boldsymbol{b})} \frac{\partial P_{-i}(y_i, \boldsymbol{a}_{-i}, \boldsymbol{b})}{\partial y_i} \equiv \Phi_{-i}(y_i, \boldsymbol{a}_{-i}, \boldsymbol{b}) \quad i = 1, \ldots, I \tag{36}$$

$$\Psi_{-j} \equiv \frac{z_j}{P_{-j}(z_j, \boldsymbol{a}, \boldsymbol{b}_{-j})} \frac{\partial P_{-j}(z_j, \boldsymbol{a}, \boldsymbol{b}_{-j})}{\partial z_j} \equiv \Psi_{-j}(z_j, \boldsymbol{a}, \boldsymbol{b}_{-j}) \quad j = 1, \ldots, J \tag{37}$$

In contrast to the two-firm, bilateral monopoly case, the price elasticity facing each firm depends directly on the indexes of market power of all other firms.

The '$i$'th seller's equilibrium output can be determined contingent on the prevailing pricing wedges by equating its own offer price with the minimum offer price that it faces from the rest of the market. Similarly, the '$j$'th buyer's wedge-contingent equilibrium input function is determined by equating its own offer price with the maximum offer price that it faces.

$$y_i = Y_i(\boldsymbol{a}, \boldsymbol{b}) \text{ solves } \frac{1}{1 - a_i} C_i'(y_i) = P_{-1}(y_i, \boldsymbol{a}_{-i}, \boldsymbol{b}) \quad i = 1, \ldots, I \tag{38}$$

$$z_j = Z_j(\boldsymbol{a}, \boldsymbol{b}) \text{ solves } (1 - b_j) R_j'(z_j) = P_{-j}(z_j, \boldsymbol{a}, \boldsymbol{b}_{-j}) \quad j = 1, \ldots, J \tag{39}$$

It is also possible to establish the wedge-contingent equilibrium market price for reference purposes.

$$p = P(a, b) = \begin{cases} P_{-i}\left(Y_i(a, b), a_{-i}, b\right) & i = 1, \ldots, I \\[2mm] \dfrac{1}{1 - a_i} C'_i\left(Y_i(a, b)\right) & i = 1, \ldots, I \\[2mm] P_{-j}\left(Z_j(a, b), a, b_{-j}\right) & j = 1, \ldots, J \\[2mm] (1 - b_j) R'_j\left(Z_j(a, b)\right) & j = 1, \ldots, J \end{cases} \tag{40}$$

Of course, the same wedge-contingent equilibrium price function arises regardless of which firm is the subject of attention.

Let us reinterpret Fig. 5.1 from the standpoint of a typical supplier. The $C'$ curve continues to represent the seller's marginal cost curve, but the $R'$ curve is now the residual demand curve facing the sample seller when all other firms set their indexes of market power equal to zero. Thus, $x^*$ is the efficient output for the sample firm. The $S(a_N)$ curve, is the seller's own supply curve for the situation where its own index of market power is equal to $a_N$. The demand curve, $D(b_N)$, is the residual demand curve facing the seller in question. Since the residual demand curve is formed by adding up the demand curves of each buyer and subtracting off the supply curves of each other seller, it is drawn contingent on the pricing wedges of all buyers and all other sellers. The intersection of the $S(a_N)$ and $D(b_N)$ curves gives the equilibrium output of the supplier in question, $x_N$, and the equilibrium market price, $p_N$, contingent on the pricing wedges of all of the players. Of course, we could also reinterpret Fig. 5.1 from the perspective of a typical buyer.

Let $f_i$ and $g_j$ denote the profits of the '$i$'th seller and the '$j$'th buyer. Of course, profits are equal to revenue minus cost.

$$f_i = p_{-i} y_i - C_i(y_i) \quad i = 1, \ldots, I \tag{41}$$

$$g_j = R_j(z_j) - p_{-j} z_j \quad j = 1, \ldots, J \tag{42}$$

Wedge-contingent profit functions of each seller and buyer can now be formed.

$$f_i = F_i(a, b) = P_{-i}\left(Y_i(a, b), a_{-i}, b\right) Y_i(a, b) - C_i\left(Y_i(a, b)\right)$$
$$i = 1, \ldots, I \tag{43}$$

$$g_j = G_j(a, b) = R_j\left(Z_j(a, b)\right) - P_{-j}\left(Z_j(a, b), a, b_{-j}\right) Z_j(a, b)$$
$$j = 1, \ldots, J \tag{44}$$

Each firm chooses how much market power to wield conditional on the market power being exercised by all of the other players on both sides of the market. The first-order conditions for profit maximization are completely conventional.

$$\frac{\partial F(\boldsymbol{a}, \boldsymbol{b})}{\partial a_i} = 0 \Leftrightarrow \left( p_{-i} + y_i \frac{\partial P_{-i}}{\partial y_i} \right) = C_i' \quad i = 1, \ldots, I \tag{45}$$

$$\frac{\partial G(\boldsymbol{a}, \boldsymbol{b})}{\partial b_j} = 0 \Leftrightarrow R_j' = \left( p_{-j} + z_j \frac{\partial P_{-j}}{\partial z_j} \right) \quad j = 1, \ldots, J \tag{46}$$

Each seller sets its index of market power so as to equate the marginal cost of its own output with the marginal revenue that it faces, and each buyer equates the marginal revenue from its own input with the marginal input cost that it faces. The second-order conditions are also standard.

$$\frac{\partial^2 F_i(\boldsymbol{a}, \boldsymbol{b})}{\partial a_i^2} < 0 \Leftrightarrow \left( 2 \frac{\partial P_{-i}}{\partial y_i} + y \frac{\partial^2 P_{-i}}{\partial y_i^2} \right) < C_i'' \quad i = 1, \ldots, I \tag{47}$$

$$\frac{\partial^2 G_j(\boldsymbol{a}, \boldsymbol{b})}{\partial b_j^2} < 0 \Leftrightarrow R_j'' < \left( 2 \frac{\partial P_{-j}}{\partial z_j} + z_j \frac{\partial^2 P_{-j}}{\partial z_j^2} \right) \quad j = 1, \ldots, J \tag{48}$$

The slope of the marginal revenue function facing each seller must be less than the slope of its own marginal cost function, and the slope of each buyer's own marginal revenue function must be less than the slope of the marginal cost function that it faces.

Let us reinterpret Fig. 5.2 from the standpoint of a typical seller. The marginal cost curve of the seller in question is $C'$, the residual demand curve that it faces is $D(b_N)$, and the marginal revenue curve that it faces is $MR(b_N)$. Both the residual demand curve and the marginal revenue curve derived from it are drawn contingent on particular values of the pricing wedges of all buyers and all other sellers. Given the values of the wedges of all other firms, the sample seller's profit-maximizing output, $x_N$, is determined by the intersection of the $MR(b_N)$ and $C'$ curves. Moreover, given the pricing wedges of all other firms, the optimum degree of market power for the seller in question will just displace its own supply curve upward to $S(a_N)$ which intersects the $D(b_N)$ curve at the profit maximizing output. We could use an analogous argument to reinterpret Fig. 5.3 from the perspective of a typical buyer.

The first-order condition of each player is its reaction function. It is helpful to simplify these reaction functions using the definitions of the Lerner indexes of market power.

$$a_i = -\Phi_{-i}\left(Y_i\left(\boldsymbol{a},\boldsymbol{b}\right), \boldsymbol{a}_{-i}, \boldsymbol{b}\right) \quad i = 1, \ldots, I \tag{49}$$

$$b_j = \frac{\Psi_{-j}\left(Z_j\left(\boldsymbol{a},\boldsymbol{b}\right), \boldsymbol{a}, \boldsymbol{b}_{-j}\right)}{1 + \Psi_{-j}\left(Z_j\left(\boldsymbol{a},\boldsymbol{b}\right), \boldsymbol{a}, \boldsymbol{b}_{-j}\right)} \quad j = 1, \ldots, J \tag{50}$$

We can use Brouwer's fixed point theorem (Gravelle and Rees, 1992) to provide a general solution for the existence of a Nash equilibrium in indexes of market power for the bilateral oligopoly. When there are a small number of firms on each side of the market, each firm will always face a non-zero price elasticity and it will have a non-negligible equilibrium index of market power. Thus, there will be an inefficiently low volume of transactions on the market as a whole.

In the case where all sellers have identical cost functions and all buyers have identical revenue functions, we can reinterpret Fig. 5.4 to show how the Nash equilibrium is determined. As a result of the existence of symmetry, all sellers will have a common index of market power, $a$ in the Nash equilibrium, and all buyers will have a common index, $b$. The substitution of $a$ and $b$ into any one of the $I$ equations in 49 yields a 'solution function' for sellers that can be graphed as a curve such as AA in Fig. 5.4. Similarly, the substitution of $a$ and $b$ into any one of the $J$ equations in 50 yields a solution function for sellers that can be graphed as a curve such as BB. Thus, Fig. 5.4 indicates that the Nash equilibrium value of each seller's index of market power is $a_N$, while that of each buyer is $b_N$.

Finally, let us consider the impact of increased competition on the bilateral oligopoly. Suppose the market is replicated such that it is $N$ times its original size. In other words, there are $N$ identical firms corresponding to each and every original seller and buyer giving a total $NI$ sellers and $NJ$ buyers. If the indexes of market power for each of the $I + J$ types of firms are kept at their original levels, replicating the economy $N$ times will have no impact on the quantity transacted by each type of firm or on the market price. Nevertheless, the absolute value of the elasticity of the acceptance price facing each type of seller will decline because the slope of the (inverse) residual demand function that it faces gets smaller (i.e. see equations 13 and 15). Similarly, the elasticity of the offer price facing each type of buyer declines because the slope of the (inverse) residual supply function that it faces gets smaller (i.e. see equations 14 and 16). As the number of replications of the market goes to infinity: (i) the slopes of the inverse residual demand and supply functions facing each seller and buyer go to zero; (ii) the absolute values of all of the elasticities of acceptance prices and offer prices go to zero; and (iii) the profit-maximizing indexes of market power fall to zero (see equations 49 and 50). Thus, as the number of firms on each side of the market gets large, the market tends towards the ideal of perfect competition.

# Notes

1. In Romania and Albania there has been a forced dissolution of state and collective farms but now the individual farms tend to be too small. In Poland, small family-based peasant farms remained the dominant form even in the communist era. Issues concerning farm size and structure are discussed further in Chapter 7.
2. If fertilizer were the state farm's only variable input, then the marginal revenue function for its input would simply be its marginal revenue product function.
3. One type of simple game is a prisoners' dilemma. Suppose that there are two prisoners. Each must either confess to committing a crime or deny it and they are unable to cooperate because they are interrogated separately. If neither confesses to the crime, they will each receive a one-year sentence. If one confesses and the other does not, then the 'confessor' goes free and the 'denier' serves a life sentence. If both confess, they both receive a twenty-year sentence. The prisoners must each consider what the other will do as they make their best choice. The first prisoner rationalizes that if the second confesses, she is better off confessing, and if the second denies, she is still better off confessing. In the same way, the other prisoner rationalizes that he is always better off confessing. Thus, both confess regardless of whether either or both of them are guilty. Notice that this non-cooperative equilibrium is inefficient because both prisoners would have been better off if they had each denied. For further details on game theory, see Gravelle and Rees (1992) and Tirole (1988).
4. For example, in a (Cournot) oligopoly where each firm chooses its output, the instruments are symmetric because both choice variables are outputs, and independent because each firm has its own output.
5. We could, however, allow the firms to simultaneously choose either quantity or price constraints. For example, both the buyer and seller could choose the maximum quantity of fertilizer that they would transact. In this case each firm would always want to set its quantity constraint just below that of its opponent in order to drive the price in its own favour. Consequently, the quantity transacted would always approach zero. While quantity constraints would generate an unduly restrictive result, price constraints would be unreasonably competitive.
6. The analysis of the non-cooperative bilateral monopoly game in indexes of market power has strong parallels with the two-country tariff retaliation model in international trade (see Johnson, 1953–1954 and Dixit, 1987).
7. It is also possible to model a non-cooperative bilateral monopoly using absolute rather than proportional pricing wedges (see Gaisford *et al.*, 1995). In such an approach, the seller's absolute wedge would be equal to $p_S - C'(x)$ while that of the buyer would be equal to $R'(x) - p_B$. The model with absolute pricing wedges generates essentially the same results, but proportional wedges are more reasonable. These conventional indexes of market power have the advantage of being comparable across markets.
8. A reaction curve is simply a best-choice curve that answers the question: 'if my rival does $X$, then what is my best choice?' for the whole range of possible $X$s that the rival could choose. Thus, the seller's reaction curve shows its best response to the buyer's actions. For this reason, reaction curves are sometimes called best-response curves.
9. Figure 5.4 shows a situation where both firms have negatively-sloped reaction functions and the indexes of market power of the seller and the buyer are strategic

substitutes (see Bulow *et al.*, 1985). Even at the other extreme where reaction functions were positively sloped and the indexes were strategic complements in the neighbourhood of the equilibrium, the quantity would still be restricted to a greater extent under the bilateral monopoly than with either a pure monopoly or a pure monopsony.

10. As it happens in Fig. 5.1, the seller is equally well off in the Nash equilibrium and the competitive position because area 3 is equal to area 5. If iso-profit curves for the two firms were drawn through the Nash equilibrium point in Fig. 5.4, the seller's iso-profit curve would cut through point 0 while the buyer's iso-profit curve would cut the horizontal axis to the right of point 0.

11. It is especially important to note that cooperation between firms on the same side of the market (e.g. cartels) typically reduces efficiency while cooperation across markets by buyers and sellers can enhance efficiency. Contract law in virtually all advanced market economies recognizes that such across-market cooperation is in the public interest. The advanced market economy countries are less uniform in their treatment of collusion between groups of sellers or groups of buyers. In the USA, Britain and Canada, contracts between groups of sellers or groups of buyers are often prevented by competition policy, but in Japan some forms of cooperation of this type are actively promoted.

12. For example, consider the cooperative game that would arise in the situation described by Figs 5.1 and 5.4, the quantity transacted would be $x^*$ but the price would not be $p_0$. Footnote 10 indicates that the seller would be as well off at the non-cooperative Nash equilibrium (the disagreement event) as in the competitive position. In the Nash bargaining solution to the cooperative game, the seller and the buyer will share the extra surplus that arises from cooperation. Since both firms must be strictly better off from cooperating, the competitive position is ruled out as a solution to the cooperative game. Rather, the Nash bargaining solution to the cooperative game will involve a positive (rather than zero) index of market power for the seller and an offsetting negative (rather than zero) index for the buyer. Consequently, the cooperative price will be higher than $p_0$ and we can say that the seller is in a stronger bargaining position than the buyer.

13. Where there are specific investments that must be made in advance by either the buyer or the seller, there are likely to be examples of opportunism where firms strategically renege on contracts (e.g. see Williamson, 1975; Klein *et al.*, 1978). Opportunism was discussed in more detail in Chapter 4. In a related vein, asymmetric information can also lead to inefficiencies (e.g. Myerson and Satterthwaite, 1983).

# References

Brooks, K. and Lerman, Z. (1993) Land reform and farm restructuring in Russia: 1992 status. *American Journal of Agricultural Economics* 75, 1254–1259.

Brooks, K., Guasch, L.J., Braverman, A and Csaki, C. (1991) Agriculture and the transition to market. *Journal of Economic Perspectives* 5, 149–161.

Bulow, J.I., Geanakopolos, J.D. and Klemperer, P.D. (1985) Multimarket oligopoly: strategic substitutes and complements. *Journal of Political Economy* 93, 488–511.

Coase, R.H. (1937) The nature of the firm. *Economica* new series 4, 386–405.

Csaki, C. (1995) Presidential address: where is agriculture heading in Central and Eastern Europe? Emerging markets and the new role for the government. In: Peters, G.H. and Hedley, D.D. (eds) *Agricultural Competitiveness and Policy Choice*. Dartmouth, Aldershot, pp. 22–24.

Dixit, A.K. (1987) Strategic aspects of trade policy. In: Bewley, T. (ed.) *Advances in Economic Theory, Proceedings of the 5th World Congress of the Econometrics Society*. Cambridge University Press, Cambridge UK, pp. 329–362.

Gaisford, J.D., Kerr, W.A. and Hobbs, J.E. (1995) If the food does not come: vertical coordination in the CIS food system: some perils of privatization. *Agribusiness: an International Journal* 11, 179–186.

Gravelle, H. and Rees, R. (1992) *Microeconomics*. 2nd edn. Longman, London, 752 pp.

Johnson, H.G. (1953–1954) Optimum tariffs and retaliation. *Review of Economic Studies* 21, 142–153. Expanded and revised as: Johnson, H.G. (1958) Optimum tariffs and retaliation. In: Johnson, H.G. (ed.) *International Trade and Economic Growth*. Harvard University Press, Cambridge, Massachusetts, pp. 31–61.

Klein, B., Crawford, R.G. and Alchian, A.A. (1978) Vertical integration, appropriable rents, and the competitive contracting process. *Journal of Law and Economics* 21, 297–326.

Myerson, R.B. and Satterthwaite, M.A. (1983) Efficient mechanisms for bilateral trading. *Journal of Economic Theory* 29, 265–281.

Spengler, J.J. (1950) Vertical integration and anti-trust policy. *Journal of Political Economy* 58, 347–352.

Tirole, J. (1988) *The Theory of Industrial Organization*. MIT Press, Cambridge, Massachusetts, 479 pp.

Williamson, O.E. (1975) *Markets and Hierarchies: Analysis and Antitrust Implications*. Free Press, New York, 286 pp.

Zedgar, J.S. (1995) The impact of economic transformation on the development of rural areas in Central and Eastern European Countries. In: Peters, G.H. and Hedley, D.D. (eds) *Agricultural Competitiveness and Policy Choice*. Dartmouth, Aldershot, pp. 387–396.

# 6 Few Sellers: Agroinputs for the Farm Sector

The entire question of the provision of agroinputs in post-command economies is extremely complex. In modern market economies there has been a slow evolutionary process whereby fewer and fewer of the technically separable stages in the production and distribution chain of agrifood systems are coordinated within the management structure of individual farms. In the past, farms performed many of the processing and distribution activities, including the transfer to final consumers. Pig farmers often slaughtered their own animals, preserved the meat, butchered it and transported it to final consumers. The farmer may have even delivered the meat from door to door. Egg rounds were common. Cream was processed into butter on the farm and sold to retailers or directly to consumers. Over time, most of these activities have been taken over by separate commercial enterprises. At the same time as farms were shedding these activities and the responsibility for their coordination, they were expanding horizontally by producing more of their primary agricultural outputs. This horizontal expansion was facilitated by the economies of size brought by mechanization.

Similarly, farms gradually shed agricultural input activities. Seed saved from crop production was replaced by seed provided by commercial suppliers. Industrially produced fertilizer replaced or supplemented manure on many operations. Tractors replaced draught animals bred on the farm. For some commodities, even the various stages of farm level production became vertically segmented. In the North American beef cattle industry, cow-calf operations now undertake breeding activities which provide calves for a separate (both geographically and in ownership) feedlot industry. A similar division is observed in the swine industry with breeding operations being distinct from fattening operations. Chick hatcheries are segmented from egg production. One farm becomes an input supplier for another farm.

Markets initially replaced managerial direction by the farmer. In recent years, coordination between input suppliers and farmers through markets has, in some cases, been replaced by contracts which tie farmers to input suppliers. These contracts may have clauses which, in effect, allow the supplier to direct many aspects of on-farm production. Hence, the distinction between farmer and input supplier becomes blurred. No standardized pattern for the relationship between input suppliers and primary agricultural production has emerged in modern market economies. Spot markets, auctions, contracts, leasing arrangements, farmer-owned input purchasing cooperatives, full vertical integration and a host of other arrangements can all be identified. Often a range of coordination mechanisms co-exist within the same industry. As a result, when one wishes to examine the liberalization of the agroinput industries in the Central and Eastern European Countries (CEECs) and the New Independent States (NISs), there is no single model which can be used to answer the question – transition to what?

The arrangements which have emerged in the initial stages of the liberalization process to coordinate the provision of farm inputs in the CEECs and the NISs are also very complex. As a result, in the agroinput sector one is faced with a transition process in which it is almost impossible to identify the starting point and where there is no clear end in sight. Given these constraints, the remainder of this chapter will attempt to provide some insights into the progress made, and the difficulties encountered thus far in the process of transition.

## 6.1. Agroinputs in the Command Economy

As discussed in Chapter 2, large-scale operations were the norm in command economies. The agroinput industry was no different. The large scale of collective and state farms allowed inputs to be supplied in bulk. Further, collective/state farms stressed centralization, including centralization of physical infrastructure such as on-farm storage for fertilizer. This naturally led to the existence of large-scale input supply facilities.

While specialized input suppliers arose from the ideological belief in large-scale production/distribution units, they also arose because the planners, following what they observed in modern market economies, moved many of the input activities traditionally undertaken on the farm into the industrial sector. In fact, the process of vertical segmentation often went much further than was the case in modern market economies. The early Soviet attempts at tractor and machinery depots, where non-fixed machinery was to be shared amongst collective/state farms over a wide geographic region, was a clear example of this principle. Centralized machinery maintenance facilities were common in the former command economies whereas in market economies, most routine machinery maintenance is undertaken on individual farms.

Separate input supply enterprises were often created within the structure of agro-industrial complexes (AICs). For example, a common feed mill would have been built to supply all of the collective farms in an AIC. Heifers were often removed from the dairy operations of individual collective/state farms and raised in centralized facilities until they were returned to the dairy herd just prior to the start of their first lactation. Artificial insemination units often served all of the collective/state farms in a particular area. The ownership of these input enterprises was (and remains) mixed. In some cases, the input enterprises were owned directly by the state. In other cases, they belonged to AICs. They may also have been owned by a cooperative made up of collective/state farms. A single collective farm may have owned an input facility and supplied other local collective/state farms. Questions of ownership and administrative control are important in the context of the privatizations initiated as part of the liberalization process.

Input industries were often subsidized as part of the overall attempt to keep food prices low. In effect, this meant low official prices for agro-inputs.

The main legacy of the command era for the process of transition is that, for all intents and purposes, agroinput enterprises were geographic monopolies. Alternative sources of inputs were seldom available to farms.

## 6.2. Transition and the Absence of Competition

Transition problems began to arise once the planner's visible hand was removed and prices were freed. Where the agroinput industries were controlled directly by the government, the freeing of prices allowed officials in the state bureaucracy to set monopoly prices for agroinputs. In its 1994 report on transition in the CEECs and the NISs, the Organization for Economic Cooperation and Development (OECD, 1994a) reported that 'the continuing monopolization and inefficiency in the supply of inputs has also pushed prices paid by producers above what would be available under more competitive conditions' (p. 14).

With the removal of central planning, government-owned agricultural input industries were faced with a combination of exogenous shocks – reduction or elimination of their subsidies, rapidly rising prices for foreign inputs and supply disruptions as a result of the break-up of the Council for Mutual Economic Assitance (CMEA). At the same time, these industries were expected to become centres of profit or at least to minimize their losses. It is hardly surprising that their management turned to monopoly pricing strategies. With the freeing of prices, farms were also faced with a monopsonistic government food processing sector which was able to depress farm gate prices:

> As a result of the transition process of the last few years, agriculture in the region was caught in a price cost squeeze between the upstream and downstream components of the food chain: within a sometimes high general price inflation, agricultural output prices have not risen as rapidly as input prices.
>
> (OECD, 1994a, p. 13)

## 6.3. Strategies for Reducing the Market Power of Agroinput Suppliers

The World Bank, in its 1994 report on the Ukraine, suggested that reducing monopoly power in the agrifood chain should have a high priority.

> In tandem with enterprise commercialization, actions to demonopolize the input supply, marketing, processing and retailing sectors must be taken to address the bulk of the system . . . which continues to be heavily monopolized . . .
> There are essentially three approaches to demonopolization: (a) actively pursuing foreign competition through open trade; (b) breaking up existing large-scale monopoly units; and (c) regulating the large-scale monopoly units through legislation and financing.
>
> (World Bank, 1994, p. 84)

While, in theory, the latter strategy is available in the former command economies, in practice it may be counter-productive. As suggested in Chapter 5, this is because the regulation of prices (or finances) would appear to be a return to state interference in the agrifood chain – a return to the command system in another guise. Further, the reason for regulating monopolies in market economies is because they represent market failures. Removing market failure requires that the competitive market solution be approximated when the price and quantity are established by the regulation. The determination of an appropriate regulated price has proved to be an extremely complex and resource-intensive process in market economies where large numbers of individuals with specialized human capital exist. The arbitrary price setting skills of former planners are not the same set of skills required to establish a regulated price for a monopoly. As a result, it is unlikely that the regulated price established by a price-setting authority in former command economies would approximate the price that would arise in the absence of market failure. As suggested in Chapter 5, inefficient quantities of output would result. Further, if the regulated price does not reflect the relative value of the agroinput, a distortionary price signal will be put in place. As a result, inappropriate investment decisions may be made.

Breaking up existing large-scale monopoly units, while theoretically possible, may prove to be extremely difficult in practice (World Bank, 1994). Further, given the physical infrastructure bequeathed by the command system, this strategy may only have limited success. In some countries, the opportunity to break up monopoly units may have already been foregone in the rush to privatization. In Hungary, for example, the state-owned companies trading in equipment and materials required for agricultural production (known as AGROKERS) were command firms enjoying regional monopolies.

> By May 1993, of the 15 AGROKERS, 13 had been transformed into joint stock companies and two into limited liability companies. Five of the transformed companies have been privatized with a majority Hungarian interest and one with a majority foreign interest.
>
> (OECD, 1994b, p. 70)

In Russia:

> The upstream and downstream enterprises with which farms interact continue to privatize; by July 1993, about half of the state food-processing enterprises were private. Yet, change in their official status has altered their behavior little. Both upstream and downstream enterprises continue to use their market power to 'price-squeeze' all types of farms – state, collective, private.
>
> (OECD, 1994a, pp. 152–153)

The World Bank (1994) suggested that prior to privatization the national enterprises should be:

> broken up into rayon-level[1] businesses operating independently and reliant on their own financial results for working capital. In the overall scope of the country's privatization program, these rayon-level enterprises could eventually be transferred from state ownership to private hands.
>
> (World Bank, 1994, p. 84)

The agroinput supply industries present a dilemma for privatization agencies. The large command firms were attractive privatization entities precisely because of the market power which they represent. Breaking these large national or regional monopolies into smaller enterprise-sized units would have considerably reduced their attractiveness for privatization. Foreign investors, in particular, find acquisition of a major ownership position in existing monopoly supply networks attractive. The existing networks provide a ready-made distribution system for their exports and their market power can be used to discourage other foreign competitors. One of the Hungarian AGROKERS was acquired by a foreign firm (OECD, 1994b), and in the Ukraine the largest private import supply company is a joint venture with a major foreign input supplier (World Bank, 1994).

Breaking up national or regional monopolies by dividing the existing infrastructure into separate administrative units which would approximate firms in modern market economies may have only limited success as a demonopolization strategy. These new input supply firms might either be owned by the state, but with an independent management, or be privatized. The major reason why this strategy will not remove the monopoly position of agroinput suppliers is the investments in large-scale facilities made in the command era. The existing large-scale facilities will simply become local monopoly suppliers. The expected success of this privatization strategy is based on the assumption that, once broken up, the new firms will begin to compete with each other.

The difficulties faced by a firm which wishes to expand into new geographic markets, however, should not be underestimated. The poor transportation and communication infrastructure, particularly between rural areas, will mean considerable costs will be associated with the penetration of lateral markets. Further, existing local import distributors are likely to take strategic actions aimed at retaining their monopoly position. If they have a geographic cost advantage they will be able to use prices strategically to discourage new entrants. The existing inputs suppliers in a region may find other ways to prevent the entry of competitors. A competitor will have to acquire land to establish its operation. The required land may still belong to collective/state farms. The existing inputs supplier may be able to use the threat of its market power to influence the land allocation decisions of its customers.

Further, as agroinputs have never been surplus commodities, without new investment existing local facilities may not have product available to expand into new market areas. There may, however, be some short-run surplus capacity in agroinput facilities due to the general decline in agricultural production.

Competing directly with each other is likely to be detrimental to both the firm attempting to expand its market share and the firm attempting to defend its market share. In the absence of an enforceable anti-trust/anti-combines policy, local monopolies may find it easier to reach an accommodation with each other whereby they agree not to compete directly. There may be some competition at the geographic margin but no concerted effort to wrest control of a local market from an existing agroinput firm. If a clear advantage exists for an agroinput firm to operate in two or more markets, the expansion into the new market area can probably be accomplished through a merger or buy-out rather than head to head competition.[2] If the expansion takes place through acquisition then larger monopolies will arise.

A third means of introducing competition into the agroinputs sector is through the entry of foreign input suppliers into the market. In some countries this has already taken place to a considerable degree. Foreign agroinput suppliers have often had considerable advantages over local suppliers when they have expanded into the CEECs and the NISs. In many cases they were able to provide new products that were simply not available when the former regime was in place. Further, even when they provided the same product, the quality of their products may have been demonstrably superior to those provided locally. Their marketing, logistics management, technical support and financing capabilities far exceeded those of the former command firms with whom they had to compete. Foreign suppliers also benefited from the collapse of the CMEA. Many agroinputs were shared out as part of CMEA cooperation and then traded. Countries often specialized in the production of individual lines of products. With the collapse of CMEA, supplies of agroinputs from former CMEA partners were disrupted and local suppliers dependent on imports were no longer able to supply their customers. Foreign firms were able to move into this vacuum.

In some cases, joint ventures have been formed between local suppliers and foreign firms. As a result, foreign firms simply became an alternative source of supply for the existing local monopolies rather than competitors. Direct entry by foreign firms seems, however, to have been the preferred method of entering markets in the CEECs and the NISs. Cooperation between foreign and domestic managers in joint ventures has always been difficult given differences in managerial approach and experience. Direct entry also reduces the problems created by the lack of infrastructure support for business in the CEECs and the NISs. The inability to enforce contracts, the absence of a commercial credit system and the lack of clear direction from governments and bureaucrats imposes costs on foreign as well as domestic firms. By being vertically integrated, foreign input suppliers were able to minimize their transactions cost. For the most part, foreign input suppliers represent long distribution chains which emanate from their home country and terminate in local distribution centres in the former command economies. As a result, they have only a limited impact on the development of the local economy.

When foreign firms have directly entered the market they have often put the former command firm out of business (or would do if governments withdrew their subsidies). There may, however, be considerable competition among foreign input suppliers in some areas. This will benefit the local farm sector. The general decline in agricultural activity and the financial difficulties experienced by many farms has, however, blunted the enthusiasm of foreign input suppliers to some extent. Their assessments of the levels of business and political risk have made them wary of making long-term investments in production facilities. As a result, they remain largely distribution networks for products originating in their home countries.

It is interesting to note that, in its discussion of demonopolizing the Ukrainian agroinput sector, the World Bank (1994, p. 84) laid out only three strategies: (i) regulation of monopoly prices; (ii) fostering increasing competition by breaking up existing state monopolies; and (iii) fostering competition from foreign agroinput suppliers. There is no discussion of competition from new local agroinput suppliers.

Except on a very small scale – individual machinery repair services, veterinary services, etc. – there is little evidence of newly constituted enterprises entering the agroinputs industry. The difficulties associated with raising capital, in obtaining raw materials and establishing a sufficiently wide distribution network may effectively discourage entry. As many of the large collective/state farms remain in being, new small-scale agroinput operations will also face, in effect, monopsonistic buyers for their products.

New domestic entrants also face competition from the existing domestic agroinput suppliers and foreign entrants. Given the size and market power of the former and the managerial expertise of the latter, it will be difficult for new entrants to survive. In some countries, the existing collective/state farms directly control a limited range of agroinput supplies. These enterprises may have an opportunity to grow into a competitive threat to the former state monopolies but the poor financial position of most former collective/state farms means that capital will be scarce, limiting their ability to expand.

## 6.4. Agroinputs for Private Farms

Some agroinputs are controlled by AICs, groups of former collective/ state farms or individual former collective/state farms rather than the government. For the most part, the range of agroinputs provided tend to be limited to those which are the direct result of agricultural production or on-farm agroservices. For example, artificial insemination services may be provided by an enterprise controlled by a number of local collective/state farms. Processed animal feeds, dairy heifers and day-old chicks may be provided by specialized enterprises within AICs. Spare parts and machinery repair services may be controlled by collective farms. This control over inputs, directly or indirectly, by former collective/state farms may act to inhibit the establishment of private (family) farms.

Collective and state farms have been allowed to retain their existing organizational structure in most countries, although they may have been privatized as joint stock companies or through some other form of shared ownership. Private farming is, however, usually allowed. Private farms require land (and possibly other resources) which must be given up by the reconstituted former collective/state farms. Hence, their management may tend to view private farmers as both competitors in the market and competitors for their resources. In some countries, specific provisions have been legislated which force reconstituted former collective/state farms to relinquish farmland to private farmers. The ability of reconstituted collective/state farms to control the flow of inputs to private farmers can, however, be used to discourage individuals from attempting to establish private farms. With no alternative supplier to the local former collective/state farm, private farmers remain dependent upon their major competitors. In Russia, for example,

> The absence of a well-developed (and preferably private) system of services and input supply continues to hinder the development of private farming . . . The inadequacy of support services and marketing outlets tends to keep private farms dependent on the state and collective farms from which they broke away.
>
> (OECD, 1994a, p. 149)

Hence, structural reform at the farm level is dependent, to a considerable degree, on the establishment of alternative agroinput supply networks.

## 6.5. The Effect of Transition Problems in Other Sectors

The transition process in agroinputs has also been affected by a large number of exogenous shocks. Energy prices, which had been kept considerably below world prices in the command system, rose quickly to world prices in many CEECs. Fuel costs for farms increased, adding to their financial difficulties. This price shock also worked itself through the energy intensive agroinput industries such as fertilizers. Import-dependent inputs also suffered from rising prices as the adjustment to world prices was made.

In some cases there were direct disruptions to fuel supplies when the CMEA collapsed. In particular, cheap Russian energy supplies were simply cut off. With refineries and transportation links dedicated to Russian supplies, it was difficult to organize alternative supply chains.

In the CMEA era, low-priced supplies of Russian animal fodder were available in the CEECs. Alternatives to this bulky, transportation-intensive input have been difficult to organize. Spare parts for machinery made in former CMEA partners have often become imposs-ible to obtain. In one case, a tractor plant in the former East Germany which had supplied a number of CMEA countries was simply shut down as uneconomic under the German reunification strategy.

## 6.6. Conclusions

The agroinput sector in Central and Eastern Europe and the former USSR remains a major bottleneck in the process of the transition of the entire agrifood system. Far more attention has been given by politicians, bureaucrats and foreign advisors to the transformation of the farming system itself. While the issues of land reform, farm privatization and stimulating production in primary agriculture have garnered a great deal of thought and attention, the handling of the transition process in the agroinputs sector has been conducted on an *ad hoc* basis. As a result, the existing agroinput monopolies have survived and, due to the freeing of prices, have been able to capitalize upon their market power. This has reduced the profitability of the farm sector and inhibited its ability to become commercially viable. As a result, the fiscal difficulty of gov-ernments has been increased as larger subsidies have been required for the farm sector.

As yet, little domestic competition for existing agroinput suppliers has arisen. Competition is being provided to some degree by foreign agribusinesses. It is not yet clear to what extent foreign firms have begun to compete with each other. At this point, foreign input suppliers are providing agroinputs which, for the most part, were simply not available in the command era.

Unless a greater degree of competition can be introduced into the agroinput sector, prices will remain high, thereby reducing the compet-itive position of primary agriculture. Given the legacy of large-scale investments in physical infrastructure, the breaking-up of former state agroinput monopolies into individually managed facilities will do little to reduce monopolization. The real challenge is to foster a competitive alternative to the former state system.

## Notes

1. Rayons were the local administrative districts in the old Soviet Union. Twelve to twenty rayons constitute an oblast, the administrative district that corresponds most closely to a province or state.
2. See McGee (1958) on this topic.

# References

McGee, J.S. (1958) Predatory price cutting: the Standard Oil (N.J.) case. *Journal of Law and Economics* 1, 137–169.

OECD (1994a) *Agricultural Policies, Markets and Trade in the Central and Eastern European Countries (CEECs), the New Independent States (NISs), Mongolia and China.* Monitoring and Outlook 1994. Centre for Co-operation with the Economies in Transition, Organization for Economic Co-operation and Development, Paris.

OECD (1994b) *Review of Agricultural Policies: Hungary.* Centre for Co-operation with the Economies in Transition, Organization for Economic Co-operation and Development, Paris.

World Bank (1994) *Ukraine; the Agricultural Sector in Transition.* A World Bank Country Study, International Bank for Reconstruction and Development, Washington DC.

# 7 Taking Farms Apart: Large Units or Small Farms?

Unlike the transformation of the wider agrifood system in former command economies, the reform of farming has received considerable attention since liberalization. A substantial literature on the topic already exists. As a result, this chapter will concentrate on the interaction of the farm sector with the rest of the agrifood system. Relationships between the farm sector and the agricultural inputs sector and between the farm sector and the processing sector can influence the effectiveness of farm level reforms. On the other hand, failing to reform primary agriculture can endanger the functioning of the entire agrifood system. As a result, the topic of farm level reforms cannot be ignored when the transformation of agrifood systems in former command economies is being discussed.

The reason why the reform of primary agriculture has received a disproportionate amount of attention throughout the liberalization process is that the issues involved are both complex and very contentious. It is a contentious issue because altering the property rights to agricultural land cannot be divorced from the distributional question of who gets those property rights. As property rights in land represents tremendous income and wealth generation potential, particularly in the long run, the initial distribution of those rights in a post-communist state will play a fundamental role in determining the future prosperity of those who may have a claim to a proportion of the resource. The legal limits put on property rights to land, such as restrictions on transferability, will affect the efficiency with which the land is eventually used. Restrictions on property rights will affect the willingness of farmers to invest in land improvements and conservation measures. The willingness and ability to enforce the set of property rights which are established will also affect the efforts and investments of farm units by altering the risks faced by farmers.

The issue of reforms in primary agriculture has also been contentious because land ownership and the organization of agricultural production were central issues in the communist era. The removal of private property rights to land was a fundamental tenet of communist doctrine. The collectivization of agriculture was only achieved at a very high cost – a horrendous cost in human life and suffering in the former USSR. Of course, it was resisted with some degree of success in the Central and Eastern European Countries (CEECs), particularly in Poland. Landlords, with their potential ability to exploit peasants – often providing the smoking gun of conspicuous consumption themselves – were particularly vilified in the Marxist press and education system. Hence, while there was widespread disillusionment with central planning, there has been much less consensus among reformers regarding the desirability of private ownership of farmland. In other words, while it was conceded that markets would be an improvement over the planners' visible hand in the allocation of resources, the reinstitution of private property in farmland was not perceived as providing the same potential for gains among those charged with implementing reforms. This was particularly the case in the CEECs where many of the former estate owners and private farmers (or their descendents) laid claim to former lands which had been confiscated.

The issue of privatization, and particularly decollectivization of farms, is also contentious due to concerns over food security. While the collective/state farm system was generally recognized as inefficient, it did provide a degree of food security for the great majority of the people in command economies. It became obvious, after even a few months of liberalization, that the process of transforming command economies into market economies was not going to be a simple process and that considerable disruptions could be expected in the economy. As pointed out in Chapter 2, while disruptions to the flow of products could be tolerated in many industries over the period of transition, this was not the case for fuel and food. Even short-run disruptions to food supplies can impose considerable hardship on consumers. In the midst of all the other turmoil in former command economies, an experiment with wholesale land reform was simply perceived as too risky by many policy makers.

Consequently, the central questions relating to the reform of primary agriculture have revolved around: (i) who should receive the property rights to agricultural land?; (ii) should large scale farming be replaced by some variant of the family farm model typical in modern market economies?; and (iii) can the reforms be accomplished in a manner which does not lead to a reduction in the general level of food security? Of course, these three questions are closely interrelated. The answers to these questions are also very much bound up with the evolution of the entire agrifood chain.

## 7.1. Farming in the Command Economy

One of the major tenets of Lenin and other Bolshevik ideologues was that Russian society should become industrialized. This broad desire for industrialization was radically applied to primary agriculture which is not generally perceived to be amenable to that process. Industrialization is typically associated with manufacturing in centralized facilities. In modern market economies, the geographically dispersed nature of most agricultural production and the non-routine nature of many farm tasks are both considered to lead to large monitoring costs being associated with centralized managerial direction – largely to prevent shirking and pilfering. These arguments were rejected by communist theorists. Of course, the reorganization of agriculture on industrial lines also served a valuable political objective, for Stalin in particular, in that it allowed for the elimination of one of the major quasi-capitalist groups in society, prosperous peasants known as *kulacks*.

The industrialization of agriculture removed two major underpinnings of capitalism in one stroke. Property rights to land were removed from individual farmers and, along with them their rights to retain any surplus. Entrepreneurship was eliminated by replacing self-management with managerial orders delivered through a hierarchy (Zhurek, 1994). Peasants voluntarily joined collective farms (*kolkhoz*) vesting their assets in the organization – at least in theory. For all intents and purposes, of course, control of land was vested in the state. The myth of vesting assets with the collective, however, does have considerable ramifications for the process of privatization in the post-communist era. State farms, *sovkhoz*, were closer to an industrial factory in organization with workers simply paid a wage and having no formal legal claim to the land. When the command economy model was extended to the CEECs after the Second World War the *kolkhoz/sovkhoz* model was applied there as well. The *sovkhoz* system was often used for large landed estates of the former gentry – even in Poland where collectivization was resisted.

The industrialization of agriculture imposed a hierarchical management structure for decision-making on the large economic units which constituted collective/state farms. With land used in common and tasks assigned to individuals by managers, over time labour specialization arose. For example, in dairy operations those workers involved in fodder production reported to an agronomist in charge of feed production who, in turn, reported to the chief agrologist in charge of all cropping activities on the farm. Milkers, on the other hand, reported to the breeder in charge of animal operations in the dairy unit who reported to the animal scientist in charge of all animal operations on the farm. Maintenance in the milking barn was the responsibility of a central maintenance unit with its own reporting channel. State/collective farms were, therefore, organized along functional lines running in parallel to each other and seldom, if ever, converging. The hierarchical system discouraged lateral communications among these reporting chains. As a result, there was little appreciation of the problems of those further along the production chain. Lack of cooperation meant that gaps developed in assigning responsibility. Problems which crossed over hierarchical boundaries could only be communicated and rectified by information and decisions moving up and down the respective chains of command. This slowed the speed of response which is often extremely important when dealing with biologically based production or perishable products. The vertical segmentation of this hierarchical system tended to foster conflict as middle level management had little or no incentive to cooperate. Furthermore, middle level management could improve their status with their staffs by defending the actions of the unit for which they were responsible.

Over time, the hierarchical system, with its poor lateral communications, meant that collective farm workers lost the more holistic perspectives over the farm business associated with farmers in modern market economies. Further, collective farm workers failed to develop a wide range of farming skills. In the case of state farm workers with no farming tradition, individuals did not develop farming perspectives and skills. Further, the hierarchical nature of the management structure meant that individual initiative was seldom rewarded but often punished, particularly if something went wrong. Workers became accustomed to acting only upon orders. With no direct incentives to improve their performance, farm workers were inclined to shirk. As predicted by critics of large scale farming, monitoring costs were very high and, as a result, shirking and pilfering became endemic. Productivity in state and collective farms was low.

The only comparatively bright spot in command agriculture was the private plots where individuals were directly able to reap the rewards of their own efforts. While very productive in a relative sense, their small size made them intensive production units whose output could only be sustained by the large degree of shirking which farm workers could get away with on collective/state farms. Farm workers could devote considerable effort to their very small plots. The average household garden plot was only 0.2 hectares for Russian *kolkhoz* families (Durgin, 1994).

The Soviet plan for industrialization of the country required that labour be released from the land to work in urban industrial enterprises. The only way to accomplish this goal was to replace farm labour with machinery (Durgin, 1994). Collectivization on its own could not provide the increases in productivity required. Consolidation of the land allowed for the use of large scale machinery. This dovetailed well with the faith in economies of size which was evident in all sectors of command economies. The machinery produced for working the land tended to be suitable only for extensive cultivation. The predilection for bigness, however, was applied to all aspects of mechanization. Barns used for milking operations were of a standardized model designed to accommodate approximately 200 animals. Swine operations tended to be large scale. Machinery repair facilities tended to be centralized in one location on collective/state farms. Storage for inputs such as fertilizer and seed, as well as outputs, was also concentrated in large facilities. Housing tended to be grouped in villages rather than on individual farmsteads.

State and collective farms were also responsible for providing most rural infrastructure. Housing, school, medical, retirement and social infrastructure was part of the farm unit's responsibility. The legacy of all these past investments has considerable ramifications for the liberalization process.

The large scale farming operations, in turn, suggested large scale input facilities and processing units. As the command system required farms to deliver their products to designated delivery points and to receive their inputs from designated suppliers, there was no need for alternative supply and marketing channels for individual farming units. As a result, individual farms faced monopolistic suppliers of inputs and monopsonistic buyers of outputs once prices were freed and processing firms and input suppliers were expected to make profits.

By the end of the communist era, primary agriculture was perceived as one of the major failures of the command system. Productivity remained low and the sector was, directly or indirectly, heavily subsidized. Farms did not, however, go broke and those who worked on farms had a job and food security, although their incomes were lower than workers in the non-agricultural sectors. As a result, a considerable proportion of farm workers have resisted, or maybe more correctly ignored, the changes brought by the process of transition. Faced with the choice of establishing themselves as private farmers with all the risks apparent in transition economies or remaining part of a large reformed cooperative farm, many have chosen the latter. Certainly, some workers from the former collective/state farms have chosen to strike out on their own but, as yet, these remain the exception. This is not the case, of course, in countries such as Romania and Albania where collective/state farms were abolished outright by law. Becoming a private farmer is extremely risky because: (i) there are few rural market support institutions (such as market based lending institutions) in place; (ii) security of title in land seldom exists; (iii) suppliers and purchasers are monopolistic; and (iv) it is difficult to compete with the still subsidized former collective/state farms. As a result, the transition to private farming has been far from smooth. The legacy of command farming has been hard to overcome.

## 7.2. The Question of Property Rights in Farm Land

While property rights to land on collective farms was legally vested in the collective, in effect the state controlled the rights to retain any surplus generated from the land. For *sovkhoz*, the legal right to the land was vested in the state.

One of the central tenets of the process of reform has been privatization. This has been extended to farm land. As ownership of land offers considerable potential as a source of future wealth, the distribution of the rights to farm land held by the state has proved extremely contentious and complex. One finds that there are three major contenders for the property rights to land in the post-command era. The contenders are: (i) those who claim rights to land on the basis of use – the members of collective/state farms at the onset of reforms; (ii) those who claim the rights to land on the basis of ownership prior to confiscation of private property rights by the communist state; and (iii) the state itself.

Those who claim rights to land on the basis of use are the existing workers on farms and other members of farms' wider social structure such as their pensioners. At the most basic, divestiture of ownership to members of collective/state farms can, in theory, be simple. The existing land assets of the collective/state farm can be equally shared among the members of the collective farm. Each member can receive a voucher equal to a share in the farm's assets. The farm's non-land assets can also be shared out in this manner. Of course, a number of variations on this theme are possible. Instead of equal division, division can be made based on years of service and/or wage levels at the time of privatization.

This method of vesting property rights is workable if the former collective/state farm is to remain in being. It is not workable if land itself is to be distributed to farmers. This method of privatization cannot answer the question – who gets what piece of land? This might be handled by apportioning individual pieces of land among the claimants based on their value. Each claimant would receive the same total value of land resource. However, in the absence of well functioning land markets there is no way to establish the relative value of individual parcels of land. Hence, allocation to individuals, when carried out as in Romania where collective/state farms were broken up, has been an extremely complex process. As a result, in most countries while moves towards vesting ownership in land have taken place, the individual's land assets remain within the organizational structure of the reformed collective/state farm. Alternative arrangements for individuals to withdraw their assets from the common pool have had to be found.

In the CEECs and the Baltic States, where removal of private property rights was relatively recent, the former owners and their descendents have come forward to demand restitution of their property rights. Governments have generally responded positively to these claims. In the case where these claimants are members of collective farms, the process of apportioning land among the farm membership was simplified. Individual parcels can be returned to their former owners.

A large number of problems have arisen, however, with the restoration of property rights. In many cases, the records from former times have not survived and verifying claims has proved difficult. Further, many of the claimants are not members of the former collective/state farms. As a result, they may have no interest in farming the land. The existing system of inheritance can further complicate the process. If inheritance is based on equal division of land among all a family's children, the number of claimants can be very large and the land area received by each claimant too small to be a viable commercial enterprise. The process of establishing the legitimacy of claims and even locating and informing potential claimants is very time consuming. As a result, while many claimants registered their interests and may have received tentative rights to individual pieces of land, few have yet received clear title to their claims.

In the case of the former owners of large scale estates, which often formed the land base for state farms, the farm workers had no claim to use of the land. In some countries, acreage limits were placed on former owners' claims while in other countries offsetting parcels of land or shares in privatized industries have been offered to claimants so that farm workers could receive rights in the lands controlled by their former *sovkhoz*.

When former owners do not wish to take up private farming themselves, their land is often rented back to the former collective/state farm. In some cases, the former collective/state farms have been transferred into joint stock companies with the former owners being the shareholders and the farm workers becoming salaried employees. When the individual allocations of land are small and, as a result, the rents received are small, stockholders have little interest in the operation of the farm. Hence, there is a separation of management and ownership. This has led to considerable latitude for management to act in the interest of the farm workers and/or themselves rather than the owners. Those owners interested in altering the managerial direction of farms face considerable organizational costs in mounting a challenge to the existing management. On many farms the pre-liberalization management has remained in place. As a consequence, there has been resistance or indifference to the process of reform.

In some countries, a portion of the collective/state farm lands were given to farm workers. At a later date, the claims of former owners were also allowed. If the former owner's land had already been distributed, alternatives had to be negotiated. If the distributions plus the claims of former owners exceeded the total land base of the collective/state farm then alternative assets had to be offered. This process has slowed down the pace at which clear title can be established.

The state itself has, in some cases, acted as a claimant. If there is land left over after the other claimants have been satisfied, the government may sell off the remaining land to increase its revenue. Foreign buyers with hard currency have been purchasers in some cases. In other cases, surplus farm land has been sold along with other state assets as part of wider state privatization schemes. As a result, another group of owners has been introduced to the farm sector. While some of these outside owners may be interested in farming, others wish to use the land for non-agricultural purposes such as housing, while still others are willing to rent out their land.

The introduction of many new owners in the rural areas has, as one might expect, led to conflicts. Where land was distributed to former owners, the members of state and collective farms have felt left out or cheated by the process. Poor records and lack of resources for land surveys has slowed the process of establishing clear title. Restrictions have been put on the use, transfer and acquisition of farm land as governments try to reconcile their socialist perceptions of a land owning class with the need to establish a market economy.

Where the question of security of tenure has not been resolved, land markets have failed to develop, consolidation of individual holdings into viable economic units has been delayed and, probably most important, investment into maintenance of the land resource or expansion of its productive capacity has been delayed. As a result, productivity is declining and, in some cases, long term damage is being done to the land resource.

With the ownership status of land unclear, in some cases in the rush to privatize, the non-land assets of state/collective farms have been separated from the farms' land assets. The ownership of the non-land assets has been vested in the members of collective farms or workers on state farms. This might be an equal division per head or based on some combination of years of service and rate of pay. The usual model is to leave the state/collective farm in being with the non-land assets held by the owners in some variant of a joint stock company. Restrictions are usually placed on the disposal of ownership certificates. In some cases, large collective/state farms may have been divided into two or three large units and then privatized. In either case, this has meant that the existing organizational structure of the farms remains virtually unchanged. Until questions surrounding the issue of land tenure can be sorted out, little is likely to change. Where actual distribution of both land and other assets has taken place, it has led to excessive fragmentation of land and underutilization of machinery and other facilities which were designed to be used on large scale operations.

The separation of the ownership of land from the non-farm assets of now privatized state/collective farms (farm corporations) has created a number of problems. Without security of tenure, most non-farm member owners have been willing to rent their land to the farm corporation which controls the other assets, at least until the problems associated with land tenure are resolved. This means, in effect, that the owners of the land have little control over the corporate farms' management. Under these circumstances, management has no incentive to manage the land in the best interest of the owners. As a result, there is underinvestment in the land resource. As productivity remains low, rents are low. Further, management has no incentive to alter its practices. This is particularly true in the case of employment. Management receives support from the farm workers who fear division of the corporate farm's land assets to private farmers. This division would lead to the corporate farm being dissolved and the livelihood of both the remaining farm workers and managers being taken away. The result is continued overstaffing, high wage bills and, hence, low land rent.

Low land rents arise whether the contracts are specified as fixed rent contracts or some form of sharecropping arrangement. As owners can only rent to the existing corporate farm, the farm can act as a monopsony and drive down rental prices. As ownership is often widely dispersed, land owners face considerable organizational costs and, hence, are not able to collude to provide a sufficient degree of countervailing power to improve their returns.

The considerable degree of inflation in many of the former command economies has led land owners to specify fixed rent in kind rather than nominal prices. This has meant, however, that owners may be faced with disposing of large quantities of produce which farms have delivered as rent payments. This leads to spoilage. Further, if harvest prices are depressed and land owners (often urban dwellers) have no storage capability, the produce received in lieu of rent must be sold off at low prices.

In the case of some former collective farms, ownership of the land has been divided between the members of the collective and former owners. In these cases, former owners will have even less control over management. Farm managers are likely to draw support from those land owners working on the farm as a result of management's ability to control wages. Land owners who are also farm workers will be indifferent as to whether their income is derived from wages or rent. If wages can be kept high, that means that workers will tend to gain at the expense of land owners as all land owners receive a low rent while only workers receive wages. Of course, control of management will depend upon the division of land assets between the two groups. Former members of collective farms are, however, likely to face lower organizational costs relative to widely dispersed former owners and, hence, may have more power than their share of assets suggests. The existing evidence suggests that few changes of management have been made as the privatization of collective/state farms into corporate farms has taken place. In part, of course, this reflects a shortage of skilled managers outside the former command system. The net result has, for the most part, been resistance to change on the part of corporate farm management. While some managers have aggressively attempted to change the production/market orientation of their farms, there nonetheless has been little change to the existing management structure.

The dual dependence of the existing management and their former collective/state farm members has meant that reforms aimed at altering manning levels, wage structures or work effort could be resisted by workers. This, in turn, has slowed the transition of corporate farms into truly commercial operations. As these difficulties are endemic in former state/collective farms, it has meant widespread insolvency in the corporate farm sector. If only a few corporate farms were insolvent, governments could safely allow them to declare bankruptcy. Large scale bankruptcy is not politically acceptable and governments have been forced to subsidize corporate farms or to intervene in markets to set prices which ensure their survival. The management of large state farms have also been much more effective than small private farmers in lobbying the government. This means that the former state/collective farms have been able to garner a disproportionate share of the available government resources. Until the vestiture of property rights in land has been completed, farming will be unattractive for investment and remain primarily in the control of corporate farms unless the collective/state farms are terminated as happened in Romania. In reality, this means that the process of transition is inhibited. The experience in Romania would suggest that compulsory abolition of former collective/state farms, in the absence of security of tenure and a functioning land market, leads to excessive fragmentation of land holdings into uneconomic units (OECD, 1994).

## 7.3. Family Farms or Corporate Farms

When privatization of the farming sector in the former command economies is discussed by those in modern market economies, the model typically envisioned is the stereotypical family farm which exists in western Europe and North America. Farmers live on a farmstead on their land and are independent business people. Family farms are also envisioned as an appropriate model for a liberalized agricultural sector by many in the former command economies. However, the family farm model is far from being universally accepted in the former command economies. In many areas of the former USSR, this mode of agricultural organization did not evolve in the pre-communist era. In the CEECs, while the model was more often observed prior to collectivization, concentration of land ownership in the hands of an elite and various forms of tenant farming were the norm. Further, as suggested above, there is considerable vested interest, both for workers and managers, in the continuance of some modified form of the collective/state farm model. Finally, the investment in large scale facilities, both on the farms and in the inputs and processing sectors, may make division of former collective/state farms an extremely complex and difficult process. Many of these problems would remain even if land could be distributed in such a way that economically viable family farm units could be created. While other models can be envisioned for a segment of agriculture – self-sufficient peasant farms, part-time farming and income supplementing garden plots – this discussion will centre around family farms versus the continuation of corporate farms.

If one considers these two models (family farms and corporate farms) as the future basis for commercial agriculture in the former command economies, the issue of privatization relates primarily to the control of resources – land but also equipment, livestock and to a lesser extent, facilities. This is because establishing private family farms requires that corporate farms give up some of their resources. As long as corporate farms remain in being, private farmers will be viewed both as competitors for resources and as competitors in the product market.

In the case of former owners who have historic title to identifiable parcels of land, removing these parcels of land from the corporate farm may threaten the financial viability or production capabilities of the entire corporate farm. If land for a state farm was assembled around a core based on a landed estate, allowing the restitution of the original estate to its former owner may leave the corporate farm with only poor land which is geographically dispersed. The corporate farm's buildings, repair facilities or social service infrastructure may be located on the land of former owners. In some cases, land swaps have been attempted but in the absence of a land market, and given the difficulties with appraisals, former owners have often been unwilling to exchange their parcels. Typically, when faced with the need to swap land, the corporate farm offers land of a lower quality in an attempt to protect its resource base.

Provision has often been made in privatization legislation for members of former collective or state farms to receive land, and sometimes other assets, from their corporate farm if they wish to become private farmers. Corporate farms are directed in the legislation to make available to private farmers a portion of their land resource. Administratively, this may be accomplished by a portion of the corporate farm's land being removed from its control for redistribution by the state or it may be a direct negotiation between the prospective private farmer and the management of the corporate farm. In both cases, the corporate farm can be expected to offer only its lowest quality land for redistribution. By doing so two purposes are served. It ensures that the corporate farm retains control over the best land and it makes it very difficult for those who wish to be family farmers to succeed.

Both the partitioning of a proportion of the corporate farm's land for redistribution and the one-on-one negotiation system works to the advantage of the corporate farm's management. In the case of one-on-one negotiation, an individual member of the corporate farm will have little or no bargaining power as he or she attempts to wrest resources from the corporate farm. At best, prospective private farmers can expect to obtain poor land, inferior livestock and obsolete or worn out equipment. Considerable overmanning existed on state/collective farms in the command economy era. When the partitioning policy is followed, the management of the corporate farm is able to retain the loyalty of the majority of the workers because there will not be sufficient land available for all members of the former state/collective farms to take advantage of the opportunity to farm privately. Those left out of the process have few alternatives but to continue to support the corporate farm's existence.

Even when a family farmer is able to acquire land as well as some livestock and equipment to start farming privately, a number of difficulties remain. As there is virtually no private distribution system for inputs, family farmers must rely on the corporate farm (their competitor) for a wide range of inputs, particularly services. As the corporate farm controls repair facilities, spare parts, veterinary services, fuel, even water, it may decide not to provide those services or only to provide them at a prohibitively high cost. If the first few members with family farm aspirations can be driven into bankruptcy, other members will be discouraged from attempting to pursue it.

The previous collective/state farms were responsible for the provision of rural social services and much of the rural infrastructure such as roads. By withdrawing from the corporate farm, family farmers may deprive their families of health care, deprive their children of an education and may not be able to secure the building of a road to their property.

Rural commercial banks are rare so the family farmer is not likely to be able to access credit. Further, as clear tenure to the land is unlikely to be established in the near future, lending institutions will be reluctant to advance loans to family farms. As suggested in the previous chapter, when the input industries such as fertilizer, seed, machinery supply, etc., have been privatized and separated from the former state/ collective farm, these suppliers are likely to represent local monopolies. Further, if a corporate farm is the largest customer of an input supplier, it may be able to put pressure on the supplier to deny inputs to family farms. Private farmers are also likely to face monopsonistic buyers when it comes time to sell their output. This problem will be discussed in greater detail in Chapter 8.

Support services for family farms also tend not to exist. Little or no research has been done on appropriate technologies for family-farm scale enterprises. While there is considerable effort being put into the development of extension services based on models from modern market economies, as yet they have not become operational on a widespread basis. As a result, family farmers lacking knowledge about production methods, prices or marketing alternatives have no low cost means of acquiring such information.

In short, the prospects for family farms are generally bleak, with poor resources, underinvestment, low productivity and poor knowledge endemic. Of course, a few exceptional entrepreneurs have been able to make a go of family farming.

Corporate farms retain control of most of the fixed farm assets constructed in the communist era. Most of this infrastructure is not suitable for small scale farming. Dairy barns are large. Feed handling systems are large scale. Storage facilities were centralized. As a result, family farms start out with little physical capital. In the absence of savings out of profits or access to credit, family farms are unlikely to have the resources to invest in farmstead centred infrastructure. Even if access to common facilities for family farmers could be arranged, sharing common facilities when biological systems are involved is extremely complex. Different herds sharing common milking facilities leads to problems of disease transmission among herds and questionable health standards for milk. If milk from a number of herds is pooled in bulk milk facilities, individual farmers may be able to cheat on their sanitary procedures leading to a classic free rider problem.

Given the constraints to family farming described above, the continued existence of large scale farming is often proposed as an alternative. Instead of subdividing collective/state farms into family farm units, their transformation into corporate farms is proposed. While, conversion to legal corporate farms has, to a considerable degree been accomplished, there is little evidence that transformation into market oriented/ capitalist farms has taken place. Instead, the existing collective/state farm system remains relatively unchanged. Directors or managers have more autonomy because the visible hand of the planner has been removed. Still, the management of corporate farms has little room to manoeuvre. The supply of inputs and the purchasing of the farms' products are often controlled by local geographic monopolies. Even if corporate farm managers wished to diversify their sources of supply and markets, alternatives often simply do not exist.

Privatization has meant that managers are, to some extent, responsible to their workers. As shares have been distributed to the existing workforce, labour force rationalization is often extremely difficult.

As the state heavily subsidized credit in the command system, many corporate farms carried considerable debt with them into privatization. In the command era this debt was never expected to be fully serviced or repaid. Governments, now short of cash, have been less willing to forego interest payments. Further, large debts give the government considerable control over corporate farms. Often the government is able to use these debts to force management to supply product to processors so that the processor's survival can be ensured for reasons of political patronage. The prices at which these products are supplied may be less than what could be obtained from the market.

Managers have a vested interest in maintaining control of corporate farms. They may have few alternatives if they are fired from their positions. Retaining the existing cohort of managers leads to the continuation of hierarchical managerial control rather than a more consultative approach to management. At the same time, management may be pandering to workers by paying high salaries, turning a blind eye to pilfering or failing to punish shirking by workers.

High debts, poor access to credit and adverse prices mean that the new corporate farms are often lumbered with poorly designed and/or outdated physical infrastructure, tractors and field equipment which are too large and other equipment which is only suitable for the production of inappropriate crops. Spare parts for the equipment may no longer be obtainable. In the wake of the collapse of the Council for Mutual Economic Assistance (CMEA) trade relationships, out of country supplies have been particularly at risk. To become competitive enterprises, corporate farms need to re-equip but they have no access to the resources required. Technical skills also require upgrading and new management needs to be trained, or existing management retrained. While some progress has been made in setting up extension systems and in establishing market based agricultural education, much remains to be done before information and training is generally available.

In short, both the family farm model and the corporate farm model are encountering considerable difficulties in their transitions into commercial operations in market systems. Where the former collective/state farms have been broken up and the land base has been fragmented, many land holders do not possess sufficient resources to become commercially viable. Until questions of tenure are finalized and a land market can develop, consolidation into viable units will be forestalled (Christensen, 1994). Even the once productive plots are no longer prospering in some areas. While their land base was small, they could often access additional resources from the collective/state farms. Fodder, fertilizer, machinery and fuel were often appropriated from the large farm to which the plot owner belonged. As these previously subsidized resources have become scarce, corporate farms have had to increase the control of their use. The problems with transition at the farm level have led to declining output and even more dramatic reductions in deliveries to the formal food distribution system.

## 7.4. Farm Level Reforms and Food Security

Commitment to reform by the governments of the former command economies varies considerably. In some countries little progress has been made while in others voters disillusioned with reform have returned former, but now reformed, communists to power. In a number of countries, agrarian political parties have considerable legislative influence. Even the most pro-reform governments, such as the Czech Republic, could not ignore the effect of declining production and rising prices on both total food consumption and the composition of diets (OECD, 1994). Those countries with good transport links with western Europe have been less affected by the decline in food security arising from falling production than those which are more isolated. Imports, and in some cases food aid, have been used to forestall food crises. While widespread famine conditions have not been observed, largely because of widespread access to subsistence plots, the erosion of purchasing power for those on fixed incomes such as pensioners has reduced their food security to near or below an acceptable minimum.

The farming systems in the former command economies are in disequilibrium. As a result, governments, given their concern for food security, have not been willing to allow the farming system to reach an equilibrium. Intervention has been universal. Intervention to forestall bankruptcy has been evidenced in all countries. Prices of many staple commodities are set below market levels. Fixed prices are liberalized, only to be reimposed. In some countries, compulsory deliveries to the formal state distribution system have had to be reinstated. The diversion of food away from the formal food distribution system is a major problem in the NISs. As distances are considerable, there are large populations in non-food growing areas which are particularly vulnerable to breakdowns in the existing distribution system. In CEECs and the food growing areas of the NISs, a large proportion of farm output has been diverted from official channels into physical marketplaces. Given poor information, transportation and banking systems, however, little of this food moves outside the local area. The transactions costs are simply too high to allow arbitrage among distant markets to take place.

Governments are faced with a dilemma. Intervention in farm gate markets slows down the pace of transition. The disruptions to farm output brought on by liberalization, however, erodes the food security of voters. Hence, in the farm, as well as the energy sector, intervention has remained. In other sectors, disruptions to supply are more easily tolerated and markets can be allowed to find their equilibria. This suggests that transition in the farm sector can be expected to be longer and more drawn out than in other industries.

## 7.5. Conclusions

Considerable progress has been made in altering the legal structure of former collective/state farms. These changes have provided the mechanism whereby the existing farms have been privatized. Legislative and legal changes have also been made to allow for the establishment of owner/operator (family) farms on the model common in modern market economies. These legislative initiatives, however, are only necessary, but not sufficient, changes to lead to the establishment of a commercial market-based farming sector.

The most immediate constraint to the transformation of farming in the former command economies is the slow pace with which property rights to land can be formalized. Conflicting claims, inadequate land surveys, bureaucratic delays and, in some cases, a reluctance to return to full private property rights in land has meant that farm owners cannot be sure of their titles. As a result, land cannot be used as collateral for loans. Investment in agriculture is reduced. Further, direct investments in measures to improve farmlands' productive capability are inhibited. While some of the delays in establishing property rights in land cannot be avoided, more efficient mechanisms for registration and the establishment of a rightful interest could speed up the process and provide sufficient security to encourage investment. Only when the property rights to land are settled can a functioning land market develop.

The problem of how land assets are to be distributed has not been satisfactorily resolved. As overmanning of collective/state farms was common, dividing the assets among all members of collective/state farms leads to fragmentation into uneconomic units. If the collective/state farms are left largely intact – albeit as reformed corporate farms – they compete for resources with those who wish to become private family farms. As private farmers pose a threat to the viability of corporate farms, managers and those collective/state farm members who cannot or will not become family farmers, have a vested interest in retaining control of land and other resources. As a result, family farmers have difficulty obtaining adequate resources and may find their corporate farm neighbours hostile competitors. Many would-be family farmers are likely deterred from leaving the corporate farm.

Both corporate and family farms also face a cost-price squeeze from input suppliers and buyers. As these tend to be large geographic monopolies, the removal of fixed prices and the guiding hand of the planning ministry has meant these agribusinesses are free to set monopoly prices for inputs and monopsony prices for outputs. As a result, farm level insolvency is high.

Little has changed in most former collective/state farms. Hierarchical management structures remain and, for the most part, managers from the previous era remain in place. Shareholders in former collective/state farms – a large proportion being the former members or workers – have considerable control over management. They have a vested interest in keeping employment and wages high. This adds to the farms' costs and reduces their ability to become profitable. As insolvency is widespread, governments are unwilling to allow formal bankruptcy. The result is either subsidization or direct intervention in farm gate markets to keep the corporate farms in business. These interventions slow the pace of transition. Intervention also takes place as a result of concerns over food security, further inhibiting the development of markets. Hence, it seems that the transition to a commercially viable and competitive farm sector will be a long and, probably, erratic process.

# References

Christensen, G. (1994) When structural adjustment proceeds as prescribed: agricultural sector reform in Albania. *Food Policy* 19, 557–560.

Durgin, F.A. (1994) Russia's private farm movement: background and perspectives. *The Soviet and Post-Soviet Review* 21, 211–252.

OECD (1994) *Agricultural Policies, Markets and Trade in the Central and Eastern European Countries (CEECs), the New Independent States (NISs), Mongolia and China.* Monitoring and Outlook 1994. Organization for Economic Cooperation and Development, Paris, 249 pp.

Zhurek, S.Y. (1994) Transforming Russian agriculture: why is it so difficult? *The Soviet and Post-Soviet Review* 21, 253–281.

# 8 To Whom Do I Sell?: Marketing Channels for Farmers

The economic success of the newly privatized primary agriculture sectors in the Central and Eastern European Countries (CEECs) and the New Independent States (NISs) will depend crucially on the availability of marketing channels through which to sell agricultural output. In modern market economies, marketing channels have evolved gradually bringing changes in the structure of industries and changes in transportation, storage and processing technologies. Many agricultural products were once sold directly to consumers in local farmers' markets. Live cattle, for example, were once walked to marketplaces in nearby towns and cities by cattle drovers. Eventually the development of rail and road transportation led to the movement of agricultural products to more distant markets via rail and by truck. Live[1] auction systems developed for some products and in recent years, electronic and video auctions have been introduced. Contracts between producers and processors have become common, with standardized agricultural contracts often being used. Governments in most modern market economies have involved themselves in the marketing of some agricultural products through the formation of compulsory marketing boards through which all the output of an industry is sold. Finally, group marketing of agricultural commodities and (voluntary) agricultural cooperatives have arisen. In many countries, a number of these marketing channels exist simultaneously. The former command economies do not have the luxury of gradual evolution and experimentation with different marketing channels over a long period of time. The rapid establishment of functioning and effective marketing channels for agricultural produce is essential.

Marketing channels perform a number of roles. First, they provide a means by which the forces of supply and demand can establish market-clearing prices. They bring together buyers and sellers for the physical exchange of products and provide the means by which the requirements of consumers are transmitted to producers through appropriate price signals. Many of the institutions which facilitate the efficient, smooth operation of marketing channels are discussed in later chapters, for example, the legal and financial institutions which form the bedrock of modern market economies. This chapter focuses on the marketing channels themselves, describing the pre-reform command economy system before focusing on alternative marketing channels and highlighting some of the issues which will be important in the marketing of agricultural products in economies in transition.

## 8.1. Marketing Channels in Command Economies

Freely operating marketing channels guided by market forces did not exist in command economies because markets did not exist. Certainly, products moved from the agricultural sector to processors, distributors, retailers and final consumers but this was the result of bureaucratic decisions rather than at the direction of market forces. To label these channels marketing channels is therefore somewhat fallacious. Rather (as discussed in previous chapters) production, processing, distribution and retailing were organized like the links in a chain. Just as large state farms or cooperatives were required to purchase their inputs from designated suppliers in the chain (see Chapter 6 for a discussion of agroinputs), they were also required to transfer all of their output to designated processors, distributors or retailers in the same chain. Sometimes, the large farms undertook processing, distribution and/or retailing functions themselves as one fully vertically integrated unit. In this case, the marketing channel was completely internal to the state or collective farming enterprise. There was no need for the price-setting features of market economy distribution channels since the allocation of resources and products was determined by bureaucratic decisions.

Information costs for farm managers were low because they did not have to spend time and effort discovering market prices or searching out and evaluating potential buyers, etc. Negotiation costs were low or non-existent because the terms of the transaction were not negotiable; delivery schedules, for example, were set by the planning process not through producer–processor negotiations. As the return earned by the farm was independent of the quality of the product, there was no need for farms to monitor the storage, distribution and handling of their products by distributors or processors. Although the system as a whole incurred high (bureaucratic) transaction costs in administering the flow of products from farms to processors, individual farm managers did not incur significant transaction costs because there was no incentive (or opportunity) to market products to processors. Without price signals and market incentives, however, resources were not allocated efficiently and considerable wastage occurred both on the farm and in the distribution system.

## 8.2. Marketing Channels in Transition

As was discussed in Chapter 7, land reform and privatization of the agricultural production sector in the CEECs and the NISs has led to a two-tier industry consisting of a number of small, private farmers and large former state-owned or corporate farms. At the other side of the farm's marketing transaction, privatization of the processing sector, in many cases, has created local monopsonies. The development of marketing channels for agricultural products must, therefore, be viewed within the context of this changing industry structure and against the backdrop of ineffective or non-existent financial and legal institutions. The effectiveness of different marketing channels in modern market economies depends on the economic and legal environment in which those channels operate. This environment will also be important in determining the choice and effectiveness of marketing channels in economies in transition.

The newly established (and, in Poland, existing) private farms tend to be relatively small. They are at a significant bargaining disadvantage *vis-à-vis* privatized processing companies which are of a large scale. A number of potential marketing channels for private farmers could emerge in CEECs and the NISs, including: farmers' markets, live auctions, electronic auctions, sales through dealers or agents, direct sales to processors with and without contracts and various types of joint marketing. Large-scale former state or collective farms also have a number of marketing channel options including: live and electronic auctions, direct sales to processors and vertical integration. The conditions necessary for the successful functioning of these marketing channels are discussed below.

Two concepts useful in evaluating the efficiency of marketing channels are operational efficiency and pricing efficiency. Operational efficiency refers to the efficiency with which products physically move through the marketing chain from farmer to consumer, or the transformation of products from one form to another. It measures input–output relationships. A change in technology or developments in the marketing infrastructure which increase the quantity of goods moved through the marketing chain with a less than proportional increase in resource use would represent an increase in operational efficiency. Alternatively, for highly perishable products, an increase in the speed with which goods moved from producer to consumer would reduce the risk of product deterioration and spoilage, representing an increase in the operational efficiency of the marketing chain. Some marketing channels have numerous middlemen between the farmer and end-user – commission agents, wholesalers, distributors, etc. These intermediaries usually earn a commission, or mark-up the price of the product, for performing various marketing services such as locating buyers, negotiating the transaction, storing or transporting the product, etc. If the services which these intermediaries perform increase the value of the final product, or match the demands of consumers to the availability of product from suppliers more effectively than would otherwise have been the case, they enhance the operational efficiency of the marketing channel. On the other hand, numerous intermediaries sometimes serve to increase the final product price without performing services of a commensurate value. They also act as a barrier between producer and end-user, impeding the flow of information from the market back to the producer, therefore making it difficult for the farmer to know what type of product is required by the end-user. This is often a criticism levelled at the Japanese food distribution system which is a complex, multi-layered system consisting of several layers of primary and secondary processors, wholesalers and retailers. The distribution system has proved somewhat of an enigma to would-be exporters (Kerr *et al.*, 1994).

The second type of efficiency is pricing efficiency. This refers to the accuracy with which market prices measure the value of products to end-users. It also refers to the effectiveness with which this price information is transmitted back through the marketing channel to producers. If prices more accurately reflect the preferences of consumers, a more efficient allocation of resources will occur.

### 8.2.1. Farmers' markets

Farmers' markets, in which farmers bring their produce to a central point (usually in a nearby town) for sale directly to final consumers have been expanding both in numbers and in importance in the CEECs and the NISs. Small private farmers, rather than large former state or collective farms, tend to use these markets because the volumes which they have for sale are (individually) relatively small. Clearly, this marketing channel is only suitable for some agricultural products, namely those which do not require further processing such as fruits and vegetables. Livestock products (with the exception of poultry), unless slaughtered, deboned and further processed by the farmers themselves, could not be sold safely in this way. Milk is highly perishable and would present numerous logistical and sanitary problems if it were to be transported to a marketplace, stored all day without access to refrigeration and sold in small quantities to individual consumers.

Generally speaking, in farmers' markets the price is set by the seller and the buyer either accepts or rejects this price, although there may be opportunities for negotiation of individual transactions. Farmers probably have to transport their products to the marketplace themselves because of the absence of an efficient or affordable private transportation industry in former command economies (transportation issues are discussed in Chapter 10). At harvest-time, when supply may outstrip demand, if products cannot be stored and sold on the market at a later date, either prices will fall considerably in order to clear the market or extensive product wastage occurs. The latter may be more likely as local farmers' markets are inefficient at matching demand with supply. Consumers may not know that in a particular market on a particular day there will be large supplies of cauliflower for sale at reduced prices. Farmers' markets are an extremely time consuming method of marketing for the farmer who must travel to and from the marketplace and often spend an entire day waiting to sell produce. The time commitment will be greater if the market is a long way from the farm and if local transportation infrastructure is inefficient.

Farmers' markets have poor operational efficiency. They are time consuming: numerous small private farmers sell from individual stalls to numerous consumers, neither of whom can be sure that the other will have/want the products in question prior to attending the market. There are few opportunities to smooth out regional fluctuations in supply or demand through storage or through transportation to other areas where there may be greater demand for the product. Farmers are unlikely to have access to information about other potential markets and would probably be unable to service more distant markets. The fact that the farmer has direct contact with the consumer does shorten the marketing channel and avoids the need to use commission agents or distributors; this would appear to improve operational efficiency. On the other hand, the seller is entirely reliant on a sufficient number of consumers attending the market that day; an agent or distributor may be able to improve operational efficiency by having access to a more dispersed group of buyers.

Pricing efficiency is enhanced because the farmer has direct contact with the consumer. If the product will not sell, the farmer must reduce the price, hence, price information should be transmitted rapidly and

effectively to farmers. However, the method of price determination –
numerous bilateral negotiations between seller and consumers – is inefficient.

The use of grades or classifications for agricultural products
increases pricing efficiency because they enable consumers' preferences
for products with certain quality characteristics to be more accurately
relayed to producers in the form of price premiums and discounts for
products of different qualities. Produce sold at a farmers' market is
rarely, if ever, graded. Certainly, in the newly developing economies of
the CEECs and the NISs, effective grading of agricultural products has
not yet occurred. This means that the consumer at a farmers' market
will incur sorting costs in trying to measure the true value or quality of
a product. A discussion of the problems of valuing product quality was
presented in Chapter 4. It was suggested that it is more efficient for the
seller to incur the sorting costs of measuring the value of the products
once, than for each individual buyer to go through the same sorting
process. For example, suppose an apple farmer were selling apples to
numerous individual consumers at a farmers' market on the outskirts of
a small town in the Ukraine. It would be in the seller's interest to sort
the apples according to certain quality characteristics of interest to the
consumer, such as the degree of bruising, size, firmness or colour. Provided the grades or classifications were consistent and reliable, consumers' product information costs would be reduced and the marketing
channel would operate more efficiently. A more effective system, of
course, would be an independent, nationally (or at least regionally)
recognized quality sorting system. The logistics of implementing this at
farmers' markets, where the produce of tens or hundreds of small private farmers is brought for sale, would probably be impractical. High
sorting costs for consumers is likely to remain a feature of farmers' markets.

From a farmer's perspective, bilateral private sales to individual
consumers can incur high transaction costs because of the need to determine an appropriate market price in the absence of market information
on supply, demand and general price trends. Although the farmer
avoids having to pay commission charges to an agent or auctioneer,
negotiation costs arise because attending the sale is extremely time consuming. The opportunity cost of attending a farmers' market for one or
more days per week to sell perishable products will be extremely high
during harvest-time when the farmer could otherwise be harvesting the
remainder of the crop. As private farms tend to be relatively small, most
farmers do not have employees to continue the work on the farm or to
take the produce to market; the use of family labour for these tasks is
very important.

The lack of an effective transportation infrastructure in most former command economies means that farmers have to transport the products to market themselves. This raises negotiation costs. Monitoring costs would be relatively low, however, as there are no intermediaries between the farmer and the consumer, hence, the farmer does not have to worry about improper handling of the product by transport companies, processors or wholesalers. Of course, if hired help is used to carry out the marketing function, the farmer would incur monitoring costs in ensuring that the worker(s) do not shirk and that the products are handled, displayed and sold properly.

Farmers' markets were once common in many market economies (and have subsequently seen a small revival but mostly on a novelty level in some urban areas). Developments in storage, processing and transportation technologies eventually made farmers' markets relatively inefficient and obsolete. The same is likely to occur in the CEECs and the NISs. The high transaction costs from using farmers' markets, both for sellers and buyers, make them a relatively inefficient marketing channel. In the absence of the financial and legal institutions to facilitate transactions through other marketing channels, however, these markets may well thrive for suitable commodities for the time being. Payment is on a cash-on-collection basis and there are no long-term contractual agreements between buyer and seller. Hence, the financial and legal institutions essential to the successful operation of other marketing channels may be less important for farmers' markets. This may explain the recent growth in this marketing channel in the CEECs and the NISs. For example, in Tajikistan: 'farmers markets are playing an increasing role in marketing and distribution' (World Bank, 1994, p. 101). As a price setting and market clearing mechanism, however, they are relatively inefficient. In modern market economies auction markets developed as an alternative to farmers' markets.

### 8.2.2. Live auctions

Schotter (1976) defines auctions as 'exchange mechanisms without a tâtonnement or recontracting provision in which the seller is relatively passive and goods are often indivisible' (p. 4). Auctions allow buyers to view the commodities prior to the sale, forming an opinion of the probable value of the commodity. Sometimes additional information is provided, this may include the farm of origin or, in the case of a livestock auction, the weight and breed of the animal and possibly information about production methods. Buyers bid on the basis of their visual assessment of the commodity.

There are two main types of auction. The English auction proceeds with the auctioneer calling out a starting price and then dropping the price until the first bid is received. Bid prices then rise until the last bid is received which determines the final price paid. Sellers may specify a reserve price below which the commodity will not be sold. During the sale, buyers gain information about the prices which other buyers are willing to bid for a commodity, although they cannot know what the upper limit will be on a buyer's willingness to bid. English auctions are used for fish and, in particular, for livestock[2] in modern market economies.

An alternative is the Dutch auction, with a large clock in the auction room or yard which begins at 2 minutes or 60 seconds and counts down the time remaining in the sale. As it does so, the auctioneer calls out a price and drops the price as the clock ticks. The first buyer to bid, usually by pressing an electronic button, is awarded the sale. This is a very rapid method of selling and is used for highly perishable products such as cut flowers in the Netherlands. Buyers have no information about the price which competitor buyers are willing to bid for a particular lot as the first bid ends the sale.

Buyers are responsible for transporting the commodity away from the auction sale. Once the auctioneer's hammer has fallen, all responsibility for, and title to, the commodity is transferred to the buyer. The auction market earns a commission on the sale which is usually paid by the seller.

Auctions operate on six basic principles:

**1.** Inspection by the buyer reveals most of the characteristics of a commodity relevant to consumer demand.
**2.** Commodities are purchased *caveat emptor* or buyer beware. There is no recontracting process if mistakes are made by the buyer in the visual inspection of the commodity. As Schotter (1976) observes: '. . . not only do mistakes occur, but all mistakes are final' (p. 4).
**3.** Successful transacting through an auction market requires that buyers understand the product characteristics demanded by end-users, and thus are aware of the characteristics for which consumers will pay a premium.
**4.** Auction markets provide sellers with a competitive price for their commodity. A sufficient number of buyers must be present to ensure that there is enough interest in the sale and to make the costs of organizing collusive activities among buyers prohibitive.
**5.** Lot sizes are a trade-off between time and price discovery. This is particularly important for livestock auction sales. Although smaller lots allow more accurate pricing of the individual commodity (or animal), they take longer per commodity (animal) to be traded and are, therefore, less time-efficient for the auctioneer, the buyers and any farmers attending the auction.

**6.** All variables except the price are decided in advance, therefore price is the sole determinant of the sale. This stems from the condition that commodities are accepted *caveat emptor*.

Auctions have a number of advantages and disadvantages which are influenced by the institutional environment peculiar to economies in transition. The chief advantage of an auction is its role as an efficient price-setting mechanism; auctions improve pricing efficiency relative to bilateral transactions between farmers and buyers. If the auction is operating effectively, sellers are not at a bargaining disadvantage relative to buyers as a number of buyers will be competing for the same commodity. This advantage depends on a key assumption: that there are a sufficient number of buyers present at an auction to stimulate competitive bidding. If too few buyers are present, there is a danger that they will collude and bid together strategically, either dividing lots among them or agreeing not to bid beyond a certain price for each lot. The privatization of the processing sector in the CEECs and the NISs has often created regional monopsonies. The sheer scale of the processing activities and equipment has, in most cases, made it impossible to privatize processing facilities into several smaller processing companies. This limits the number of buyers likely to be present at any given local auction market. If there are insufficient buyers, prices are not determined competitively and the auction system fails.

The auction system also acts to clear the market because all commodities offered for sale can be sold at some price (assuming that farmers do not withdraw the commodity from sale and return it to the farm if it does not reach the reserve price). Again, this advantage of the auction system depends on a sufficient number of buyers being present at the sale, something which simply cannot be assumed in the CEECs and the NISs.

An important role of the auction system is in the generation of publicly available price information. In modern market economies, auction prices are usually published in newspapers and in farming magazines and may be broadcast on local radio and television stations. Farmers (or buyers) can also attend an auction sale to obtain price information. Auction prices give farmers and processors a guide as to the market value of a commodity, even if they do not intend to sell/buy through an auction. In modern market economies, auction prices are often used as the basis for pricing agricultural commodities sold directly from farmers to processors, hence, auctions reduce price information costs for sellers and for buyers. The extent to which these prices can be used as a guide to the market value of a commodity depends on whether a representative proportion of a region's or a nation's commodities are sold through the auction. If only a small percentage of cattle or sheep are sold through the auction system, then the prices received may not be representative of the wider market; the auction market is said to be thin. This will be important in former command economies where auctions are a new phenomenon. A sufficient proportion of farmers must be willing to sell their commodities through newly established auction systems for the prices to be representative.

Another advantage of the auction system for farmers is that once the sale has taken place the farmer's return is known because it is the final bid price. Any uncertainty over the perceived versus actual value of a commodity is borne by the buyer.[3] Farmers receive payment quickly, often collecting payment from the auction company within a day of the sale. In many cases, the processor does not pay the auction company until much later (two to three weeks later in some modern market economies). The auction company finances the lag between the buyer taking possession of the commodity and when payment is made and assumes any risk of the buyer defaulting on payment. This protection against default on payment is important to many farmers in modern market economies, particularly in times of recession when processing companies can be taken into receivership before a farmer has been paid for commodities sold to the processor. Although auction companies provide this service, buyers only rarely default on payment (otherwise the auction company would be put out of business). The indemnity protection provided by auction markets becomes crucial when one considers the situation in most CEECs and the NISs. Much of the former state-owned processing sector is in debt and is financially unstable. Newly established auction companies would therefore face a considerably greater risk that processing companies would default on payment to the auction company. This aspect of auction markets in modern market economies must be addressed for the successful development of auctions to take place in the CEECs and the NISs. Presumably, if auction companies were to perform this role, in order to offset the risk, they would charge higher percentage commissions than do auction companies in modern market economies. Furthermore, processors are legally obliged to pay for commodities purchased at an auction sale; enforcement requires an effective commercial legal system – something which is either in its infancy or non-existent in many CEECs and the NISs.

Auction sales may be less operationally efficient than direct sales to processors since they involve an additional transportation leg (commodities must be transported from the farm to the auction and from the auction to the processing plant). The additional handling and transportation of commodities can cause increased stress in livestock which can lead to carcass damage and weight loss; it can lead to bruising and product spoilage in the case of other commodities, such as fruit and vegetables.

The auction system has been criticized for its failure to effectively transmit information back to farmers on the quality characteristics preferred by consumers. Commodities are sold on the basis of a visual assessment; for example, livestock are sold on a liveweight basis, i.e. a price per kilogram (or pound). Farmers rarely know how the commodity was graded nor its final destination. Without this information, it is more difficult for farmers to adjust their production practices in response to changing consumer demands.

Another constraint to the use of an auction system in former command economies is the need for farmers to transport (or arrange transportation of) commodities to the auction market. This either requires that the farmer has the transportation equipment or that a private transportation industry exists which can perform this function. Small private farmers in the CEECs and the NISs are unlikely to have the capability to transport their own cattle to the marketplace. The transportation industry in most modern market economies is very competitive, particularly in the case of livestock haulage. Large numbers of small firms exist and farmers have access to competitively priced transportation services. This is not the case in former command economies where a separate transportation industry did not exist prior to liberalization. The transportation function was vertically integrated within collective or state farms or within processing units. When these units were privatized, if the transportation function was sold off at all, it tended to be on a large scale, often creating a regional monopoly in transportation. These monopolies are able to charge high prices to small private farmers for transporting their commodities to a weekly auction market. Reports from Albania typify the problems faced by farmers in many transition economies:

> Lack of market information and lack of transport has caused many products to go bad before they can be sold. Price liberalization has meant that transport rates are exorbitant and private farmers cannot afford them.
> (World Bank, 1995, p. 78)

Chapter 10 discusses transportation and distribution issues in more detail.

### 8.2.3. Electronic auctions

Electronic or video auctions were first used in the US in 1962 to market pigs and are now used in many countries including Canada, Australia, the Netherlands and the UK. Electronic auctions operate on many of the same principles as do live auctions and are subject to many of the same advantages and disadvantages. Products are sold on the basis of a written description or video of the commodity. This means that the characteristics of the products of interest to buyers must be easily summarized in a verbal description or visually detectable in a video transmission. Participants (buyers) access the sale through remote electronic telecommunications, with bidding carried out competitively through a centralized computer system. Buyers bid from their offices, either via telephone or through a modem link to the central computer of the auction company. Operating from the centralized location, the auctioneer begins by dropping the price until a bid is received from a buyer, either over the telephone or by pressing a computer key. The sale then usually follows the rules of an English auction, with the final bid determined after there have been no subsequent bids for a set period of time, e.g. ten seconds.

The commodity remains on the farm if and until it is sold. In some cases the buyer is responsible for collecting the commodity from the farmer, alternatively, the farmer must deliver the commodity to a central collection point. The farmer usually receives payment within one to two weeks of the commodity being collected. The auction company receives a commission from the sale. Final price determination differs, some systems allow the bid price to be the final price, hence, for livestock this would be a liveweight price based on the written description or video of the commodity. Some products (e.g. fruits and vegetables or grain) may already have been graded and this information included in the written description. In the case of livestock, it is more common to determine final payment by the deadweight grade of the carcass because the true value of the carcass cannot be known until after slaughter. In this situation, bids are made on the basis of an average grade with a predetermined schedule of price premiums and deductions awarded for carcasses which deviate from this average standard. Farmers receive a price which reflects the carcass characteristics of the animal.

One of the advantages of an electronic auction system is that it improves the operational efficiency of the marketing chain. Commodities do not leave the farm until after they have been sold, thereby reducing transportation and handling. This lowers costs and reduces the risk of product damage. In the cases where farmers are paid according to the final grade of the commodity, rather than on the estimated value of the commodity, price signals as to the preferred quality characteristics are transmitted more effectively back along the marketing chain from buyers to sellers. Published electronic auction prices also serve as a source of market price information for the whole industry. Electronic auctions can give sellers access to a wider pool of buyers than is the case with live auctions or local sales directly to processors, although an insufficient number of buyers is still likely to be a constraint in former command economies. Buyers benefit from having access to a large number of commodities without having to send procurement staff to widely dispersed regional auction markets; therefore buying costs are reduced.

The implementation of an electronic auction system in former command economies faces many of the same obstacles as a live auction would face. The primary limitation would be an insufficient number of buyers. Furthermore, electronic auction companies play the same bridging role between the seller receiving and the buyer making payment as is the case with live auction sales. This role would carry with it magnified risks in former command economies, probably resulting in higher percentage commissions than in modern market economies. Difficulties with transferring funds between regions may significantly limit the potential buying audience as will limitations on long distance transport. Unfamiliarity with the auction system may also impede the adoption of electronic auctions. Even in countries in which farmers and buyers have a long familiarity with the auction system, electronic auction companies have had to convince farmers and processors of the merits of using this system. In rural areas, communications systems may not be available to support the technology embodied in electronic auctions.

In conclusion, live or electronic auction systems could be used by either small private farmers or by large former state or collective farms, provided that there are a sufficient number of buyers. For live auctions, larger farms would probably not face the same transportation constraints as small private farmers as they are more likely to own trucks, etc. Larger farms would be able to reap economies of scale in transportation as they would have more livestock to ship per sale whereas smaller farmers may be shipping only a handful of livestock to the market at any one time. Electronic auctions free both types of farmers from the need to transport livestock (assuming that buyers collect the commodity from the farm). The rules of an electronic (or live) auction concerning minimum lot sizes, may affect the viability of this channel for smaller farmers if they only market small quantities of a commodity (e.g. one or two cattle) at a time. The appropriateness of logistical rules such as minimum lot size, delivery/collection arrangements, etc. for both small private farms and corporate farms should be thoroughly investigated before live or electronic auction systems are set up so that the systems can be tailored to the particular industry environments in which they must operate. Furthermore, the rules over price determination, delivery or collection responsibility, etc. affect the division of transaction costs such as grade uncertainty between the buyer and the seller and will therefore affect the success with which auction systems are adopted.

Both live and electronic auctions are governed by a widely accepted set of rules concerning the operation of the sale and the liabilities and responsibilities of the three parties: sellers, buyers and auctioneer. These rules are essential for the auction system to perform and, importantly, to be seen to perform its role. For either auction system to be effective in the economies in transition: (i) an enforceable set of rules must be established which protect and uphold the basic principles of an auction system; and (ii) a sufficient number of buyers must be present for prices to be determined competitively.

### 8.2.4. Sales through dealers

Another option for small private farmers is to market their produce through an independent dealer or agent. The dealer either purchases the commodity from the farmer and then re-sells it to a processor, or sells the commodity on behalf of the farmer earning a commission on the sale. Dealers act as a go-between in the marketing channel when farmers or processors have poor access to market information or transportation and storage facilities. Provided that the service which the dealer performs contributes more to the final value of the product than the mark-up or commission charged, then the operational efficiency of the marketing chain is improved. Sometimes, however, dealers act as a buffer between producer and processor, hampering the flow of product and price information along the marketing chain. Given economies of scale in marketing, large-scale former state or collective farms would probably find it more efficient to internalize this function and employ their own marketing staff.

### 8.2.5. Sales directly to processors – without a contract

Sales directly to processing firms could occur either with or without a contract. In many countries, farmers develop long-term relationships with processing companies, delivering their commodities to them on a regular basis but without a formal written agreement between the two parties. In the case of livestock, the farmer either receives a liveweight price per head (in which case the buyer assumes all of the uncertainty that animals will not grade as expected) or a deadweight price based on the carcass grade (the seller assumes all of the grade uncertainty). The latter option enhances pricing and operational efficiency, enabling the appropriate price signals to be sent to the producer. The value of other agricultural products (crops, fruit and vegetables) can usually be assessed before they are processed, such that farmers can be paid according to the quality of the product delivered. If there are a large number of potential buyers in an area, farmers may believe that they do not need to lock down a buyer (or price) for their produce through a contract, instead preferring to find a buyer when the crop is ready for sale. The farmer depends on there being a sufficient number of buyers to ensure a competitive price for the product. If the product can be stored, and current prices are low, the farmer may be able to market the commodity strategically at a later date to obtain the best price. Direct sales to processors without a contract would probably not be a feasible option for large former state or collective farms, given the sheer volume of produce which they would need to market.

Small private farmers are likely to find themselves dealing with large monopsonistic or oligopsonistic buyers. This means that they have little bargaining power *vis-à-vis* the buyer and would be forced to accept the price offered by the buyer. This leaves them in a vulnerable position. Highly perishable products such as milk and soft fruit have to be sold immediately; any delays in the marketing channel would result in rapid product deterioration and spoilage. This puts the farmer in an even more vulnerable position if there are few buyers and inadequate storage facilities. It is unlikely that small individual producers would have storage capacity. Any storage capacity which previously existed belonged to large state or collective farms. Where these farms are still in existence they would probably be unwilling to share this capacity with competing private farmers. If these large farms have been divided up the existing storage facilities are likely to be too large for most individual private farmers.

A further problem with marketing directly to processors is price determination. Not only are individual farmers at a bargaining disadvantage, they lack a reliable source of information on prices. A World Bank study of the CEECs found that in Romania, for example:

> Many managers of the cooperatives say they receive too little information about market prices, availability of suppliers, etc. . . . Individual farmers do not have sufficient information either about the domestic and international market situations to adjust their production to the demand.
>
> (World Bank, 1995, p. 81)

In modern market economies, farmers selling directly to processors also incur information costs in finding out about representative market prices but because of the existence of a range of information sources these costs are less important (Hobbs, 1995). Where auction prices for similar commodities exist, they provide sellers with a guide to representative market prices. Industry associations often collate and publish average prices for direct sales to processors, either collecting this information from the processors (as the Meat and Livestock Commission does in the UK) or from producers (in Canada, CANFAX collects and disseminates price information on deadweight slaughter cattle prices as a service to member ranchers); wholesale markets for fruit and vegetable products in many modern market economies provide a source of price information, etc. Without these (usually national) sources of price information, a farmer's price discovery costs when selling directly to a processor would be extremely high. Considerable time and effort would be spent determining whether the price offered by a buyer was representative of average prices for that quality.

In the CEECs and the NISs, direct sales to processors (usually former or existing state enterprises) do occur. Often commodities are simply sold on a per head, weight or volume basis rather than priced by quality. This means that producers have little incentive to produce commodities which suit the preferences of end-users. This quality issue will be discussed in Chapter 9. In addition, in many modern market economies, the farmer receives payment for the commodity several days or even up to three to four weeks after delivery to the processing plant. The customers of the processing plant (retailer, wholesalers, secondary processors, etc.) operate a similar delayed payment system. This system would be untenable in most CEECs and NISs given the current financial instability of most enterprises. Many transactions are conducted on a cash-on-delivery basis and this extends to farmer–processor transactions. A World Bank report, based on a series of farm sector surveys in the CEECs, observed that in Hungary:

> Another problem is the reliability of buyers since some have gone bankrupt . . . Because of this uncertainty about the survival of private enterprises, farmers are said to prefer to deliver to state enterprises or to other enterprises that pay in cash. They might pay lower prices but the risk is less.
>
> (World Bank, 1995, p. 80)

Marketing products directly to processors without a contract leaves farmers in an extremely vulnerable position. The effectiveness of this marketing channel depends upon a competitive processing sector and adequate storage and information infrastructures.

### 8.2.6. Sales directly to processors – with a contract

Farmers can sell to processors according to pre-agreed terms established in a written contract. This method of marketing is often used for grain commodities in modern market economies. Contracts could be of interest to large former state-owned or cooperative farms since it enables them to lock down a buyer for the large volume of commodities produced on the farm.

When farmers sell their commodities through a contract with a buyer, they are producing for a forward market, making an agreement in the current time period for delivery of goods at some future date. The relationship between two independent firms in a contract is closer than in the case of auctions or uncontracted sales because each party has made commitments to aspects of the transaction in advance (Mighell and Jones, 1963). Under a contract, a producer devolves control over various aspects of marketing and/or production of the product to a buyer.

Contracts can be classified into three broad groups (Mighell and Jones, 1963).

**1.** Market specification contracts represent an agreement by a buyer to provide a market for a seller's output. The seller transfers some risk and the decisions over when the product is sold and how it is marketed to the buyer. Control over the production process, and the risks associated with it, remains with the seller. Provided that the contract is honoured, the producer is assured of a market for the commodity; the price, or the basis for establishing a price, is detailed in the contract.

**2.** A production-management contract gives more control to the buyer than a market-specification contract. The buyer participates in production management through inspecting production processes and specifying input usage. This will be important when specific quality characteristics of the output are of importance to the buyer and can be influenced by production practices.

**3.** Even more control rests with the buyer in the case of resource-providing contracts in which the buyer provides a market outlet for the product, supervises its production and supplies key inputs. Often the buyer owns the product, with the seller paid by volume of output. This is the closest contractual arrangement to full vertical integration. For example, a feedstuffs manufacturer might contract with pig producers, supplying feedstuffs, overseeing production methods and marketing the finished pigs. The producer is paid per kilogram of pig. The contractor (buyer) assumes all of the production risk and keeps a tight control on production methods.

Contracts have a particular set of risks, however, when considered within the context of an economy in transition with a poorly developed system of commercial law. While contracts increase pricing and operational efficiency by allowing closer coordination of controllable product characteristics with end-user preferences, the monitoring costs of contract production can be particularly high in these economies. If there are few (or only one) buyers for a commodity, having committed resources to producing this commodity under contract, the seller faces the risk that the buyer will act opportunistically and renege on the contractual agreement. It may be that conditions in the processors' marketplace have changed and the processor stands to make a loss if the product is purchased at the pre-agreed contract price, or it may be that the processor no longer requires the pre-agreed quantity or quality of products, or the processing company may simply act opportunistically to increase its profits at the expense of the vulnerable private farmer. Whatever the reason, as Teece observes:

> Even when all of the relevant contingencies can be specified in a contract, contracts are still open to serious risks since they are not always honored. The 1970's are replete with examples of the risks associated with relying on contracts . . . open displays of opportunism are not infrequent and very often litigation turns out to be costly and ineffectual.
>
> (Teece, 1976, p. 5)

In economies with a well-developed system of commercial law, sellers may have recourse to contract law and could sue for damages if the contract were reneged upon. The seller could incur legal costs and would face uncertainty over the outcome of any litigious action. The decision to apply the law of contracts would probably depend on whether the benefits to be gained from forcing the other party to uphold its contractual obligations outweighed the costs of legal action, allowing for the uncertainty of the judicial process. Often, the potential for opportunistic behaviour remains, despite the existence of contract law. Firms in former command economies are usually without the protection of even the most basic contract law to govern the establishment and enforcement of contracts. As a result, both parties to a contract incur considerable monitoring and enforcement costs in ensuring that the contract is adhered to. These high transaction costs will prevent the widespread adoption of contracts for the supply of agricultural products, given the current financial and legal instability of economies in transition. If farmers have made asset-specific investments in the production of commodities tailored to the quality specifications of particular buyers, they are especially vulnerable to buyers acting opportunistically. A study of fruit and vegetable distribution systems in Russia revealed attitudes among farmers which are typical of the problems associated with agricultural contracts in the current economic and legal climates of economies in transition:

> There is no trust of contracts, which are largely unenforceable . . . State farms and newly privatized farms are uncertain what to plant because they feel vulnerable without guaranteed outlets. They do not trust the vegetable bases to honour contracts and the feeling is mutual.
>
> (Jones, 1993, p. 22)[4]

The reluctance to enter into contracts and reluctance to produce commodities to the specifications of particular buyers will hamper the ability of agrifood chains to evolve into efficient systems, responsive to the preferences of end-users.

Those entering into contracts for the supply of agricultural commodities to processors should be aware that contract terms which influence the division of risk between seller and buyer are critically important. Lang (1980) reports the results of a nationwide survey of fruit and vegetable bargaining associations in the US. The author analyses the effect on the organizational slack and X-inefficiency[5] displayed by processors engaged in collective bargaining with a group of producers, as opposed to independent contracting by individual producers. The survey found that the increased bargaining power provided to farmers by collective bargaining resulted in changes in marketing contracts which forced processors to take up organizational slack, reducing X-inefficiencies and leading to a Pareto improvement[6] in terms of reducing joint costs in the marketing chain.

Lang (1980) provides examples in which collective bargaining led to contract terms which improved product handling, storage, scheduling of delivery or collection, etc. In all cases, the processor was in a better position to control the risk of product damage than the seller. Shifting the risks from farmers to processors enabled a minimization of joint costs, therefore leading to a Pareto improvement.

The terms of payment and the point at which the buyer takes title to a commodity significantly affect the allocation of risk among sellers and buyers. This is particularly important for perishable agricultural products where the timing of delivery, storage and handling methods can have a considerable effect on the final value of the commodity. To maximize economic efficiency, contracts between farmers and processors in the CEECs and the NISs should be written so that the allocation of these risks reflects the ability of either party to minimize joint risks, in other words, so that organizational slack and X-inefficiencies are minimized. Contracts without an appropriate incentive structure will fail to generate products of the desired qualities and will impede the development of a modern, efficient market-based agrifood chain. Fundamental to the use of contracts, however, is an effective system of commercial law for the establishment and enforcement of contracts.

### 8.2.7. Joint marketing

As the above discussion suggests, sometimes there are advantages to farmers acting jointly to market their products. Three types of joint marketing can be identified: marketing boards, marketing cooperatives and group marketing schemes.

## Marketing boards

Marketing boards are quasi-governmental organizations which control various aspects of the marketing of agricultural products within a country. Some marketing boards in modern market economies have a strictly advisory and promotional role and do not have a distortionary effect on market forces. Often, however, marketing boards have powers to control the supply of a commodity. The ability to control supply means that a higher price can be secured thus improving the incomes of the board's producer members. It is usually compulsory for all producers in the industry to sell their products to these marketing boards. Supply management marketing boards have been criticized as inefficient because they lead to increased food prices for consumers and because they protect existing farmers from competition by restricting the entry of new producers into the industry (Veeman, 1982). Table 8.1, adapted from Abbott (1987) summarizes some of the key features of marketing boards.

Recent policies in many countries in the European Union and in North America have reduced government involvement in the agriculture sector. This has included policies to disband existing marketing boards and to discourage the establishment of new boards because of the budgetary cost to taxpayers and the distortionary effect on market forces, including higher prices for consumers. In the CEECs and the NISs, domestic monopoly trading marketing boards would be unpopular and other marketing boards which require government subsidization would probably not be feasible given the limited financial resources of these governments. Furthermore, financial aid from foreign governments and transnational organizations such as the World Bank is usually dependent upon the CEEC or NIS adopting policies which encourage the development of a market economy, allowing market forces to operate relatively unfettered. The creation of monopoly supply management marketing boards as part of agricultural policy initiatives would be contrary to the conditions set out by donor governments or organizations. Marketing boards which serve a promotional function or carry out research into improved methods of storage and distribution may, however, be an option.

Although often initially established to protect agricultural sectors for food security or countervailing power reasons, monopoly marketing boards in modern economies created powerful vested interests in the *status quo* among the agricultural community. As economies evolved and the marketing environment changed, the initial reasons for establishing the boards often disappeared. Once established, however, the boards have proved difficult to disband. The long-term economic impacts of creating a domestic monopoly supplier of an agricultural commodity need to be fully understood before such a policy is contemplated.

**Table 8.1.** Features of marketing boards.

| Type of marketing board | Responsibility | Nature of government commitment | Financial implications | Effect on free market structure | Marketing and administrative skills required | Benefits for producers | Implications for domestic consumers | Conditions of implementation |
|---|---|---|---|---|---|---|---|---|
| Advisory and promotional | Undertake or commission promotion and research | Authorize compulsory levy on sales of products | Self-supporting as long as levy accepted | Continues undisturbed | Low; specialized service can be hired | Improved sales returns should exceed cost of levy | None | Marketing centralized at processing plant |
| Regulatory | Control quality, form and quantity of product offered on particular markets | Delegate powers of compulsory membership and authorize collection of levy | Self-supporting as long as levy accepted | Continues, subject to controls; no discrimination between enterprises | Skills needed to assess benefits of actions envisaged and taken | Improved unit sales returns; quantities sold may be restricted on historical basis | May be lower-quality products at higher prices | Producers few and identifiable or products pass through limited processing points |
| Stabilizing prices without trading | Establish prices paid to producers. May operate a reserve fund for this purpose | Delegate powers of compulsory membership and authorize collection of levy | Self-supporting as long as levy accepted | Wholesale operations continue free; purchasing from farmers subject to licence | Skills needed to estimate appropriate price to pay farmers | Wholesaler purchasing price clearly established; may be lower than on free market | Minimal | Wholesale purchases relatively few and easily identifiable |

**Table 8.1.** contd.

| Type of marketing board | Responsibility | Nature of government commitment | Financial implications | Effect on free market structure | Marketing and administrative skills required | Benefits for producers | Implications for domestic consumers | Conditions of implementation |
|---|---|---|---|---|---|---|---|---|
| Stabilizing domestic prices through buffer stock on a free market | Trading monopoly on imports and exports; implement pre-announced minimum price to farmers | Provide initial capital; guarantee for commercial credit; support import or export controls | Ulikely to cover own costs – government must subsidize the difference | Domestic trading continues subject to competition with board; prices kept within band by buffer stock policy | Ability to estimate supply and demand; judgement in buying into and selling buffer stock and skill in stock management | Announcement of guaranteed minimum price before planting facilitates production planning | Protected against very high prices by releases from buffer stock, but price floor prevents benefits from lower prices | Product easily storable |
| Export monopoly trading | Sole exporter of designated products; implements pre-announced minimum price to farmers | Provide initial capital; guarantee for commercial credit; support for export monopoly | Operating costs normally covered from margin allowed by price paid to farmers | Former exporters eliminated or required to act as agents for the board | Routine management skills required; initiative and judgement to exploit monopoly power and to see new opportunities | Announcement of guaranteed price facilitates production planning but price may be low | Minimal or none | Exports limited to a few border points that can be policed |

| Domestic monopoly trading | Sole trader in designated products in defined domestic markets and implements minimum prices to producers | Provide initial capital; guarantee for commercial credit; support for monopoly | May incur trading deficits if capital and operating costs excessive or if badly managed | Former traders eliminated or required to act as agents for the board | Routine management skills required; initiative and judgement to see new opportunities | Pre-announce guaranteed price but it may be low | Price of products sold likely to be higher than on a free market | Feasible to control evasion of monopoly |
| --- | --- | --- | --- | --- | --- | --- | --- | --- |

Source: Adapted from Abbott (1987), p. 20.

*Marketing cooperatives*

An alternative to compulsory marketing boards are voluntary marketing cooperatives. These are significantly different from the agricultural cooperatives of command economies. This difference is important because the term cooperative is often met with considerable suspicion among farming populations in former command economies, who naturally associate the word with the compulsory collectivization of land under the communist system. In modern market economies, a cooperative is a voluntary association of producers with a common commercial objective, such as securing cheaper inputs or securing a greater market share and higher prices for outputs. Cooperatives can be of two types: supply cooperatives enable farms to reap economies of scale in the provision of inputs, such as bulk-buying of fertilizers and seed or machinery rings which share the costs and usage of agricultural machinery. Marketing cooperatives focus on agricultural outputs, with farmers cooperating to improve the marketing of their produce. For example, a cooperative may be vertically integrated downstream in the marketing chain, undertaking processing, packing and distribution of agricultural outputs or it may simply perform marketing functions such as providing market information, locating and negotiating with buyers, etc. A primary aim of a marketing cooperative is to increase the bargaining power of its members *vis-à-vis* downstream firms and to reduce the costs of marketing by exploiting economies of scale.

Cooperatives operate on a number of principles which may vary between countries but generally include: (i) open membership (any producer can join, or leave); (ii) one member-one vote (regardless of the degree to which the member uses the cooperative); and (iii) the interest on a member's share and loan capital is limited and any profits which are surplus to requirements are distributed among the members.

Voluntary marketing cooperatives could play a useful role in helping small private farmers in former command economies to market their produce. To be successful in counteracting the bargaining power of local monopsonistic or oligopsonistic processors, marketing cooperatives would have to handle a significant proportion of local produce. This will depend on the proportion of agricultural production accounted for by small private farmers and the extent to which these farmers join and participate in the cooperative. This is one of the weaknesses of the voluntary cooperative system. The cooperative has little control over the supply of the product by the cooperative because membership is open to any producer and members are often not obliged to sell all of their produce through the cooperative. Fluctuations in the supply of products through the cooperative occur if farmers perceive that the open market temporarily offers them a better return.

Large-scale former state-owned and collective farms would be less likely to participate in a voluntary marketing cooperative because they may have enough market power by themselves (or they may still be vertically integrated with processing facilities). The one member-one vote principle means that these large farms would have a disproportionately small say in the management of the voluntary cooperative relative to their throughput. This voting principle, although ostensibly democratic, leads to problems in modern market economies when a small number of producers supply a large proportion of a voluntary cooperative's output, yet a large number of marginally active cooperative members have considerable influence over the running of the cooperative. This leads to dissatisfaction among the loyal members. Cooperatives often have difficulty raising capital from their members if they follow the principle of limited return on share capital. The lack of a functioning agricultural credit market in most former command economies means that outside sources of financing would also be difficult to obtain.

Probably the biggest constraint to the creation of voluntary marketing cooperatives among private farmers will be a cultural aversion to this type of organization because it could be associated with the compulsory collectivized cooperatives of the command system. Those individuals who have become small farmers are likely to be more independent by nature. They will be far less willing to become involved in a voluntary cooperative than those who have elected to remain with an old-style cooperative farming complex. Furthermore, a functioning commercial legal system is necessary to enforce the rules of voluntary marketing cooperatives.

### Group marketing schemes

An alternative form of producer cooperation is a group marketing scheme. This is a voluntary association of farmers with a common commercial objective and is in many ways similar to a voluntary cooperative; however, membership is restricted. Members must agree to adhere to specified production methods such as breeds, feeding regimes or sowing and fertilizing practices, etc. Marketing groups improve farmers' bargaining power relative to downstream firms. Group marketing schemes may be better placed to meet the specific requirements of downstream firms because they have more control over product quality. As a result, they improve the operational efficiency of the marketing chain.

As with individual contracting, if group marketing schemes tailor production to the needs of specific buyers, they are subject to the risk of opportunistic behaviour on the part of the buyer, particularly if the producers have made an asset-specific investment. The risk may be offset to some extent by the bargaining power of the group. This depends on the proportion of a processor's inputs represented by the marketing group. Conceivably, a non-cooperative bilateral monopoly situation could develop. The effect of bilateral monopolies was discussed in Chapter 5.

Opportunistic behaviour is, of course, a double-edged sword. Buyers also face the risk that producers will renege on a supply agreement if market conditions change. The problem is more serious, however, if one party is more dependent on the other as a source of supply or as a market outlet. If processors have a number of suppliers to choose among, the effects of opportunistic behaviour being practiced by one of their supplier groups, whilst undesirable, can be mitigated. On the other hand, if a producer group contracts to supply all of its output to one buyer, the effects of opportunistic behaviour on the part of that buyer are potentially far more serious.[7] In former command economies, the extent to which agricultural production has been privatized into small independent farms will influence this power relationship. If private farms do not represent a significant proportion of local produce, a private farmers' group marketing scheme has less bargaining power *vis-à-vis* a monopsony buyer than if private farming represents the major proportion of local produce (assuming most farmers join the scheme). As with voluntary cooperatives, group marketing schemes are not likely to appeal to large former state or collective farms as they may already have sufficient market power locally. Observations from the Hungarian agri-food chain are typical:

> The processing industry prefers to deal with co-operatives because they can supply large amounts of products of a constant quality on a regular basis, whereas the private farmers can only supply in small quantities.
> (World Bank, 1995, p. 80)

The risks of opportunistic recontracting on the part of the buyer notwithstanding, group marketing schemes and voluntary marketing cooperatives can reduce transaction costs for farmers by providing market price information, providing information about the product characteristics demanded by buyers, reducing negotiation costs through economies of scale in negotiating and reducing monitoring costs through collective bargaining agreements which encourage processors to improve the storage and handling of products.

### 8.2.8. Vertical integration

A final option is for the farm enterprise to vertically integrate forwards into processing and distribution. This may occur if the transaction costs of coordinating economic activity through the price mechanism outweigh the costs of within-firm coordination. It is an option more likely to be pursued by large-scale former state or collective farms given economies of scale in processing; some may have been privatized as vertically integrated units. It could also be an option for some small producers able to service local markets with further processed products. The lack of a functioning credit market, however, will seriously impede any investments of this kind. Assuming the farm has not previously been involved in further processing or distribution, a new set of skills will probably be necessary, requiring an investment in human capital by the farmer or by farm workers. Additional new transaction costs which would arise in this situation would include the costs of monitoring workers to prevent shirking. The effects of transition on the processing sector are discussed in Chapter 9.

## 8.3. Conclusions

A key problem facing agricultural production sectors in most former command economies is the regional monopsonies in processing and distribution created by liberalization of the downstream industry. This leaves small private farmers at a considerable bargaining disadvantage *vis-à-vis* buyers and creates a problem of bilateral monopoly between existing large-scale former state or collective farms and buyers. Small privatized farms and large-scale farms face many of the same problems in marketing their produce, although large farms have the additional problem of disposing of large volumes of product. On one hand, this gives them considerable bargaining power in the marketplace but on the other, if the processing sector is monopolized, leaves them vulnerable with large quantities of product to dispose of to one (or only a few) buyer(s).

Assuming that the liberalization process continues and small-scale private farming eventually emerges as the dominant form of agricultural production in most CEECs and NISs, a number of marketing channel options present themselves. These include farmers' markets, auction markets, dealers, sales directly to processors with and without a contract and different forms of joint marketing. While all of these marketing channels have their strengths and weaknesses, to different degrees they are all dependent on a functioning and efficient system of commercial law, on the availability of storage and transportation technology for agricultural commodities (particularly those which are highly perishable) and on a system of disseminating market information among producers. Without these fundamental market economy institutions, the marketing channels will fail to generate representative market prices and therefore fail to allocate resources efficiently. Liberalizing the agricultural production sector and agrifood processing and distribution is a necessary but not sufficient condition for the successful transition of agrifood systems. The underlying institutions which facilitate the operation of marketing channels must also be in place. These institutions are discussed in more detail in Chapter 16.

# Notes

1. This means the physical presence of buyers, and often sellers, in the place where the auction was taking place.
2. The exception being pigs which are less likely to be sold through auctions because of problems with disease transmission.
3. This is quite different from the situation in which farmers are paid according to a quality assessed grade, where the farmer bears the risk that the commodity will not grade as expected.
4. Vegetable bases, a remnant of the centrally planned system, are storage centres located in cities in order to guarantee a supply of fruit and vegetables to the city population. They originally provided a link between state farms and state shops.
5. Cyert and March (1963) observed that organizational slack occurred in concentrated industries and argued that firms would forego pecuniary efficiencies in order to achieve other managerial goals such as maintaining good intra-firm relationships among management or between management and employees, etc. Liebenstein (1966) argued that, where organizational slack existed because of the absence of external competitive pressures, a firm displayed X-inefficiency, i.e. it did not operate to its maximum productive efficiency (at the outer-bound of its production possibility frontier). In other words, people preferred to maintain a working atmosphere which entailed less pressure and better interpersonal relationships. Internal changes such as a change in management or increasing costs, or external pressures such as falling demand, increased competition or a change in bargaining relationships within a marketing channel may force a firm to take up some of this organizational slack and reduce X-inefficiencies.
6. A Pareto improvement is a change which leaves at least one person better off and nobody worse off.
7. In other words, the buyer's Lerner Index of monopoly power is higher than the producer group's, see Chapter 5.

# References

Abbott, J. (1987) Alternative agricultural marketing institutions. In: Elz, D. (ed.) *Agricultural Marketing Strategy and Pricing Policy*. A World Bank Symposium, pp. 14–21.

Cyert, R.M. and March, J.G. (1963) *A Behavioural Theory of the Firm*. Englewood Cliffs, New Jersey, Prentice-Hall, 332 pp.

Hobbs, J.E. (1995) A transaction cost analysis of beef marketing in the UK. PhD thesis, Department of Agriculture, University of Aberdeen, Scotland.

Jones, S. (1993) The future for fruit and vegetable distribution in Russia. *British Food Journal* 95(7), 21–23.

Kerr, W.A., Klein, K.K., Hobbs, J.E. and Kagatsume, M. (1994) *Marketing Beef in Japan*. The Haworth Press Inc., New York, 201 pp.

Lang, M.G. (1980) Marketing alternatives and resource allocation: case studies of collective bargaining. *American Journal of Agricultural Economics* 62, 760–765.

Liebenstein, H. (1966) Allocative efficiency vs. X-efficiency. *American Economic Review* 56, 392–415.

Mighell, R.L. and Jones, L.A. (1963) *Vertical Coordination in Agriculture*. USDA ERS-19, Washington DC, 90 pp.

Schotter, A. (1976) Auctions and economic theory. In: Amihud, Y. (ed.) *Bidding and Auctioning for Procurement and Allocation*. Proceedings of a conference at the Center for Applied Economics, New York University Press, New York, pp. 3–12.

Teece, D. (1976) *Vertical Integration and Vertical Divestiture in the US Oil Industry: Economic Analysis and Policy*. Stanford University, 141 pp.

Veeman, M. (1982) Social costs of supply-restricting marketing boards. *Canadian Journal of Agricultural Economics* 30, 21–36.

World Bank (1994) *Tajikistan; the Transition to a Market Economy*. A World Bank Country Study, International Bank for Reconstruction and Development, The World Bank, Washington DC, 240 pp.

World Bank (1995) *Farm Restructuring and Land Tenure in Reforming Socialist Economies: A Comparative Analysis of Eastern and Central Europe*. World Bank Discussion Papers no. 268, Euroconsult, Centre for World Food Studies, The World Bank, Washington DC, 148 pp.

# 9 Quality Control and Reliability: The Challenge for Processors

The food processing sectors in the Central and Eastern European Countries (CEECs) and New Independent States (NISs) face a number of challenges which arise as a result of both the pre-liberalization structure of these industries and the market environment in which firms must now operate. Internal challenges and constraints stem from the general level of indebtedness of newly privatized enterprises, concomitant problems of financing and investment, obsolete technology and equipment, quality control problems and the need to develop managerial and marketing expertise. Developing relationships with suppliers, retailers and foreign competitors presents food processors with a series of external challenges. This chapter begins by discussing the nature of food processing under the command system. The alternative privatization options pursued by the economies in transition are then outlined and the internal and external challenges facing food processors are assessed.

## 9.1. Food Processing Under the Command System

In the command system, food processing was organized on a large scale, often with only one processing unit serving a region with a particular product. Sometimes food processing facilities were an extension of state or collective farms. Processors received raw agricultural product from designated suppliers and transferred their output to designated distributors or (state operated) retailers within the same command economy supply chain. Decisions over the quantities and qualities produced were made by bureaucratic decree in accordance with the wider economic plan of the region or country, rather than in response to market forces. Prices had no role to play in the allocation of resources to the production of different food products (see Chapter 3). Food processing units did not incur information costs in locating and assessing suitable suppliers or buyers (see Chapter 4). They did not incur monitoring costs in ensuring that raw materials were of a consistent quality because they were required to process all agricultural products from their designated suppliers, regardless of quality. A free-market pricing mechanism which would have evened out surpluses and shortages did not exist, hence, considerable wastage occurred in the system. Quality was generally poor because the processors could not provide producers with appropriate price signals to induce them to supply high quality inputs. The processors themselves did not have an incentive to improve the quality of their products. Food processing units received frequent government financial transfers or credits and were not expected to be financially self-sufficient. These credits did not have to be repaid, hence, the system did not encourage fiscal responsibility. Considerable restructuring has therefore been necessary in the food processing sectors of all CEECs and the NISs, particularly given the large-scale, regional monopoly nature of food processing facilities.

## 9.2. Food Processing in Transition

Privatization of food processing has occurred at different rates and by different means in the various CEECs and NISs. Common to all of these countries, however, has been the problem of the pervasive monopoly in food processing and distribution. Privatizing processing facilities in their existing form invariably creates a regional monopoly and, when the same privatization process is applied to other stages of the agrifood chain, a situation of bilateral monopoly occurs. This leads to an inefficient outcome with higher prices and lower output than would be the case in a competitive industry situation (see Chapter 5). Breaking up large processing units, however, is far from easy. Often the equipment is indivisible and can only operate efficiently on a large scale. It is, therefore, inappropriate for small-scale processing. Of course, the lack of investment capital also hampers privatization.

A number of alternative privatization models have been tried. The Czech Republic, Slovakia and Lithuania have used a voucher privatization system for the majority of enterprises. In some cases, a small percentage of processing enterprises were returned to their original owners, where these individuals could be identified and where the processing enterprise had been kept as a separate entity under communist rule. In the Czech Republic there were two waves of privatization, the first in 1992 and the second, and last, in 1994. By the end of 1994, all food processing enterprises had been privatized as joint-stock companies. All Czech residents were issued with vouchers which they could exchange for shares in a joint-stock company. In this way, the food processing sector was transferred from state to private ownership and the state withdrew from the running of these industries, although it still plays a part in the enactment and enforcement of anti-monopoly legislation. The privatization process in the Slovak Republic followed the same model.

Lithuania applied a similar privatization process to most key sectors of the economy over a three year period. Small processing firms were privatized and became part of agricultural partnerships, although the privatization of most processors in urban areas occurred through the issuing of shares. Privatization of these larger firms proceeded at a some-what slower pace because much of the equipment was old and obsolete and most vouchers had already been used in the privatization of the smaller firms. At a later stage, in an effort to encourage closer integration between agricultural production and food processing, the Lithuanian government allowed farmers to purchase shares in food processing companies. At least 50% of the shares of these enterprises were to be distrib-uted to farmers (OECD, 1995). While foreign investment was allowed (and in some cases encouraged) in the Czech Republic and Slovakia, the Lithuanian privatization process relied almost entirely on internal investors. This has led to problems for the Lithuanian industry as it has tried to modernize and update its equipment. Many food processing firms are seeking joint ventures with western firms to provide an infu-sion of investment capital and technological know-how.[1]

Much emphasis has been placed on attracting foreign investment into the Hungarian economy. The aim is to have all state companies privatized by 1997. This procedure was well underway by late 1994, with no food processing sector more than 50% state owned. In Nov-ember 1994, the dairy industry had the greatest state presence (49%), whereas the brewing, confectionery and vegetable oil industries were completely privatized (OECD, 1995). There is a significant share of for-eign capital in many Hungarian agrifood industries – one extreme being the vegetable oil industry which is 100% foreign owned. The degree of foreign ownership differs across industries. In 1994, the brewing indus-try was 51% foreign owned, whereas there was no foreign ownership of milling or baking. There has been little foreign investment interest in the dairy, meat and poultry industries, all of which are major industries. Reasons for the lack of interest include the debt and liquidity problems of these sectors, in addition to excess capacity problems.

The high level of foreign investment in some countries invites the danger that a foreign investor will achieve a position of monopoly power in an industry. This would occur if the investor was able to purchase the main processing facility plus any additional smaller processing enterprises in the region. Economic theory suggests that it may be more efficient to buy up (and either amalgamate or close) competitors than to try to engage predatory or limit pricing tactics to force them out of business (McGee, 1958). Foreign investors are more likely than domestic entrepreneurs to have access to the necessary financial resources to allow them to achieve a position of monopoly power in an industry.

A different privatization procedure has been adopted in Estonia. Food processing industries have been sold off by auction through written (sealed) bids and by the public sale of shares. Enterprises may either be purchased for cash or with privatization vouchers which are issued to all permanent residents of the country with the distribution weighted by the number of years in employment. Although public share issue is used, often a single investor purchases an entire enterprise. This can lead to problems of monopolization. By early 1995, 90% of the Estonian food processing sector had been privatized, although the main sectors of the food industry, meat, dairy and distilling were still partly state owned (OECD, 1995).

In Poland, the emergence of new small and medium-sized private enterprises has made a more significant contribution to the share of private activity in food processing (which had reached 50% by 1995) than the privatization of former state food industries. For example, by the end of 1993, the number of state enterprises had been reduced by only 21%, whereas the number of private firms had increased by 250%. In some sectors, enterprises were turned into joint-stock companies which were initially 100% Treasury (government) owned. The shares were then sold to the public, with the workers (in practice, the managers) of former state and cooperative enterprises, having first access to the shares. This has created a conflict of interest during the transition period. When initially sold to the public (i.e. managers and workers of the firm), the share price is determined by an assessment of the value and profitability of the enterprise. It is therefore in the interests of managers to mismanage the firms, allowing organizational slack and X-inefficiencies,[2] in order to reduce profits while the firm is in the transitionary phase between treasury ownership and public sale. If the firm records a low level of profit, initial shares will be available to managers at correspondingly low prices. Problems of this kind have been observed in Polish industries. Investment capital has, for example, been available for updating obsolete equipment, with the potential to improve the profitability of the firm but has been left in a bank account while the firm was in the transition phase. Strategic behaviour by enterprise managers has clearly affected the pace of transition.

The privatization process has been undertaken with less zeal and enthusiasm in other countries, particularly those of the NISs. These countries face much greater macroeconomic instability and a poorer developed legal and market infrastructure and have had a longer history of state ownership than some of the CEECs. In the Ukraine, for example, the privatization of food processing began hesitantly in 1993 but was subsequently halted in 1994, at which time there were only four fully privatized enterprises in the country (OECD, 1995). The privatization program involved each enterprise issuing stock, with 51% to be sold to the firm's workers and suppliers at a nominal price and the remainder offered for public sale. As the OECD observes, the process was met with much suspicion and criticism:

> This programme was widely criticised for its slow pace, corruption and mismanagement, especially the lack of clear procedures for valuing property to be privatised and awarding it to particular individuals.
>
> (OECD, 1995, pp. 152–153)

Clear potential exists for opportunistic behaviour on the part of administrators (rents to be extracted in the valuing of property and allocation of share rights) and enterprise management (mismanagement to facilitate purchase of the enterprise at below its true market value). Following the suspension of the original privatization programme, a new pilot programme of voucher privatization was begun in five Ukrainian provinces in early 1995.

The creation of private market monopolies is an endemic problem in Russia where food processors were built as state monopolies with one or occasionally two plants of each type in an area. Compounding this problem is the fact that the transportation infrastructure was invariably designed around these monopoly collection and processing points, fanning out to farms from the central processing location. The monopoly situation is exacerbated by restrictions on the movement of products between districts, so that producers face a monopoly purchaser for their products. Where these processing facilities have been privatized, the purchasers have often been outside interests. Alternatively, plants were sold to their employees. Agricultural interests were either excluded from the privatization of food processors or could not raise sufficient capital to invest in the downstream industry. As pointed out in Chapter 6, this has left the farm sector facing a cost-price squeeze as it must purchase inputs from privatized (often monopolized) suppliers and sell its outputs to separately owned monopsony processing firms. In some cases, this had led to farms establishing their own processing facilities. These tend to be small scale and face severe investment constraints. Furthermore, as the downstream industry (including the retail sector) privatizes, these small-scale farms/processors may have to deal with large-scale distributors and retailers; this will raise their negoti-ation (transaction) costs of doing business. Alternatively, the farm-processors could sell directly to end-consumers, as was discussed in Chapter 8, however, this is time-consuming as well as operationally inefficient from a system-wide perspective. Small-scale integrated farm-processors are unlikely to remain competitive in the long run.

Very little privatization has occurred in the food processing sectors of Belarus and Albania. In Belarus, only 4% of state property had been privatized by the end of 1994, with much opposition to further privatiz-ation from the country's parliament. The situation in Albania is perhaps typical of that which has faced many more progressive CEECs and is still facing a number of other CEECs and the NISs as they embark along the road to a market economy:

> The entire state-owned food processing and marketing sector is in a serious condition. The subsistence orientation of farmers has cut deliveries to most plants, which are now working at very low capacity; most poultry, eggs and pork plants are closed or operating at a very low level of activity. This situation is also reflected in the high levels of processed food imports. In addition, these industries are also seriously hampered by obsolete equipment, out-of-date technology and a lack of knowledge on the part of management of how to operate in a market environment.
>
> (OECD, 1995, p. 22)

These, and other problems, are discussed in the sections below.

## 9.3. Internal Challenges

Food processors in the CEECs and the NISs face a number of internal restructuring challenges, not the least of which is the need for invest-ment. Without the discipline of market forces, industries were ineffi-cient, excess capacity had developed and there was little incentive to update technology or invest in product improvements. As a result, con-siderable re-investment in food processing technology will be required. Many food processing facilities, for example, produce canned food. The technology exists to extend considerably the shelf-life of foods through freezing and through rapid chill and vacuum-packaging. Consumers in modern market economies increasingly demand these more convenient, nutritious and tasty methods of food preservation and presentation. Consumers in the CEECs and the NISs will be no different, particularly as incomes increase. The domestic industries of these countries will therefore face stiff competition from foreign firms who already utilize these technologies.

In addition to often being obsolete, food processing technology tends to be large scale and inefficient. Processing lines are far less automated than those in modern market economies, relying on manual labour for many tasks which can be performed more efficiently, more consistently, and often more hygienically by an automated process. Existing state owned processing facilities tend to be over-staffed. Furthermore, governments may wish to break-up large processing and distribution units into numerous smaller, more competitive firms to avoid creating situations of bilateral monopoly. The existing processing equipment was designed to capture economies of large-scale operation and may be unsuitable (and too expensive) for small firms.

Most former state or collectively owned processing enterprises were privatized with substantial debt-loads. Under the command system, the state commonly issued industries with credits to finance their operations which, in reality, were grants as they were rarely repaid. These credits appeared as debts on the balance sheets of enterprises to be privatized, with repayment required after privatization. Processing firms in transition economies are faced with a dilemma which has significant financing and investment implications. They must decide whether to invest in new technology or whether to modernize and improve upon existing technology. The first option could put the firms on an equal technological footing with competing foreign firms, enabling new and improved products to be produced more efficiently. It is, however, a very expensive option, requiring access to considerable investment funds. These funds are often not available to firms in transition economies because of the high cost of credit, the depleted asset base of these firms (the absence of sufficient collateral) and, in many cases, a poorly developed commercial credit system.

The second option, to utilize, and try to improve upon, existing technology would require less investment capital but may leave the firm in a vulnerable position *vis-à-vis* competitors (particularly those from modern market economies). The extent to which this latter option is possible will depend upon the nature of the product. Investment in simple packaging technology could lead to product improvements without requiring large amounts of capital. It may be possible to alter packaging size to better meet the needs of end-users and retailers or to improve the presentation or functionality of the package (e.g. longer shelf-life, easier to store, easier to open, etc.). New product developments, such as the production of yoghurt or ice-cream at a milk processing plant, will require a more substantial investment in new technology. The lack of internal sources of credit and the difficulty of obtaining external financing will severely hamper the development of independent private processing industries in transition economies. A priority for policy makers should be to facilitate the development of reliable and functioning financial institutions. This issue is discussed further in Chapter 17.

Another difficulty associated with modernizing, or even maintaining, a processing firm's existing set of machinery is the lack of replacement parts. Command economies had enormous difficulties in allocating resources to the production of spares. With many heavy machinery industries either shut down or short of raw materials, replacement parts are difficult to obtain. This situation has been exacerbated by the collapse of Council for Mutual Economic Assistance (CMEA) trade relationships. Obtaining spare parts from firms in other CEECs or NISs is often no longer possible. As a result, technology from modern market economies often appears as an appealing option even if it is more expensive because spare parts are easily accessible.

Processors also face considerable challenges in maintaining their quality control laboratories and in modernizing them so that they can export. Often, equipment from modern market economies is not compatible with the requirements of domestic regulations. Existing testing and monitoring equipment is difficult to keep at acceptable levels of operating efficiency due to shortages of spare parts and technical support from manufacturers. If equipment from modern market economies can be used in quality control laboratories, staff need to be trained in its operation. Given the differences in education systems, training of quality control personnel must often be undertaken out of the country at high cost. This training process may require considerably more time than is typical in modern market economies to compensate for differences in basic levels of training.

Managerial skills pose another internal challenge for food processing industries. In the command system, managers responded to bureaucratic directives rather than market forces. The managerial skills required tended to be organizational, logistical and often political in nature. Managers could not seek alternative sources of supply and could rarely try new product formulations; market outlets were designated and assured. Profit-making was not an objective, so there was no incentive to manage the enterprise in an economically efficient manner. Command firms played a larger role in the lives of their employees than is common in most market economies, often performing social functions. For example, many companies operated vacation hotels and resorts for their employees. Most employees could expect to have a job for life with the same state firm, hence, complacency often developed as there was little threat of redundancy or dismissal to discourage shirking. The managerial system tended to be very hierarchical in nature, with innovation often discouraged. Promotion was usually on a seniority or political patronage basis rather than by ability to manage the enterprise efficiently.

A very different set of managerial skills are required in a market economy. Managers must understand (often implicitly) the functioning of a market economy. They must be able to assess the needs of the marketplace and how best to serve those needs, including what to produce, how to produce it and to whom to sell it in the most cost-efficient, revenue maximizing manner. They must be able to identify input supply needs and assess alternative suppliers. Included in all these aspects of market-enterprise management will be at least a basic knowledge of marketing, finance, economics, consumer behaviour, personnel management and processing logistics. None of these skills were necessary in the command system.

In some cases, processing enterprises have been privatized in their entirety with the existing management in place. Where these individuals happen to be innovative and market-economy oriented, meaning that they will quickly learn the skills necessary to adapt to a competitive situation, the enterprise will be well managed. These are not the type of individuals, however, who were likely to have been in senior managerial positions under the previous regime. Where previous managers who have not acquired the necessary managerial skills are still in control of newly privatized firms, these firms are likely to remain inefficient, particularly if they have monopoly power.

In addition to management skills, a change in management philosophy may be required. Often, at best, a product-oriented and at worst, an internal company-oriented philosophy pervades many former command economy firms. In other words, managers are maximizing an objective function based on pre-liberalization modes of conduct: 'how can I best manage this enterprise to enhance relationships with the workforce or to produce output with minimum effort or to supply only those products which best suit my production technology?' A successful firm in a market economy has a consumer-oriented approach: 'how can I produce a product which best suits the needs of my customers?' This consumer-oriented attitude pervades all aspects of the processing function, from product design to packaging and delivery schedules in modern market economies.

An illustration of the change needed in management philosophy is provided by a typical newly privatized Polish dairy enterprise near Wroclaw which had diversified into the production of yoghurts. A number of different flavours of yoghurt were produced, with the flavour indicated on a label fixed to the lid of the yoghurt pot. This style of packaging, however, meant that a consumer could not see the flavour label when the product was displayed in the retail store because the yoghurt pots were normally stacked one on top of another on the retail shelf. A market or customer-oriented company would have realized that this was a problem and would have changed the package design so that the information which the consumer required when making a (usually split-second) purchase decision was clearly visible. The Polish firm, displaying a typical product or internal company-oriented attitude responded that it was the retailer's problem, who they argued was at fault for not stacking the product so that the flavour label was visible to the consumer. Foreign imports of yoghurt, with clearly visible labels, were selling alongside the Polish product and, while more expensive, were capturing market share from the Polish product.

A further challenge facing the food processing sectors in the CEECs and the NISs is the need for quality improvements. This has both an internal and an external aspect. The internal aspect of food quality is the need to improve within-plant processing practices. There is a great deal of wastage, particularly of highly perishable products because of the inefficiency of the storage and distribution systems. Processing plants are often old and lack adequate refrigeration or chilling facilities. For example, modern meat processing plants in market economies operate at controlled cool temperatures to minimize the risk of bacterial contamination. In fact, the Australian meat processing industry has been plagued with labour disputes in past years because workers objected to the cold temperatures in which they were required to work. Processing plants in all modern market economies operate according to strict food hygiene standards. The European Union introduced meat hygiene standards to which all meat processing plants were required to conform by January 1996. Those which could not meet these new stricter standards closed. The issue of food safety and consumer protection regulations is discussed in more detail in Chapter 19. Many older processing plants in former command economies lack adequate chilling facilities and operate at food safety standards well below those acceptable in modern market economies.

Within-plant quality includes elements not only of food safety and hygiene but also the ability to tailor products to the needs of particular customers. The Danish pork industry is particularly adept at this. Although a small country with no obvious natural advantages in the production of pork (land is scarce and relatively high priced, labour costs are high, feedstuff costs have been inflated in the past by EU policies to protect its cereals sector), Denmark accounts for approximately 26% of global pork and pigmeat exports. There are many reasons for the Danes' success, one of which is the efficiency and attention to quality of the processing sector (Hobbs, 1996). Pork processing plants tend to be highly automated and have different production lines or different sections of the plant dedicated to producing products for different markets. For example, there may be a cutting line dedicated to the Japanese market or a room dedicated to producing bacon products for the UK market. Employees are trained to prepare products to the standards and specifications of different markets. The whole processing system is well thought out and designed to give managers maximum control over the types and qualities of product produced. Research into product quality improvements is continuous. The processing industries of the CEECs and the NISs could learn much from the Danish model.

The ability to control and adapt product quality will be essential to the long-term success of the processing sectors in the CEECs and the NISs. Consumers in modern market economies are demanding higher quality value-added food products. Functions which were previously performed in the household kitchen are increasingly being performed by food processors. The increased ownership of microwaves and freezers has facilitated the growth of the convenience, ready-meal market. Although consumer incomes in the CEECs and the NISs are well below those of modern market economies and ownership of modern kitchen appliances is far less common, the situation will gradually change. As consumers' purchasing power increases, they will begin to demand higher quality further processed food products. Firms in modern market economies are well placed to meet these changing demands and could pose a significant competitive threat to domestic firms. This may already be occurring to some degree; Denmark exports (low-quality) pork products to Russia. Quality improvements will be essential if food processors in the CEECs and the NISs are to meet the demands of consumers in foreign market economies and to withstand competition from foreign firms. In the future, the region has the potential to become an important exporter of food products to western Europe and other markets; one indicator of this potential is the increase in Polish exports of soft fruit to western Europe.

## 9.4. External Challenges

An external aspect of the need to improve quality involves the interaction between processors and their suppliers. The quality of the final processed product is highly dependent upon the quality of the inputs. In the command system, there was no incentive for farms to improve the quality of their produce. They simply transferred their output to the designated processing or distribution point with no premiums or deductions for high or low quality produce. Hence, there was no incentive for dairy operations to alter feeding practices or experiment with new breeds to affect butterfat levels in milk or improve milk storage and handling techniques to reduce bacterial contamination; there was no incentive for cattle producers to change production practices to improve the yield or alter the fat content of carcasses; fruit and vegetable producers had no incentive to improve harvesting and handling techniques to reduce bruising and product damage.

A further, and perhaps related, challenge facing processors is in establishing reliable and regular volumes of supply. In the command system this was less of a problem as farms were required to supply all of their output to a designated processor or distributor. In a market economy (assuming that a monopsony processor does not operate in a region) farmers have a choice of whom they supply. This means that a processor does not have a guaranteed access to a continuous source of supply. In some transition economies, uncertainties over food security and the development of processing monopsonies have resulted in many rural families becoming nearly self-sufficient in food. For example, the OECD reports that an estimated 80% of food consumption by rural families in Romania comes from own production (OECD, 1995). This further reduces the supplies available to processors. In the initial stages of transition, the novelty of choice may have caused new farms to switch processing outlets randomly, just as they switched input suppliers randomly, simply because they could. Over the longer term, obtaining regular supplies of a consistent quality will require that processors establish a price incentive system which rewards producers for supplying the type of products required by the processor. It may also require a more formal relationship between suppliers and processors.[3]

A grading or classification scheme is one method by which processors could secure a reliable supply of consistent quality inputs. Grades enable products to be sorted according to a pre-determined set of quality characteristics. Payment should be linked to the grading system, so that farmers are rewarded for achieving an above average grade and penalized for a below average grade.

A grading system has four main requirements:

**1.** The characteristics which make up the grading system should be capable of accurate measurement. For example, some beef grading systems include a yield grade which measures the proportion of usable meat in a carcass, most include a measure of fat thickness at a designated point on the carcass, some include a measurement of internal marbling (intra-muscular fat); however, this is often a visual assessment and is criticized for its apparent subjectiveness. Grain grading systems usually include a measure of the foreign material content and the moisture content of the grain; pulse grading systems may include the size and colour of the pulses.

**2.** A grading system must also be seen to be accurate and reliable. Both parties (buyers and sellers) must have faith in the honesty of the system. The use of an independent third party to carry out the grading function may be essential in this regard. Disputes over the accuracy of grades reduce the efficacy of the grading system as an incentive for high quality production.

**3.** A grading system should be simple. It is essential that the system be understood by all participants in the supply chain. This will be particularly important in the CEECs and the NISs where this method of remunerating farmers will be unfamiliar. Processors should understand how to translate the requirements of their customers into product grade characteristics. For example, if consumers require a meat product which is more tender or is to be cooked in a manner which requires more intra-muscular fat, the processor can select carcasses with the necessary characteristics using features of an accurate grading system and reward farmers for producing animals with these characteristics. Farmers must also understand the grading system. If the system rewards them for producing an animal or a crop with certain quality characteristics, this must be clearly understood by the farmer at the outset, so that production methods can be tailored accordingly.

**4.** The grading system should be relatively easy to apply. It must usually be carried out rapidly at some point during processing. Subjective grading measurements (e.g. visual assessment of colour) can vary between graders and over time and are subject to criticism. The Danish pork industry has developed a fully automatic carcass grading system which eliminates human error and which is integrated into the processing line. Ideally, the system should be relatively inexpensive. This will be important in encouraging financially constrained firms in the CEECs and the NISs to adopt grading systems.

There is a role for the government, or an independent industry association, in establishing and operating a system of grading for agricultural products to ensure the honesty of the system and to reduce measurement costs for all participants in the food chain (farmers, processors, retailers and consumers). Chapter 4 discusses the measurement costs which can arise when each consumer visually assesses a product before making a purchase decision. Retailers also face similar measurement costs when choosing which processor to purchase from. Assuming the characteristics of the product of importance to consumers are part of the grade, a reliable and accurate grading system reduces consumers' measurement and sorting costs because products are sorted and measured (by the grader) just once rather than by each individual consumer (or retailer). Similarly, information costs for processors and farmers are reduced because grades provide them with a means of comparing products and prices. Farmers receive information on the types of product demanded by downstream participants in the food chain. A national (or at least regional) grading system will be far more effective than a series of individual grades particular to individual companies, as this would still leave the retailer, consumer or farmer with measurement costs in determining whether Grade A from company X is equivalent to Grade 1 from company Y, etc.

An external opportunity (or challenge) facing processors in former command economies is their relationship with foreign firms. On the one level, these firms could be competitors, either by exporting to the country or by overseas investment in competing processing facilities. On the other hand, foreign firms may prove to be a source of investment capital. Links with foreign firms can take the form of the outright sale of processing facilities, joint ventures, licensing or franchising. The sale of processing facilities to foreign firms was discussed in section 9.2 and courts the danger that foreign firms achieve monopoly power in a particular sector. Furthermore, although foreign ownership brings employment to a region and provides an outlet for agricultural produce, any profits can be repatriated to the foreign country and do not necessarily remain in the transition economy. Joint ventures, on the other hand, allow a proportion of the profits to be retained in the home country.

A joint venture is an agreement between two or more firms whereby a separate legal enterprise is established through which the firms cooperate for mutual benefit. It is different from a merger, in that the original firms remain as distinctly separate legal entities. If there is no longer a mutual benefit to be gained from cooperation, a separation can be achieved with relative ease.

There are three main types of joint venture, the most common being an investment joint venture where the partners commit investment funds to establish a new venture or where there is an exchange of stock. Investment joint ventures usually involve both parties in more risk because there is a greater commitment of resources. The division of profits between the two parties is subject to negotiation but is usually commensurate with the division of risk. Firms from modern market economies have been hesitant to enter into investment joint ventures in former command economies because of the considerable risks present. These risks stem from the unstable macroeconomic environment, the lack of effective banking and legal systems, and political and legislative instability. For example, in Russia:

> In addition to all the usual problems of doing business anywhere,
> Western investors in Russia face three unfamiliar obstacles to success.
> The first is currency convertibility. The second is supply shortages.
> Companies have trouble sourcing materials of almost every kind and often
> have a terrible time finding basic services like banking, transportation,
> communications, security, and insurance. The third obstacle is the
> constantly shifting regulatory and legal environment. Quite simply, the
> ground rules change with maddening frequency, as do the nature and
> locus of governmental authority.
>
> (Lawrence and Vlachoutsicos, 1993, p. 44)

The other two types of joint venture are marketing and technical joint ventures; these involve less risk. Under a technical joint venture, a foreign partner could provide technical knowledge to a firm in a former command economy and, in return, receive a share of any profits. Under a marketing joint venture, a firm from a modern market economy provides the product and marketing expertise to a former command economy firm who markets their product line or brand. This does not require as significant an investment outlay on the part of the foreign firm as an investment joint venture. Furthermore, former command economy firms would not have to commit significant quantities of their limited capital resources to a technical or marketing joint venture. These types of joint ventures would allow the firms to upgrade the skills and expertise of their management and employees who could learn marketing skills through hands-on experience, guided by the foreign partner. They could upgrade their technological processing capabilities without the need for extensive research and development expenditures into new product development or processing line improvements. In return for providing marketing and technical expertise to transition economy firms, western partners benefit from the local knowledge and expertise of the firm and could also use the association with the local processor to give the product a local image. Joint ventures may also be more politically acceptable than outright foreign ownership of processing facilities.

A further advantage of a joint venture for both parties is that it reduces their contracting costs. Both partners have a stake in the profits of the joint venture, hence, it is in their interests to maximize revenues and minimize joint costs. If they were to remain separate and allow a contract to govern their transactional relationship, the costs of writing a contract which covered all contingencies, given the high state of business uncertainty in former command economies, could be prohibitive. Both parties could incur substantial information and monitoring costs in determining the contingencies which should be covered in the contract and in monitoring the behaviour of the other party to ensure that the contract was adhered to. This is all the more problematic in societies in which contract law is poorly developed or non-existent.

One disadvantage of entering into a joint venture arrangement for a processing firm in a former command economy is the risk that, in doing so, a future competitor is created. This is a double-edged sword, as it is a risk which both partners face. Joint venture partners are privy to a great deal of sensitive information about the financial status of their partners, about technological developments and about the market environment in which the partner firm operates. This information could provide the partner with a considerable competitive advantage should the joint venture fail.

The managerial costs of coordination and control are probably the most difficult aspect of a joint venture arrangement. This may be particularly problematic for joint ventures between firms in former command economies and those from modern market economies because ingrained management philosophies may be fundamentally different. Many joint venture partnerships in modern market economies, even between firms from similar philosophical backgrounds, end in divorce because of disagreements between the two partners. Paliwoda (1986) cites the results of a survey of 1100 US joint ventures of 170 multinational corporations which found that over one-third of these partnerships were unstable, leading to divorce. A common source of disagreement is over the division of risks and profits. Joint venture agreements can end in divorce because the initial organizational choice was wrong, hence, extensive preliminary negotiations between the two potential partners are needed so that both firms can get to know one another and can understand fully the business philosophies and objectives of their potential partner. Joint ventures also end because circumstances change and the partnership is no longer of mutual benefit. Firms need to review periodically the objectives, organization and progress of the joint venture arrangement to ensure that it is adapting to changing circumstances and that it is still the appropriate strategy for both partners. This is particularly important in the rapidly changing macroeconomic, political and legal environments of economies in transition (Middleton *et al.*, 1993).

There may be opportunities for processing firms in former command economies to produce products under licence or through a franchise from a foreign firm. The processor would gain marketing and product knowledge from the foreign firm. For example, a brewing company in Poland could produce American beer under licence, or a dairy processing company in the Czech Republic could produce Dutch ice-cream under licence. In either case, the Polish or Czech firm would have to adhere to stringent quality control regulations and product specifications because the firm from a modern market economy will have invested considerable resources in developing a brand image for a particular product. This imposes high monitoring costs on the foreign firm. If these costs are too high, the firm is likely to internalize the transaction and invest in an overseas subsidiary directly rather than relying on franchise or licensed production.

A further external challenge facing processors in the CEECs and the NISs is learning to deal with retailers and wholesalers. In many market economies, this imposes considerable negotiation costs on processors. The nature of the relationship between the two sectors will depend upon the balance of market power between them. If a monopolistic or oligopolistic processing sector faces a highly fragmented and competitive retail sector, the advantage will lie with the processors. This was the case in many market economies at one time, when food manufacturers dominated the food supply chain and were able to dictate resale price maintenance rules to retailers, thereby controlling the prices at which products were sold. On the other hand, if a competitive processing sector faces a highly oligopsonistic retail sector, the prices received by processors will be lower. Over time, a powerful oligopsonistic food retail sector has emerged in many countries (e.g. the UK, Austria, France, Canada, the USA). The retailing sector in these economies is in a constant state of flux with frequent merger and acquisition activity. Of course, if both processing and retailing are monopolistic, the non-cooperative bilateral monopoly outcome described in Chapter 5 can result. Processors in CEECs and the NISs will need to understand the economic implications arising from the structure of their industry and the retailing industry in order to develop appropriate strategies for dealing with retailers. One of the issues which will influence this relationship is the nature of wholesaling. In some CEECs and NISs, the wholesaling function, although privatized, is slow and inefficient. This leads to considerable product wastage and high monitoring costs for processors and retailers. If this situation continues, either processors will vertically integrate forwards into the wholesaling function or retailers will vertically integrate backwards. This will give one or other party much greater control over the timing and scheduling of deliveries from processing plants to retail outlets and will enable more efficient management of inventory levels.

One further problem which is endemic in the former command economies is the absence of effective business organizations. While this is true at all levels of the food supply chains in former command economies, it is particularly detrimental for food processors who require close contacts with the bureaucracy because of food safety concerns. In modern market economies, business or industry organizations serve a crucial role in lobbying government officials to reduce regulatory costs. Firms, through their associations, can point out difficulties which may arise in conforming to regulations. Often, once a problem is pointed out, small changes can lead to large reductions in the cost of regulatory compliance and no decrease in food safety. In a well functioning system, industry organizations are actively consulted by governments when regulations are being drafted.

As yet, effective industry organizations have not developed in the former command economies. Instead of attempting to work with government, businesses tend to resent (and often ignore) the government regulatory regime. In part, this stems from the experience with heavy-handed bureaucracy in the command era. Primarily, however, the important role that industry organizations can play is seldom communicated to newly privatized firms. They are often too busy surviving in the novel world of competition to consider the benefits of industry level cooperation on issues relating to government regulations. Until effective industry organizations come into being, compliance costs will remain high and efficiency levels will be reduced.

## 9.5. Conclusions

Privatization of food processing is occurring at differing rates and with differing degrees of success in the CEECs and the NISs. Considerable restructuring is required before the processing sectors can be competitive and efficient. Most processing facilities are large scale, many are operating with obsolete equipment and technology and considerable retraining of management is required. In some cases, processing enterprises have been privatized in their entirety to create regional monopolies. The farming sector is left facing a cost-price squeeze – monopolistic input suppliers on the one hand and monopsonistic buyers on the other. This has exacerbated the decline in agricultural output.

Processors require a reliable supply of inputs which are of a consistent quality. This will require the establishment of commodity grading systems to provide farmers with the appropriate price signal incentives. There is a role for government policy to establish and operate (or at least monitor) such a system. Quality improvements and the development of an efficient, competitive processing sector is essential in ensuring the smooth functioning of the food chain. This is important if consumers are to have access to adequate supplies of food at affordable prices and, over the longer term, if the food processing sectors of the CEECs and the NISs are to become competitive on domestic as well as global markets.

## Notes

1. Joint ventures are discussed further in section 9.4.
2. Chapter 8 discusses the concepts of organizational slack and X-inefficiency.
3. A discussion of different supply chain relationships (e.g. contracts, group marketing schemes, etc.) can be found in Chapter 8.

## References

Hobbs, J.E. (1996) *Danish Pork in Asia-Pacific Rim Markets: A Culture of Excellence.* EPRI Study No. 96-01, Excellence in the Pacific Research Institute, University of Lethbridge, Canada, 74 pp.

Lawrence, P. and Vlachoutsicos, C. (1993) Joint ventures in Russia: put the locals in charge. *Harvard Business Review*, January–February, 44–54.

McGee, J.S. (1958) Predatory price cutting: the standard oil (N.J.) case. *Journal of Law and Economics* 1, 137–169.

Middleton, J., Hobbs, J.E. and Kerr, W.A. (1993) Poland's evolving food distribution system: joint venture opportunities for British agribusiness. *Journal of European Business Education* 3, 36–45.

OECD (1995) *Agricultural Policies, Markets and Trade in the Central and Eastern European Countries, Selected New Independent States, Mongolia and China: Monitoring and Outlook 1995*. Centre for Co-operation with the Economies in Transition, Organization for Economic Co-operation and Development, Paris, 232 pp.

Paliwoda, S.J. (1986) *International Marketing*. Butterworth Heinemann, Oxford, 308 pp.

# 10 Moving to Market: Transportation and Distribution

The food transportation, distribution and storage system was notoriously wasteful under the command system. For the former Soviet Union, transport and storage were cited as major factors in an estimated loss of 30–40% for agricultural products (OECD, 1991). Assessments of the future export potential of the former Soviet Union (e.g. Johnson, 1992) and the Central and Eastern Europe Countries (e.g. Tangermann, 1992) have pointed to the potential for reduced post-harvest losses in areas such as transport and storage. In this chapter, we examine the state of the transportation infrastructure and equipment that has been inherited from the command system, the growing problems associated with maintenance and repair and the need to revamp the economic organization of the distribution system.

## 10.1. Existing Infrastructure and Equipment

The Central and Eastern Europe Countries (CEECs), the Baltic countries (i.e. Estonia, Latvia and Lithuania), Belarus, the European portion of Russia and the Ukraine, have all been bequeathed extensive road and rail networks. In Siberia and the Central Asian New Independent States (NISs), the relative importance of rail versus road networks is largely driven by geography. The vast distances in Siberia and Kazakhstan have tended to favour rail transport especially for long-haul freight (see World Bank, 1993b). Within the mountainous republics of South-central Asia such as Tajikistan and Kyrgyzstan, roads are predominant for intra-regional transport (see World Bank, 1994 and 1993c). While the extent of the transportation networks that the CEECs and the NISs inherited from the command system tends to be adequate, there are still a wide variety of problems with the networks themselves.

The planning system coupled with the philosophy of self-sufficiency at the regional, national and Soviet bloc levels tended to distort the development of the transportation infrastructure. Not surprisingly, the entire transport system of the Soviet Union, and to a lesser extent that of the Soviet bloc as a whole, developed with Moscow as the key hub or focal point. While Moscow will certainly remain a key centre, its overall influence is likely to decline as total foreign trade grows relative to trade within and between the NISs.

While the transportation system is adequate to serve the large state and collective farms handed down from the command system, difficulties will arise when and if these farms are divided into smaller family operations. In reference to Russia, Pockney reports that:

> [t]he infrastructure for the successful development of small farms is lacking. Whether we have in mind roads (or the lack of them) to link farms with markets, or storage facilities in the countryside or at the market depots in the towns (they are large and reflect the development in the last 30 years of rings of state farms around the cities . . .), or the reluctance of banks to advance more than short-term loans (usually 3 months at the most) we find that the infrastructure is not there. Even in those villages where there are roads the peasant farmer lacks his own transport or a cooperative pool from which he can draw.
>
> (Pockney, 1994, p. 335)

Thus, the incentive to establish family farms typical of much of agriculture in the developed market economies is limited by the transport and distribution system as well as factors such as the market power of processors and the sellers of farm inputs.

In some regions, the over-development of economic activity, especially in agriculture, was accompanied by a related over-development in the transportation infrastructure. In some cases, development that was unwarranted at the time that it was undertaken has probably reached a point that it is now basically self-sustaining but in other cases development may be at least partially reversed. The forced opening of the Virgin Lands that was undertaken in Kazakhstan in the 1950s is probably an example of the former type. Despite the current state of depression in Kazakhstan, it seems likely that agriculture will remain viable in the sense that revenues have the potential to cover production plus shipping costs.[1] Of course, this is not to say that the current extent of cultivation will remain unchanged, nor even that it would be economic today to open all of the land that will remain under cultivation if the associated costs were not sunk.

Nonetheless, a major consolidation of the transport system may not be needed in Kazakhstan. In the short run, however, the transportation system appears over-built. This means that the system must be operated at sub-optimal levels which increases per unit costs for shippers. Further, maintenance costs are spread over too few loads, again increasing customer costs or the required subsidies. Given the financial difficulties of the government, maintenance is likely to be foregone leading to a deterioration in transportation capital. Rebuilding the system in future may require large capital expenditures. On the other hand, in the remote mountainous Republic of Kyrgyzstan, the World Bank (1993c) recommended that, in the likely event that funds could not be made available for the proper maintenance of the whole road network, 'plans should be formulated to concentrate existing resources on a smaller network of priority roads while substantially lowering the standard of low volume roads' (p. 155).

As well as regional anomalies in economic activity and the related transportation system caused by the command system, the central plans de-emphasized the movement of some types of goods. For example, there is currently very little in the way of capacity to distribute or store refrigerated products as a result of the command legacy. The underdevelopment of so-called cold chains is discussed further in Chapter 12.

In spite of the generally adequate overall extent of transportation networks, the quality of these networks is often low. Particularly in the NISs, the low quality of the transport networks reduces speed and requires transport equipment that emphasizes durability at the expense of capacity. For example, a World Bank study states that:

> [e]ventually, Latvia will want to upgrade to modern and efficient equipment, but existing fleets must be operated for some time to come. In the case of trucking, the practical approach would be to continue to operate the more robust Soviet vehicles, despite their lower fuel efficiency and poorer payload/gross weight ratios, until roads have been improved.
>
> (World Bank, 1993d, p. 133)

With the dissolution of the former Soviet Union, the fragmentation of transportation systems has created significant infrastructure problems. Separate road and rail administrations in each independent republic have created additional transaction costs that restrict the movement of agricultural products and commerce in general. These problems are exacerbated by the presence of enclaves that are separated from the main territory of various republics. The most important example is the Baltic enclave of the Russian Federation which is cut off from the rest of Russia by Lithuania and, depending on the route, either Latvia or Belarus.[2]

Similar problems have arisen with regard to transshipment for international trade. The Baltic countries have a natural transit role for the sea-borne trade of many of the other NISs. The Baltic countries, Belarus and the Ukraine have a similar role with respect to overland trade with the CEECs and western Europe. For example, the port at Tallinn in Estonia has been bequeathed modern facilities for handling grain imports and refrigerated foods that were bound for the entire former Soviet Union (World Bank, 1993a). It is, therefore, extremely important to adopt policies and, in some cases make new infrastructure investments that minimize the transaction costs associated with transit. For example, World Bank studies on each of the Baltic countries have emphasized the importance of maintaining the transit function with respect to Russian trade such as wheat imports. In their pricing practices, port and rail authorities in Estonia, Latvia and Lithuania have to be cognizant not only of competition from each other, but also of the possibility of diverting traffic to Russia's own Baltic Sea port at St Petersburg, especially in the long term. It is also important for all of the NISs, and indeed the CEECs as well, to adopt and maintain market-based policies relating to transit, such as the UNCTAD Multi-Modal Convention, in order to ease customs difficulties and provide certainty and streamlining for financial matters such as insurance and document collection (see World Bank, 1993a, p. 148). The World Bank (1993a,e) has also emphasized the potential for joint ventures in the operation of ports and cooperative arrangements with regard to railroad operations that would eliminate the need to change locomotives and train crews at borders. Joint ventures in the agrifood system are discussed further in Chapter 15.

There are important barriers to transport between the NISs and other adjacent countries. Of course, there are obvious geographic barriers relating to distance and/or mountainous terrain to developing and upgrading rail and other land-transport linkages with countries such as China, Iran and Turkey.[3] A less obvious problem is that the railroad networks in the NISs and parts of the networks in the CEECs are fundamentally incompatible with those in western Europe. In the nineteenth century, Russia adopted a broader railroad gauge than the rest of Europe (i.e. a larger distance between the two rails), and the broad gauge was utilized throughout most of the Soviet Union. As the costs associated with complete gauge conversion would be prohibitive, there is now a significant barrier to through freight and passenger traffic between the NISs and Europe.[4] For bulk products such as grains, the costs of reloading or using an alternative mode of transport will have a negative impact on the development of trade flows with Europe. Containers do hold promise for reducing the costs associated with reloading for many manufactured products, although they are not yet widely used. In the case of perishable products, the delays associated with reloading adds to the risk of product deterioration. Reloading can lead to the bruising of fruit and other high quality products.

For the Baltic countries there are some advantages, as well as the obvious disadvantages, of the broad gauge since Russia cannot easily diversify its bulk transit trade to Polish or German ports. International trade matters are discussed in depth in Chapters 13 and 14.

Telecommunications networks suffer from many of the same problems as transportation networks. In fact, the problems in telecommunications are often somewhat more extreme than those in transportation because central planning was much less dependent on reliable, advanced communications and information transmission than a market system tends to be. The telephone system in the CEECs and the NISs is underdeveloped in terms of the extent of service, equipment and technology. Consequently, significant new investment is required. Privatization of telecommunications enterprises, or at least arms length commercial operation, has the potential to encourage joint ventures which would facilitate access to improved technology and training (World Bank, 1993c). The fragmentation of the former Soviet Union has created administrative linkage problems in the telecommunications system as well as for transportation. Further, Moscow's role as a communications hub for the former Soviet Union was even more prominent than its role as a transportation hub. As a result, many international calls originating in NISs other than Russia will continue to be routed through Moscow for the foreseeable future (World Bank, 1993b).

## 10.2. Repair, Maintenance and Investment

Lack of spare parts for existing transport equipment has become an acute problem. The fragmentation of the former Soviet Union has meant that it is often necessary to engage in international transactions to obtain spare parts. Transaction costs for the acquisition of spare parts have soared because much of the international trade of the NISs is still conducted by republic-level state trading agencies, often on a barter basis. Consequently, shortages of spare parts have become endemic in the CEECs as well as the NISs. Further, even when the required automotive parts could be manufactured within the particular country in question, the factory is frequently unable to obtain the required raw materials for production.

The spare parts problem threatens to become a long term fact of life in the CEECs and the NISs. As trucking fleets are gradually upgraded, a high degree of diversity in truck models and suppliers may foster competition. Nevertheless, there is a serious danger that parts and servicing may be lacking if too many new suppliers enter the market too rapidly. Even apparently benign donations of equipment from developed countries may have the undesirable side-effect of creating repair and servicing difficulties later on (World Bank, 1993d). Consequently, both enterprises and governments in the CEECs and NISs will have to consider carefully the servicing capabilities as well as product features and selling prices offered by new foreign suppliers. Reliability and parts availability is particularly important in the agrifood sector where disruptions to supply chains can reduce food security to unacceptable levels. Agribusiness firms dealing with perishable products will suffer losses if the vehicles they acquire cannot be kept operational.

The deterioration of the transport infrastructure during the transition has also become a critical concern, particularly, in the NISs. Financing maintenance is proving to be extremely difficult. Government budget problems and the failure to adequately charge for the use of road systems through fuel taxes and vehicle registration fees has directly undermined ability to maintain transport networks. Further, for the smaller countries such as the Baltic States, the development of joint programmes for the maintenance and even the operation of the transportation system are extremely important.[5]

> A program to alleviate the pending crisis in construction material supply and equipment maintenance should be devised [by Latvia] jointly with the other Baltic nations. Reliance on the combined resources and production facilities already available in each country is clearly necessary for the rational management of road works . . . Latvia and its neighbours should [also] avoid attempting to be completely autonomous in matters such as port dredging, ice breaking, air traffic control, and facilities for heavy maintenance of rail and aviation equipment where it is more economic to have common facilities and equipment. Adequate access can be assured through joint ownership arrangements.
>
> (World Bank, 1993d)

Many maintenance functions could and should be provided under contract by private firms. For the smaller countries, reciprocal access to the bidding process for firms in neighbouring countries would help promote needed competition.

A host of problems have contributed to a decline in the use of transportation systems that makes maintenance and especially new public investment more problematic. The general slump in economic activity is of course a key contributing factor in the low traffic volumes that have become the norm for much of the transport network. Other factors include: (i) shortages of spare parts for trucks, buses, locomotives and rolling stock; (ii) new barriers to trade that have arisen with the political fragmentation of the former Soviet Union and the Council for Mutual

Economic Assistance (CMEA) trade area; and (iii) the elimination of some *de facto* transport subsidies under the command system. Nevertheless, there is a need for some new public investment in the transport system. Where such public investment may be warranted, there is still a need for appropriate financial controls and cost-benefit calculations that were foreign to the command system. For example, even on high-use road routes such as the via Baltica through the Baltic countries, up-grading needs to be appropriately sequenced and prioritized so that important projects can be completed quickly and begin to generate benefits (see World Bank, 1993a,d,e). Throughout the CEECs and the NISs there is also a need for investment in roadside facilities such as service stations, restaurants and the provision of accommodation but such investment should be the domain of the private sector (World Bank, 1993a).

## 10.3. Economic Organization

The large state enterprises that had been charged with a general trucking function under the command system have often been privatized as single entities. As a result, fleets tend to be unduly large, often consisting of 1000 vehicles (World Bank, 1993a,d). Such privatization practices are anti-competitive and do not promote maximum operating efficiencies.

> Experience in market economies suggests that small fleets of fewer than 10 vehicles are usually more efficient, principally because it is easier to exercise effective ownership control and management of the fleet. Moreover trucking is an excellent area to encourage small scale entrepreneurs, as the capital investment is minimal, especially if the vehicles are leased. Thus, although it will be more cumbersome administratively, privatization should be on a smaller scale.
>
> (World Bank, 1993a, p. 146)

Large scale privatization simply adds another potential bilateral monopoly link in the agrifood chain. While privatized trucking enterprises often do lease their equipment to independent drivers, the fact that these enterprises have local market power over the terms of the leases serves to at least partially negate the competitive benefits of having many independent drivers. In the agrifood sector and throughout the economy, large state enterprises have also tended to be privatized with their own trucking fleets left intact. The fact that privatized processors and/or state and collective farms often have retained their own trucking fleets leads to the overall under-utilization of vehicles as well as peak-load capacity problems. Further, the fact that trucking firms are tied to enterprises serves to entrench the bilateral monopoly structure that was discussed in Chapter 5.

Especially in the NISs, much of the transportation sector continues to receive large indirect subsidies. During the command era, the artificially low price of oil in the Soviet bloc constituted an effective subsidy to the transport sector. Particularly in the NISs, the slow transition to world oil prices has meant a very gradual reduction in this effective subsidy during the early 1990s. While there may be a case for temporary fuel subsidies of limited and declining magnitudes to ease adjustment problems, fuel prices should definitely not be subsidized indefinitely. In fact, there is a need to move rapidly towards a regime of appropriate taxes. For trucking and road users in general, it would be desirable to set fuel taxes and vehicle registration fees so as to cover the cost of maintaining the road system.[6] Similarly, rail-freight tariffs should cover both operating expenses and maintenance of infrastructure. Further, it is very important that the road and rail systems both be placed on a full cost recovery system to avoid artificially distorting the mode of transit selected by users. Concurrent phasing-in of user pay systems is also important.

Privatization at the level of wholesalers and distributors should be subject to careful scrutiny so as to minimize the opportunities for the exercise of market power but still allow for greater technical efficiency (i.e. a reduction in the extent of waste) in storage and marketing (see World Bank, 1995). Especially within large urban markets, it is important that retail outlets not be forced into inefficient relationships with wholesalers on a geographic basis. Rather, competition should prevail between distributors as well as retailers. Food retailing is discussed in detail in the next chapter.

## 10.4. Conclusions

An efficient transportation system is a key element in dealing with three of the major problems faced by agribusinesses in the process of transition. The transportation of food from farms to consumers, regardless of the number of intermediaries involved, is at the heart of the food security question. This is particularly true in the far reaches of Siberia and the Central Asian NISs. Given the waste of food products which has traditionally arisen as a result of problems in transportation and/or storage, improving the efficiency of the distribution system may be as, if not more, important than increasing farm level production.

In modern market economies, the major evolutionary force in the agrifood chain has been towards the provision of a larger proportion of foodstuffs which are either fresh or refrigerated. If agribusinesses in the former command economies are to be able to capitalize on any natural advantages they have in food production and enter foreign markets, they must be able to provide high quality products. Transportation is an essential link in the export chain.

The ability to broaden markets is the most important element in reducing the problems associated with bilateral monopolies (or oligopolies). While institutions to reduce transaction costs are a necessary condition for broadening markets, transportation must also be relatively low cost. The ability to identify prospective business partners and negotiate transactions will come to nothing if the cost of transport prohibits a profitable business deal.

While considerable attention has been given to the problem of farm production throughout the process of transition, as yet transportation in the agrifood sector has received little attention or resources. Given the large public goods aspect of transportation systems, the failure to give proper attention to maintenance and new investments in transportation infrastructure may thwart the recovery of the agrifood sector.

## Notes

1. In 1994 Kazakhstan attempted to complete its harvest without what is now foreign migrant labour but traditionally it has relied on such outside labour (OECD, 1995). It seems likely that such seasonal labour imports will continue to be in Kazakhstan's national interest.
2. The break-up of the Soviet Union has also led to political conflicts within Chechnya and Georgia, and between Armenia and Azerbaijan that have impaired the functioning of the transportation infrastructure.
3. Transshipment to India (or Pakistan) through Afghanistan is an even more dubious proposition both because of the terrain and the on-going political upheaval between competing factions that arose after the Soviet withdrawal.
4. The incompatability of the rail gauges had some military and strategic advantages for the Soviet Union in that real or potential invaders could not easily make use of the rail network.
5. Fully common rail systems for the Baltic countries and perhaps some of the central Asian republics would allow more efficient operation and also reduce the transaction costs associated with transshipment. The principle of exploiting joint economies goes beyond the components of the transport system that are most central to the agrifood system. Separate state-owned airlines and telecommunications firms in each of the small NISs may not be viable and will certainly lead to unnecessarily high prices for users (see World Bank, 1993a).
6. As late as 1995, some NISs such as Kyrgyzstan had no fuel tax whatsoever (World Bank, 1993c). In principle, external costs relating to congestion, accidents and pollution should also be covered by users (World Bank, 1993d).

## References

Johnson, D.G. (1992) World trade effects of the dismantling of socialized agriculture in the former Soviet Union. Paper presented to the International Agricultural Trade Research Consortium, St Petersburg, Florida, December.

OECD (1991) *The Soviet Agro-food System and Agricultural Trade*. Centre for Co-operation with the Economies in Transition, Organization for Economic Co-operation and Development, Paris.

OECD (1995) *Agricultural Policies, Markets and Trade in the Central and Eastern European Countries, Selected New Independent States, Mongolia and China: Monitoring and Outlook 1995*. Centre for Co-operation with the Economies in Transition, Organization for Economic Co-operation and Development, Paris, 232 pp.

Pockney, B.P. (1994) Agriculture in the New Russian Federation. *Journal of Agricultural Economics* 45, 327–338.

Tangermann, S. (1992) Central and Eastern European reform, issues and implications for world agricultural markets. Paper presented to the International Agricultural Trade Research Consortium, St Petersburg, Florida, December.

World Bank (1993a) *Estonia; the Transition to a Market Economy*. A World Bank Country Study. International Bank for Reconstruction and Development, Washington DC, 330 pp.

World Bank (1993b) *Kazakhstan; the Transition to a Market Economy*. A World Bank Country Study. International Bank for Reconstruction and Development, Washington DC, 234 pp.

World Bank (1993c) *Kyrgyzstan; the Transition to a Market Economy*. A World Bank Country Study. International Bank for Reconstruction and Development, Washington DC, 230 pp.

World Bank (1993d) *Latvia; the Transition to a Market Economy*. A World Bank Country Study. International Bank for Reconstruction and Development, Washington DC, 305 pp.

World Bank (1993e) *Lithuania; the Transition to a Market Economy*. A World Bank Country Study. International Bank for Reconstruction and Development, Washington DC, 397 pp.

World Bank (1994) *Tajikistan; the Transition to a Market Economy*. A World Bank Country Study. International Bank for Reconstruction and Development, Washington DC, 240 pp.

World Bank (1995) *Armenia; the Challenge for Reform in the Agricultural Sector*. A World Bank Country Study. International Bank for Reconstruction and Development, Washington DC, 198 pp.

# 11 Catering for Consumers: Retailing in Transition

Food retailing is a segment of the agrifood chain which has seen a flurry of private sector activity in the former command economies. Retail has been a fertile environment in which small private entrepreneurs have experimented with capitalism. While often viewed as a strength of the transition process, the fact that many small firms have entered this market may in fact be the Achilles heel of retailing. This chapter discusses the nature of food retailing prior to transition, compares the transition process in different Central and Eastern European Countries (CEECs) and Newly Independent States (NISs) and highlights some of the problems faced by this segment of the agrifood system. Strategies are suggested for alleviating the problems identified.

## 11.1. Food Retailing in the Command System

Food retailing was the final link in the agrifood chain through which food products passed from farms to consumers. Retail outlets were stocked according to a specific regional or sectoral plan. The only role for consumers in this system was reactive or responsive – they could choose whether or not to purchase the products which happened to be available on any given day. Clearly, in a modern market economy, consumers are far from passive. They are presented with a range of products and exercise their preferences through their purchase decision. This information is passed back down the supply chain from the retailer to the producer. Hence, the retailer acts as the interface between the entire agrifood chain and the consumer. In a modern market economy, this interface is crucial, being the point at which information about consumer preferences can be collated and transmitted to agents further upstream in the chain. In a consumer-oriented, demand-driven system, it is the engine which drives the agrifood chain.[1] The role of a retail outlet in the command system was far less important. Since decisions over what and how much to produce were made by the central planners, information did not flow back down the chain from consumer to producer and the retail outlet did not need to serve as a conduit for this information.

Retail outlets in former command economies were generally of two types: state-owned/managed chains of retail stores and stores run by consumer cooperatives. Often, the state-run stores served urban areas while consumer cooperatives served rural areas.[2] Allocations of food supplies to these stores were determined by the central planners based on an administrative evaluation of consumers' needs, with production and distribution being a priority rather than the provision of products in the quantities and qualities required by consumers. In order to facilitate this system in the Ukraine, for example, officials from each district were required to estimate consumption for every product in every store for the coming year. This data was assembled on regional and national levels and production quotas were assigned to plants in each region which would, in theory, meet these estimated consumption needs (World Bank, 1994). The system was highly inaccurate, with frequent shortages and severe limitations on the choice of products available to consumers; witness the once-familiar food queues which, to observers from modern market economies, became perhaps the most poignant visual representation of the failure of central planning.

Retail prices were fixed according to the central plan, hence, they often bore no relation to the actual value of the product. Without the price system as a rationing mechanism, products were allocated on a first-come first-served basis to those willing and/or able to queue long enough. The role of prices was discussed in more detail in Chapter 3.

The retail shopping experience under the command system was vastly different from that experienced by consumers in modern market economies, as is indicated by the following description of shopping in Poland's state-run food shops prior to the economic reforms:

> Most goods were stacked on shelves located behind service counters that were usually operated by brusque middle-aged attendants outfitted in food-stained smocks. Customers formed queues and ordered from attendants, who then retrieved items from the shelves. Shoppers had no opportunity to inspect products before making a purchase and often believed that clerks wanted them to know exactly what they were buying before they entered the store. Customers also were expected to bring their own shopping bags.
>
> (Johnson and Loveman, 1995, p. 46)

In those CEECs and NISs which have yet to liberalize food retailing, or in the state-run stores which remain elsewhere, little has changed. Hence, considerable opportunity has existed for private entrepreneurs to establish retail outlets with a customer-friendly lay-out. Changes to store design and closer attention to staff training in the provision of service are relatively low-cost investments which can differentiate considerably a private retail store's product (i.e. shopping experience) relative to the remaining state sector stores.

## 11.2. Retailing in Transition

In common with other segments of the agrifood chain, privatization and liberalization of the food retailing sector is occurring at vastly differing rates in the various CEECs and NISs. Since retailing is an activity which can be entered into with relatively low levels of capital investment (compared with food processing, the production of agricultural inputs or primary agriculture), considerable scope exists for the establishment of new enterprises once the government creates the conditions necessary for a private sector to develop (including making private retail food stores legal). Poland is an example where the private retail sector grew rapidly in the early stages of transition. One had only to visit Poland during the early 1990s to witness the rapid development of this sector over a relatively short period. Private retail stores offer their customers a number of advantages over the former state system The variety of products on offer is far greater, enabling consumers to exercise choice. Consumers can also exercise this choice in a more favourable environment. The contrast with previous state-run stores is striking. Johnson and Loveman describe the retail outlets of a new Polish supermarket company established in 1990:

> The interior furnishing for the stores – including modern shelving, promotion racks, and conveyor-belt-driven checkout counters – were all imported from Sweden ... The ... stores were more attractive and convenient than Poland's state-run alternatives. The company eliminated the customary service counters that separated clients from products. Instead, customers could walk unhurriedly through grocery-stocked aisles, inspect products, and make their own selections ... Furthermore, the supermarkets were open 24-hours a day, a service concept that contrasted sharply with the more restrictive hours usually associated with Poland's food retailing industry ... (the) supermarkets were enormously successful. Customers flocked to the stores to take advantage of the more leisurely shopping environment, the flexible hours and the Western groceries.
>
> (Johnson and Loveman, 1995, p. 46)

Where levels of government are still involved in food retailing, wastage and inefficiency permeate the system. In the Ukraine, for example, the process of transition has been slow, with food still sold through state food shops and consumer cooperative stores. *Kolkhoz* (farmers') markets are becoming more common. Factory shops and some private shops, however, have begun to emerge. The World Bank (1994) estimates that there were approximately 52,000 mostly small stores in the Ukraine in 1992–1993, with a total retail area of 5.2 million square metres and employing 230,000 people. This is a far lower density of retail stores (i.e. 1000 square metres per 10,000 inhabitants) than is common in modern market economies. Germany (prior to reunification) had 10,000 square metres of foodstuff sales area per 10,000 inhabitants. Poor management and inadequate investment has allowed considerable wastage of foodstuffs to continue in the former command economies.

The privatization of food retailing appears to be at a more advanced stage in Romania, for example. In 1993, it was estimated that the private sector accounted for approximately 50% of food retailing (Morton, 1993a). The contrast between the newly emerging private sector and the remaining state shops is also marked in Romania:

> ... the private shops tend to have a much better selection and often higher quality products than the state shops. Furthermore, the state shops, even those which have been converted to the private sector, are often typified by a lack of enthusiasm, charm and initiative displayed by managers and shop assistants alike.
>
> (Morton, 1993a, p. 19)

Furthermore, different retail industry sectors have privatized at different rates. In Poland, for example, product specific retailers have declined and all-purpose, self-service grocers have taken their place. This is particularly true for dry goods. These grocers sell a full range of food products as well as basic household and personal care products. In contrast to the dry grocery sector, the meat sector in Poland has successfully maintained much of its network of (former state-, now processor-owned) meat shops in local towns and cities (Morton, 1993b).

The growth of private sector food retailing has occurred in two ways: either shops once belonging to the state system have been sold off to private investors or new retail outlets have been established. Privatization has followed various paths in different countries and the process has been similar to that adopted for other segments of the food chain discussed in earlier chapters. The main difference is the relatively small size of the investment required (compared to a meat processing plant for example). This has allowed individual retailing enterprises to be sold to single investors in the private sector. In Poland, the privatization laws allowed for two alternative methods of privatizing state-owned enterprises: liquidation and commercialization (OECD, 1995a).

Liquidation involved the sale of all or part of the enterprise to another enterprise or transferring the enterprise to a joint venture between the Treasury and investors, or it involved leasing the assets to a newly-formed employee company. An alternative fast-track privatization approach allowed an entire enterprise to be sold as a whole to Polish investors at the initiation of privatization negotiations, with the value of the enterprise based on the book-value and annual profits. Foreign buyers could only participate if a Polish buyer could not be found. On the other hand, commercialization involved the formation of a State Treasury corporation from the capital of the enterprise which was then: (i) sold to private investors through share offerings; (ii) sold to one or more persons involved in the enterprise; or (iii) sold to management/ employee consortiums (OECD, 1995a).

In the Czech Republic, privatization of the retail trade began at an early stage of the economic reforms. The relatively recent appropriation of private property by the State (i.e. within one lifetime) enabled parts of the retail network to be returned to its original owners. Most retail outlets, however, were sold to new owners through public auctions as part of the small-scale privatization process (OECD, 1995b). At the same time, a small number of subsidiaries of international supermarket chains were established. The role of international companies in the Czech food retailing sector has become more prominent since 1993.[3] In addition, some food processing companies operate their own retail stores through which they sell 10–15% of their output. In 1995, it was estimated that small private shops accounted for over 80% of total retail trade in the Czech Republic, with (private) supermarkets accounting for the remainder (OECD, 1995b).

The privatization of a retail sector could therefore be accomplished in one of two ways: either state shops are sold to the private sector as a chain of stores or individual stores are sold separately. The former option runs the risk of creating bilateral monopolies between processors and retailers, leading to many of the problems discussed in Chapter 5. The second option has proved more popular. Retail food stores have very little physical capital – perhaps cold storage facilities for perishable produce and check-out counters. As was suggested above, state-run stores were, in any case, stark run-down edifices which required considerable re-design and re-investment upon purchase. State-run stores did not compete with one another through shopping ambience or the development of a reputation for good quality, so there would be little advantage in an investor (or investors) purchasing a chain of retail stores from the state. The one possible advantage of doing so could be location. If old state retail outlets are in prime shopping locations, then the land itself will be valuable. The infrastructure, however, would have little value.

This leads us to the second method by which the food retailing sector in many CEECs and some NISs has expanded – the establishment of new retail outlets in new locations by private entrepreneurs.

## 11.3. Challenges for the Private Retail Sector

One of the acute problems facing the private retail sector in many former command economies is that the sector is characterized by a large number of small firms. The preponderance of small-scale retail firms has two effects: first, the firms are often undercapitalized and second, they have very little ability to ensure supplies are of a consistent quality and quantity. Although the capital investment required in retailing is far less than that required in other food industry sectors, an investment, nonetheless, is required. The large number of small retail outlets may be symptomatic of excessive or destructive competition. If the prices received by the first small retailers in an industry in transition are above total costs, then potential entrants with access to the same (in this case, low) level of technology and small capital requirement will be enticed into the market. These new entrants create a wave of destructive competition[4] where there are too many firms and, hence, few are profitable (Schumpeter, 1950). The absence of profits means that further investments cannot take place. As a result, retail outlets remain undercapitalized and modernizing improvements are not made. In many cases, retail outlets do not even have the financial resources to hold sufficient inventory to satisfy consumers' needs. They are unable to build the crucial core of loyal clientele.

Retail sectors in transition economies are in a state of disequilibrium, the solution of which will probably require that a number of these new enterprises be allowed to go bankrupt in the near future. Even in modern market economies, there is a relatively high turnover of small-scale retailers.

Small businesses are usually single-owner enterprises or partnerships and therefore must rely on an efficient banking system for financing in excess of retained earnings. As is discussed in Chapter 17, the fledgling commercial banking systems in CEECs and NISs are often inadequate (or in some cases, non-existent).

Retailers in most modern market economies operate within a customary credit system with their suppliers whereby payment can be made from 30 to (in the case of large multi-store supermarket chains) 90 days. This means that they do not usually pay for supplies on a cash-on-delivery basis but can allow a period of time to elapse before making payment. The 90-day credit system is very important to supermarket companies who can invest the retained financial resources on international money markets during the 90-day period. Suppliers are assured of payment, albeit delayed, by the commercial legal systems of these countries. As with the banking system, the commercial legal systems of most CEECs and NISs are in their infancy or are non-existent.[5] Hence, supplier-credit is usually not available and most transactions are carried out on a cash-on-delivery basis because of the problems of enforcing non-cash payment methods. This compounds the financial problems facing private retailers and has resulted in severe under-capitalization of the sector.

A further consequence of the small-scale nature of private food retailing is the lack of bargaining power which retailers have *vis-à-vis* suppliers. This manifests itself in two ways. First, retailers are largely price-takers, particularly when faced with monopolistic or oligopolistic food processors (see Chapter 9). The inability to counteract the monopoly or oligopoly bargaining power of the processing sector leads to an inefficient outcome with prices higher and quantities lower than would otherwise be the case.

Second, private retailers will attempt to respond to the preferences of consumers in the new market-place. They will strive to offer goods in the quantities and qualities which consumers require. These improvements to the agrifood chain are difficult to achieve if the retailer cannot exert any bargaining power on an inefficient upstream supplier. Initiating changes which cater to consumers' requirements has become particularly problematic in the process of transition when upstream suppliers have yet to be privatized or when the privatized firm can act as a monopoly. Using the example in Chapter 9, it was very difficult for a small private retailer to persuade a former state-owned milk and milk products processing enterprise in Poland that it should package its yoghurts so that the label could be read by consumers. A small retailer cannot force a state-owned meat processing plant to produce higher quality hams and loins rather than lower quality pork cuts in response to a rise in consumer demand for high quality cuts. Ultimately, the pressure for an efficient economic outcome in terms of product pricing and quantities (and in terms of providing products with the required quality characteristics) comes from the pricing signals created by consumers exercising their sovereign right to choose. It is essential that these pricing signals be transmitted back down the food chain for the appropriate pricing and output responses to be made by upstream firms. If retailers are powerless to pass on these price signals, then there is no pressure for the system to transform and the transition to a market-based economy will be thwarted.

Another challenge facing the retail sector is the lack of managerial (and staff) skills and experience in market-based retailing. Problems with low levels of managerial skills and poor staff training are found at all stages of the agrifood chain but the retail–consumer interface is particularly important as it is the consumer's contact point with the entire agrifood chain. Managers of state-run retail outlets did not manage in the modern market economy sense, rather they administered a unit which formed part of a particular chain. They were told what they could sell, how much they could sell; at what times of the year the product(s) would be available to sell and the price(s) they could charge. Their main focus was backwards along the food chain to their supplier(s) or to the bureaucrat who determined their store's allocation. The customer was largely ignored. Hence, the switch to a market-based system in which consumer sovereignty is a central tenet requires a considerable reappraisal of priorities. The poor performance of government stores in the Ukraine is indicative of the problems with which these managers are having to grapple:

> The management of a large number of stores by Government, even with decentralization, has proven to be exceedingly cumbersome and resulted in enormous amount (*sic*) of food losses. Prices are set by the state and local store managers have no authority to deviate from the price guidelines. If products are not sold they are eventually returned and destroyed. Investment allocation, transportation and other necessary expenditures are made without regard to price signals, and are therefore not consistent with market conditions.
>
> (World Bank, 1994, p. 82)

Jones describes a sense of bewilderment in most newly privatized shops in Russia, commenting that they:

> . . . are unfamiliar with procuring and selling on their own behalf, and for their own profit, and there is little evidence of radical changes in practices being introduced. The critical shortage of purchasing power among the population at large is a disincentive for shops to be innovative, but eventually the lesson will be learned that shops offering quality produce, consumer service and competitive prices will not only survive but also prosper.
>
> (Jones, 1993, p. 22)

Clearly, there is a role for training in basic business management skills, including stock management, financial management and supplier or customer interaction. Marketing skills and market monitoring skills are also of paramount importance and often lacking. Through cooperation with academic institutions in modern market economies, a number of initiatives are currently being undertaken to provide basic marketing training for food and agribusiness managers in the CEECs and NISs. As the command economy system did not regard the preferences of consumers as important, managers are often unsure of how to identify and explore these consumer preferences, i.e. how to measure the demand for product(s). Experience with teaching courses in basic marketing to Polish food and agribusiness managers suggests most managers were unfamiliar with the concept of a consumer survey and were extremely keen to experiment with this tool for collecting and collating market information. Basic market research skills should, therefore, form a core part of any marketing course to food business managers in CEECs and the NISs.

Embracing new marketing techniques requires managers to take on board a new philosophy which places the consumer and consumer requirements at its core. Kerr (1996) found that sales management was an aspect of marketing with which a group of Russian marketing trainees were totally unfamiliar:

> They had no concept of what should be expected of either commercial sales staff or retail sales staff or how to monitor their activities . . . They had no ideas regarding how those involved in sales should be trained or motivated . . . the lack of awareness regarding these sales related activities can be traced back to the command economy system which did not have to be concerned about negative feedback from consumers because there were consistent shortages. As monopolies, Soviet firms had no need to be concerned about quality. Customer service is still virtually non-existent in Russia's state-run retail stores.
>
> (Kerr, 1996, pp. 46–47)

Many of the most basic aspects of sales management may not even be included in formal marketing courses given in modern market economies and are often left to on-the-job training. Clearly, if the fledgling retail sectors of private command economies do not have on their staff individuals with these skills, then much of the potential gain from applying the tools of strategic marketing will be lost because of poor sales techniques (Kerr, 1996). Managers need to be trained in marketing and sales management skills and employees need training in basic customer service. Pleasant, knowledgeable and helpful shop assistants can enable a retail outlet to differentiate its service offering from competing firms at little or no cost.

The concept of service is extremely important in modern market economies. The competitive strategies of major retailing groups have focused on environmental aspects of the shopping experience to compete for customers. Examples include, packing the customer's groceries into bags (even carrying the shopping to the customer's car or delivering it to their home), the provision of play areas for children, wide shopping aisles, soft music, extended shopping hours, etc. These aspects of modern supermarket retailing are part of a concerted strategy to differentiate one retail chain's stores from another. These developments are occurring at differing rates in CEECs and NISs and may be a long way off in some of the countries. However, the comparative success of private retail stores in the more progressive CEECs which have copied western European or North American supermarket models, albeit scaled-down versions, suggests that these are facets of food retailing which will also be important as the transition to a market economy progresses. In investigating the success of a new Polish supermarket company, Johnson and Loveman comment that the company:

> . . . completely reorganized the way groceries were sold. Although it didn't cut prices below those of the state stores, it offered a far better selection of items, a more pleasant environment, and superior service.
>
> (Johnson and Loveman, 1995, p. 46)

Finally, foreign competition in food retailing represents a growing threat to the domestic retail sectors of former command economies. In many cases, the lack of domestic capital has allowed foreign subsidiaries of international supermarket companies or foreign investors to attain an important role in the retail sector. This has been the case in the Czech Republic where international companies have created large retail units with a trading space of over 1000 square metres in industrial centres (OECD, 1995b). Poland and Hungary have also seen foreign investment in the food retail sector, for example, by the European retailer Tengelmann (Lyle, 1993). While foreign investment may threaten domestic retailers, clearly it benefits consumers in the short-run, giving them access to a wider range of products and modern retail facilities and, through competition, forcing other retailers to improve their product and service offering. Arguably, in the long-run, domination of the food retail sector by oligopolistic international supermarket chains could lead to higher prices for consumers and thwart the goal of creating a modern domestically owned agrifood sector. The problem of transnational firms is discussed in Chapter 15.

## 11.4. Strategies for the Retail Sector

One answer to the weak bargaining position which many small private sector retailers find themselves in is to form horizontal retail buying alliances. This is occurring, on an informal and somewhat *ad hoc* basis, in some countries. In Russia, for example, where the food distribution system has been described as 'in total chaos' (Morton, 1993a, p. 18), small groups of retailers are banding together to buy their supplies direct from processors (rather than through state wholesalers), although 'the system is disorganized, highly fragmented and confused' (Morton,

1993a, p. 18). A cohesive horizontal alliance may enable retailers to achieve economies of scale through bulk purchasing of supplies, to establish their own distribution and wholesaling facilities as a group and to exercise bargaining power over suppliers. If successful, this increased bargaining power should result in lower prices, higher quality and more consistent supplies of products. Even in Poland, which is recognized as one of the more advanced CEECs, the lack of consistency of product offerings in small retail stores is all too evident to the casual observer, with products available one day but not the next. Food retailers in modern market economies make consistency of supply (both quantity and quality) their paramount requirement of a supplier.

Horizontal retail alliances are common among small retailers in modern market economies. Often, these small retailers operate under a common chain name and logo, with an umbrella organization providing wholesaling and distribution functions. The SPAR group of retailers in Europe is one example. Given the large number of small retail food stores in most transition economies, horizontal cooperation among a few small retailers is unlikely to lead to an anti-competitive outcome. Furthermore, any collusive pricing disadvantages from the retailer alliances may be outweighed by benefits from economies of scale and greater bargaining power. Small chain stores and foreign supermarket companies will provide sources of competition for small-scale retailer alliances. The establishment of competition regulations (anti-trust/anti-monopoly regulations) in the future will be important to ensure that, if larger supermarket chains come to dominate food retailing in former command economies, they behave competitively (see Chapter 18).

A large number of small-scale private retailers increases the transaction costs of monitoring the quality of food supplies in the system as a whole. Once consumers are able to exercise the right to choose, quality becomes important. As discussed in Chapter 4, it is often difficult to measure the quality of a food product prior to purchase through visual inspection; for example, if the quality characteristic of interest to consumers is the tenderness of a meat product.[6] This means that consumers will often rely on the reputation of the retailer in making a purchase decision. If a consumer discovered that products purchased from a retailer were sub-standard, a repeat purchase is unlikely (assuming an alternative exists). Consistency of quality becomes of paramount importance to a retailer in attracting and retaining customers. The retailer incurs monitoring and enforcement costs in ensuring that suppliers provide goods of a consistent quality. In a system in which a large number of small retailers are monitoring the products of a few processors, considerable duplicate monitoring occurs. This could be reduced if retailers formed retail buying alliances or if the transaction costs are internalized through vertical integration, i.e. if processors establish their own retail outlets. The growth of factory shops has occurred in some industries.

Government(s) can play a role in facilitating the development of a private retail sector. The importance of effective legal and banking sectors are discussed in Chapters 17 and 18. Of particular importance to the retail sector is the provision of accessible and targeted marketing and small business management training courses. Usually, managers/owners of private retail stores are small entrepreneurs. They may have little time available for a long marketing training course, so a highly focused, intensive training programme would be required. Emphasis on consumer market research skills, basic sales management, service skills and staff training are particularly important for this segment of the agrifood chain because it acts as the direct interface with the consumer.

Location is critical to the success of a retail outlet. Retailers require locations which will offer a sufficient volume of traffic through their stores. Often in CEECs and NISs, this means a central high-street or downtown location because car ownership is not as common as in modern market economies. Consumers will shop more frequently and buy smaller quantities if they are shopping on foot or by public transport rather than by car. Food retailers are often positioned in rented high street locations. To encourage the development of a private retail sector, governments should make a sufficient quantity of high-street locations available to small-scale food retailers and should be aware that retailers will be vulnerable to high or fluctuating rents. Until land markets are fully developed and prices approach equilibrium levels, uncertainty over the cost of rents in the future will deter small-scale entrepreneurial investment. This raises questions over whether prime locations should be sold to private developers or retained by the (local) government and leased. If the government maintains control, then it is essential that these properties are managed in a way which provides the stability required to induce investments in improving retail facilities.

## 11.5. Conclusions

Food retailing is the segment of the agrifood chain that has seen the greatest private market activity in the former market economies. As legal restrictions on private businesses were gradually lifted, the entrepreneurial spirit of many citizens showed itself in the rapid expansion of private food retailing. As in any economy in which new business development occurs, some of these businesses failed or will fail. Others have survived, often because they grasped more quickly the concepts of convenience, service and choice and were thus able to differentiate themselves from the offerings of the previous state system and from their competitors. Entrepreneurial spirit will only take a business so far, however, and the lack of financing and credit has meant that many of these businesses are seriously under-capitalized and lack the necessary funds for further investment. Furthermore, many private retailers face a processing sector still under monopoly state control. Hence, they have found it extremely difficult to meet their customers' needs in terms of product qualities, quantities and prices. As the remainder of the agri-food system moves through the process of transition, these problems may ease. They will not ease of themselves, however. As has been discussed in previous chapters, if the elements of the former command agrifood chain are privatized as monopolies, there will no competitive pressure within that industry sector and very little competitive pressure which individual retailers can bring to bear on these firms and so the inefficiencies will continue.

A competitive private retail sector is critical to the successful transition of the entire agrifood sector because of the importance of the consumer–retailer interface. Through this interface, consumer purchasing decisions create price signals which direct economic activity throughout the agrifood chain. It is therefore in the interests of policy makers to ensure that the private food retail sector operates in an environment conducive to its further development.

# Notes

1. The transmission of information from consumers operates strongly in spite of the oligopolistic power of supermarket retailers in many modern market economies.
2. For example, in the Ukraine, the Ministry of Trade operated the *Gastronom* chain of food stores in urban areas, while the national cooperative *CoopSoyuz* operated food stores primarily in rural areas (World Bank, 1994).
3. The role of foreign investors is discussed below.
4. Schumpeter (1950) called this phenomenon 'creative destruction' (p. 87).
5. See Chapter 18 for a full discussion of the problems which this creates.
6. We can distinguish between three different types of goods. The quality of search goods can be determined prior to purchase through inspection, for example, a pair of jeans. The quality of experience goods can be determined only after the product has been bought and consumed, for example, a can of tuna fish; the reputation of the seller is important in the purchase decision. Credence goods are those whose quality cannot be determined even after consumption without expert knowledge, instead the consumer must rely solely on the reputation of the seller to determine the quality of the good (or service), for example a visit to the doctor or reading this book (Nelson, 1970).

# References

Johnson, S. and Loveman, G. (1995) Starting over: Poland after communism. *Harvard Business Review* March–April, 44–56.

Jones, S. (1993) The future for fruit and vegetable distribution in Russia. *British Food Journal* 95(7), 21–23.

Kerr, W.A. (1996) Marketing education for Russian marketing educators. *The Journal of Marketing Education* 18(2), 39–49.

Lyle, I. (1993) How can East European food industries learn from those in the West? *British Food Journal* 95(7), 40–43.

Morton, C. (1993a) Food distribution in Eastern Europe. *British Food Journal* 95(7), 16–20.

Morton, C. (1993b) The development of Poland's food marketing system. *British Food Journal* 95(7), 24–29.

Nelson, P. (1970) Information and consumer research. *Journal of Political Economy* 78(2), 311–329.

OECD (1995a) *Review of Agricultural Policies: Poland.* Centre for Co-operation with Economies in Transition, Organization for Economic Co-operation and Development, Paris, 285 pp.

OECD (1995b) *Review of Agricultural Policies: Czech Republic.* Centre for Co-operation with Economies in Transition, Organization for Economic Co-operation and Development, Paris, 298 pp.

Schumpeter, J.A. (1950) *Capitalism, Socialism and Democracy.* Harper and Brothers, New York, 437 pp.

World Bank (1994) *Ukraine: the Agriculture Sector in Transition.* World Bank, Washington DC, 148 pp.

# 12 Treating Agrifood as a System: Dealing with Perishables

While it has been useful to examine each of the components of the food production, distribution and marketing chain, for certain products it is essential to approach the process of transition to a market-led economy from the point of view of the supply chain as an interrelated system. Although the concept of a complete agrifood chain is important for all agricultural products, for those which can be stored or significantly transformed, the individual components of the chain can be the focal point for analysis. In the case of perishable products, however, the system itself is of paramount importance.

In modern agribusiness, a systems approach to perishable commodities is highly developed and concepts such as cold chains[1] are well understood. Focusing on the problem of systems management in the case of perishables in the Central and Eastern European Countries (CEECs) and the New Independent States (NISs) is extremely important because it was highly ineffective in command economies. Poor systems management in command economies had two major effects: (i) excessive waste; and (ii) a predilection for turning perishable products into non-perishables. While the latter may seem a rational strategy when faced with the former, in the post-liberalization era it means that a considerable proportion of high value fresh products will be supplied from abroad and many potentially lucrative exports will be uncompetitive in the high value, fresh produce segments of modern market economies. When markets in the command economies remained closed and, for example, the only form of fruit available to most consumers was domestic canned fruit, developing an efficient cold chain was not a necessity. With a state distribution monopoly there was no competitive incentive to reduce waste or to improve quality by providing fresh products. As a result, losses from spoilage in the dairy sector were large, fruit and vegetables were processed rather than sold fresh and fresh meat was turned into sausage, smoked, canned or transformed in other ways. The lack of cold chain facilities also distorted the production mix, as emphasis was placed on storable crops such as cabbage or potatoes rather than lettuce or strawberries. Variety in diets was restricted, and in many cases nutritional balance was difficult to attain.

It will be argued in this chapter that past investments in large-scale facilities and the hierarchical management structure of the former command economies will constrain the ability of the food chains for perishable commodities to adjust to a holistic systems approach. The problems associated with instituting a new systems approach are best illustrated by an example from the dairy industry. After the dairy example is examined in depth, the discussion is broadened to include red meats, perishable vegetables and fresh fruit.

## 12.1. A Dairy Industry Case Study

To understand the problems associated with the transformation of dairying in the former command economies, the mental picture of a dairy farm in a modern market economy must be put aside. Instead, it is necessary to envisage dairying as a number of independent stylized production activities which are combined together to convert basic raw materials such as sources of nutrients for plants, plant and animal genetic material, etc. into milk, as well as its various derivative products and by-products which are ultimately purchased by consumers. Approaching dairying in this way is necessary because the planners in the former command economies approached the provision of milk and dairy products to consumers from this perspective.

From their perspective, the dairy industry consists of a number of activities which are primarily constrained by biological parameters – forage production, forage storage, breeding, ration formulation, feeding, protection of animal health, milking, pasteurizing, etc. These activities were considered individual tasks which, if completed, would lead to a successful dairy industry. Of course, the same basic activities are the basis of the dairy industry in modern market economies. The difference is that the emphasis is focused on the system as a whole rather than on the individual tasks.

Improvements to dairy efficiency arise from increasing the total net value of the outputs which can be obtained given the numerous biological constraints on production. At any point in time, these improvements can come from three sources: (i) new scientific knowledge;

(ii) better applications of existing scientific knowledge; and (iii) more effective coordination of the activities which comprise dairying. Leaving aside the provision and application of new knowledge, the other two factors are not independent as poor coordination can inhibit the effective application of scientific knowledge and the state of scientific knowledge can determine the form of coordination which will be the most effective. Long run efficiency gains come from the development and application of better production technology and management systems. The approach to dairy industry coordination in most command economies has shaped the pattern of investment in their industries in ways which are making it difficult for an efficient dairy industry based on markets to evolve.

### 12.1.1. Dairying in a modern market economy

In a fully developed market system, the industrial structure of the dairy production, processing and distribution system is determined, everything else equal, by the method of vertical coordination which proves to be the most efficient. The combination of governance instruments which minimizes total production, distribution and transaction costs will be the most efficient system. As pointed out in Chapter 4, where managerial orders are the most efficient mechanism for vertical coordination, firms will be the form of governance, i.e. transactions will be coordinated within a vertically integrated firm. Where markets prove more efficient, firm-to-firm transactions will take place.

It follows then, that the size of dairy farms in a market-based economy is determined by the efficiency of managerial orders relative to between-firm transactions.[2] Dairy operations in the CEECs and the former Soviet Union were arbitrary administrative units within a command economy's vertical coordination system. As a result, privatization of command economy dairy operations may produce firms which are too large to coordinate efficiently, leading to higher levels of transaction costs.

In modern market economies, there is fairly consistent evidence that the most efficient dairy farms have herds of between 60 and 300 cows, depending upon the situation in the particular area. Dairy operations are centred on a family farm with the owner-operator facilitating vertical coordination through managerial orders. In some cases this entails only self-management as the farmer is the only labour input. Often, however, managerial orders will be given to family members and/or a small number of hired workers. For the most part, the herd's forage requirements are provided from the farm's land base. The dairy farm interfaces with other firms for supplies of other inputs and downstream disposal of milk and surplus livestock. These farm-to-firm interfaces are likely to be conducted through contracts (milk, machinery, nutrient additives and special feeds, veterinary services, artificial insemination services, accounting services, haulage) or spot markets (replacement heifers, bull calves, surplus heifers).

The downstream milk processing industry in modern market economies is often organized through a voluntary farmer-owned cooperative or a compulsory government sanctioned marketing board administered by farmers. The relationship between the individual dairy farms comprising the processing cooperative/marketing board is, however, kept at an arms length arrangement with the movement of milk from a farm to a processor governed by a contract. While herds under 60 animals and in excess of 300 animals do exist and forms of organization other than family farms can be found in the dairy industries of modern market economies, the structure described above can certainly be considered the norm over a very diverse set of countries. This might suggest that a similar system would evolve over time in the former command economies. It will be argued that the institutional constraints arising from the organizational structure of the dairy industry in command economies will act to inhibit the development of a dairy industry based on the family farm model.

### 12.1.2. Dairying in the command system

Contrast this stylized family farm dairy unit with that of a stylized Agrokombinate in the former Soviet Union.[3] The Agrokombinate (which we will call Red Wind) was a large coordinating organization which oversaw the production and processing of a wide range of agricultural commodities. Agrokombinate Red Wind had a milk plant, a meat plant, maintenance facilities and retail outlets in addition to its farming operations. The organization had a workforce of approximately 55,000 people.

Agrokombinate Red Wind's dairy herd was spread over 38 collective/state farms. Herd sizes on the individual farms ranged from 200 to 2000. In total, there were approximately 16,500 milking cows with total cattle numbers approaching 50,000 head. Calves were centrally raised on specialized farms, one farm raised heifers as dairy cow replacements while another fattened bull calves for meat production.

As a result of the current reforms, the ownership structure is no longer well defined. Certain operations, such as feed milling and the artificial insemination unit will remain government owned and operated while the state and collective farms within Agrokombinate Red Wind's structure have been privatized as Joint Stock Associations (JSA). Under the JSA privatization, all workers received shares based on a combination of salary and seniority. Each JSA farm still operates within the overall structure of Agrokombinate Red Wind but has its own management. As with most privatized collective/state farms, the managers from the command economy era remain in place. While there appears to be only limited change to the actual functioning of the collective/state farms, there is interest in private family farming among some members of these farms.

Agrokombinate Red Wind was, and still is, responsible for vertically coordinating all the stages of the production–distribution–marketing chain from farm level activities through to local distribution for the entire range of crops and livestock outputs produced on its farms. In some cases, including milk, final sales to consumers were also the responsibility of Agrokombinate Red Wind. Overall managerial authority was vested in Agrokombinate Red Wind's General Director who had line authority over the managers of the individual collective/state farms. There was little, if any, delegation of authority to the staff of Agrokombinate Red Wind (as opposed to the staff of individual farms) who have expertise in specific aspects of farm production or processing. For the most part, the staff of Agrokombinate Red Wind played only a monitoring and reporting role.

The managers of the individual farms were responsible for both the economic and social welfare aspects of their farm. Not only were farm managers responsible for managing a large work force, they were responsible for the provision of food, medical care, and housing for this work force and their families. Well over half a manager's time was typically taken up with social welfare related activities. The large proportion of time spent on non-income producing activities was a direct result of the size of the units for which a manager was responsible. For example, Agrokombinate Red Wind's feed mill had a capacity of 1500 tonnes per day and employed 300 people. One of the former state farms (with greenhouse, field vegetable, potato and dairy/meat enterprises, as well as maintenance facilities) had a staff of 2000, with 100 tractors, 600 other farm machines and 130 trucks and cars. The manager of the artificial insemination unit had a work force of 318 people. Managers now admit that they did not have staff to whom they felt comfortable in delegating tasks associated with either production coordination or social welfare. Basically, the large size of the units, along with the shortage of middle managers, meant that while charged with the task of accomplishing vertical coordination through managerial orders, managers were unable to carry out their task efficiently. While essential, managing dairying as a holistic system became impossible.

The large size of the dairy operations on most of Agrokombinate Red Wind's farms required that vertical coordination of milk production be divided among a number of individuals. As discussed in Chapter 7, the management system was hierarchical so that individuals responsible for the various activities which comprise milk production reported to the manager of the farm. There was limited communication or cooperation among those responsible for the various activities which comprise dairying. The primary responsibility for milk production lay with the breeder who selected the genetic material for the herd. The breeder (somewhat analogous to a herdsman in large dairy farms in modern market economies), along with Agrokombinate Red Wind's veterinary staff, was also responsible for animal health. Hence, milk production was cow centred with animal breeding and health considered the critical aspects of dairy cow management. The objective of the breeder was simply the production of milk to meet production targets provided by the farm manager. The breeders were given their production targets, the feeds they had to use, the staff they had to work with, the barn designs they had to milk in, the milking equipment they had to use, etc. There was little room for the breeder to exercise independent initiative.

The problems arising from the constraints on independent initiative were compounded by the compartmentalization of management. For example, the agronomists reported through a different channel than the breeder and had a different goal – maximizing forage production. They were often geographically separated from those working with the animals. Further, they had no incentive to seek them out. On the other hand, with the single goal of maximizing milk production, forage quality was not the concern of the breeder and his staff. The Chief Engineer was responsible for the maintenance of milking equipment. As a result, sanitary control had to be dealt with through hierarchical channels rather than by the milkers who were in the position to identify problems.

While the breeder may have been responsible for genetic selection, the important step of calf rearing was the responsibility of another person. Milk processing was carried out far from the farms. With no incentives or penalties for milk sanitation or quality on the farm, the shelf-life of milk was reduced to thirty-six hours compared with the two weeks common in a modern market economy. After dairy products left the processing facility, milk handling was shoddy and equipment such as refrigeration units in Agrokombinate Red Wind's retail outlets was poorly maintained.

In contrast, in the smaller dairy farm operations of modern market economies, each stage (e.g. breeding, heifer rearing and lactation) is based on the previous stages and receives feedback from the following stage. The processing sector, for example, provides incentives for the maintenance of high sanitary standards through differential pricing specified in purchasing contracts. As a result, on-farm milk line sanitation efforts reduce the bacterial count in milk, extending shelf-life in the retail store, resulting in increased marketing flexibility.

There are many individual factors in each production activity which must be monitored, assessed and acted upon as needs dictate. Inherent in the dairy system of modern market economies is independent decision-making and, if the farm is going to be successful, a solid understanding of the objective of the system, i.e. to produce a very high quality product to meet the requirements of very demanding consumers.

The key to the success of the dairy industries in modern market economies is the farmer as an owner-operator. He or she has a direct incentive to produce the market driven mix of quality and quantity which the farm's resource base can provide. Herds and farm operations are of a size that can be directly monitored. A small labour force consisting of family and/or hired labour keeps supervision activities to a minimum. Price premiums or penalties embedded in milk supply contracts provide incentives to produce the desired levels of sanitation and quality. Market competition guides input and secondary output markets such as those for bull calves. Unless their output must go to a compulsory milk marketing board, farmers are free to change who they sell to. Consistently poor management and/or insufficient resources will eventually lead to business failure.

## 12.2. Problems of Transition in Dairying

While some changes have taken place in Agrokombinate Red Wind, the organizational structure of the command era remains largely intact. The major question is how to reform the dairy industry in a manner which will lead to the adoption of a systems approach to dairying (McNeal and Kerr, 1995). Two avenues appear open. First, the family farm model, common in modern market economies, could be fostered. Second, the existing dairy industry could be reformed leaving its basic structure intact. In both cases there are important questions as to how a systems approach can be encouraged.

### 12.2.1. Prospects for family dairy farms

To focus the discussion on the problems associated with encouraging a systems approach in family farms, it will be assumed that crucial questions regarding herd and property division have been solved in a satisfactory fashion. In other words, the farm family unit has left the former collective/state farm and has acquired a herd of approximately 100 cows from the command farm's herd. Further, it will be assumed that the family has acquired sufficient land to provide all, or at least a significant proportion of, the herd's forage requirements. Along with the difficult problems associated with the distribution of land and livestock belonging to the former collective/state farms, there is a personnel question. There are no obvious farmers. There are unlikely to be many individuals with the complete set of skills required to succeed in dairying. The command system's investment in specialized human capital meant that there are no obvious generalists with an operational appreciation of dairying as a unified system. The hierarchical organization of activities discouraged individuals from acquiring a working knowledge of the entire system. Even if the difficult problems associated with the division of assets and personnel can be satisfactorily resolved, the existing infrastructure may inhibit the development of an efficient market based dairy sector.

Past investments have lumbered the former collective and state farms with large-scale milking and milk handling facilities. As a result, it will not be possible for most family farmers to have individual milking facilities when they leave their former organization.[4] This means that the milking and milk handling facilities will have to be shared among farmers. As milk from a number of individual farms must be held in the existing large bulk milk tanks, a sophisticated and expensive system of monitoring will be required to prevent individual farmers from free riding by cheating on quality standards and sanitation procedures.

The existing large-scale animal handling facilities will mean that dairy cows from different farms may have to be mixed. The mixing of herds can lead to health problems. Existing animal feeding facilities will make it difficult to allow for separate feeding of an individual farmer's stock. Silos at forage storage facilities are large scale, making it virtually impossible for individual farmers to control the quality of their forage. Arrangements will have to be made for sharing the costs of cleaning and maintaining the milking machines, pipe lines and milk tanks. As milking activities will be centralized, farmers are likely to have to travel to the fields where they will produce their forage. Hence, they will be physically separated from their animals for long periods each day in the growing season. The demands of twice-a-day milking will require considerable travelling time.

Investment in family farm level milk handling equipment, animal facilities and silage equipment may be possible over time, but given the low levels of profitability which can be expected in the short run, combined with inadequate capital markets, the rate at which a change-over can be effected may be very slow. In addition, as more family farms are able to acquire their own equipment and facilities, the number of farmers continuing to use the existing facilities will decline. As a result, the burden of fixed cost and maintenance will increase for those who continue to use the shared facilities. This will lower the probability of sufficient profits being generated to acquire separate facilities, and their economic viability may even be threatened. Hence, milk supplies may be disrupted before a sufficient number of farms with individual facilities exist to satisfy the demand at a price which is acceptable to those who perceive milk as a necessity.

If individual dairy farms are privatized separately from the processing system, then farmers will face a monopsonistic purchaser of milk. The large-scale investments of the communist era means that milk processing and distribution systems are regional monopolies. A privatized milk processor which is the sole purchaser of milk in a region will have unequal bargaining power when milk contracts with family farmers have to be negotiated. This problem could be handled by individual milk farmers combining to purchase the milk processing facilities – which is the case in many market economies. Overall management of processing would then be in the hands of the farmers, with the movement of milk between farmers and processors handled by arms length contracts. This model, however, took a long time to evolve in modern market economies. In the initial stages of privatization, it may be questionable whether the farmers who own processing facilities would effectively enforce provisions relating to health standards and quality. Without strong incentives for quality, and penalties for breaches in sanitation procedures, the systems approach to milk production will break down.

Input suppliers may also represent a problem for small-scale family farms. Agrokombinate Red Wind, and others like it, have made investments in large-scale artificial insemination (AI) facilities, maintenance depots and feed mills. If they are removed from Agrokombinate Red Wind and sold off, they will become monopolistic suppliers of those services in the local area. According to Gady and Peyton:

> Many of the large state monopolies still exist. Consequently, there is little competition to bring about many of the operating efficiencies that need to occur. Furthermore, many of the existing monopolies have been able to raise prices to punitive levels because of lack of competition, making life difficult for businesses having to buy or sell to these monopolies.
>
> (Gady and Peyton, 1992, p. 1180)

As suggested in Chapter 6, alternative input suppliers are likely to evolve only slowly. In the absence of competition, bilateral contracts rather than spot markets will be the method of organizing transactions between input suppliers and farmers. This will create a dependency between buyer and seller (Hobbs *et al.*, 1993). This dependency leaves farmers open to opportunistic behaviour by sellers, meaning that farmers may not be able to earn sufficient returns on investment (Weleschuk and Kerr, 1995). When farmers cannot expect to recover the full value of their investment, they will tend to reduce the level of investment (Anderson *et al.*, 1992). As a result, there will be underinvestment in family dairy farms. To the extent that underinvestment inhibits the development of a holistic system, gains in dairy efficiency cannot be expected.

One solution to the difficulties caused by monopolistic input suppliers might be to vest the ownership of the Agrokombinate's input enterprises in a dairy farmers' cooperative. Doing this, however, suggests a fundamental contradiction with the transition to a market-based system because Agrokombinate Red Wind's structure would simply be reinstated as farmers would control both the downstream and upstream components of the dairy system. The prices that farmers would receive for milk and the prices paid by farmers for inputs would be set bureaucratically, as opposed to the monopsonistic/monopolistic alternative, and markets would play no part.

It seems clear that the previous investment in large-scale production systems in the command economy era will inhibit the opportunity to create a dairy system modelled on those in modern market economies. While it might be argued that family farmers should be encouraged to operate totally outside the existing dairy structure, initially farmers will have to rely on Agrokombinate Red Wind for both inputs and access to the processing and distribution system. Agrokombinate Red Wind is likely to perceive these family farmers as competitors and act to limit their success. Abandoning the infrastructure in the existing system is not a feasible alternative in the short run as milk supplies would be disrupted. Hence, as creating a dairy system which is modelled on the dairy industries in modern market economies is likely to be fraught with difficulties, what then are the prospects for reforming Agrokombinate Red Wind to create an efficient, holistic dairy system?

### 12.2.2. Large-scale corporate dairy farms

Why have large-scale dairy operations not evolved in modern market economies? In a competitive system with market discipline, transaction cost theory suggests that such large operations must be less efficient than family farm units when it comes to organizing farm level activities. The biological operations of dairying are fundamentally the same, regardless of the size of the operation, suggesting that technical economies of scale are quickly exhausted.[5]

The major difference between a dairy operation the size of a typical family farm (60–300 cows) and the large-scale former state/collective

farms is that more people are involved in the operation. The fundamental biological operation is simply duplicated by adding more cows.
This suggests that the transaction costs associated with supervising
labour activities and monitoring the factors controlling output quality
rise steeply when the herd exceeds a size that is directly manageable by
the owner-operator and trusted family members (Eggertssen, 1990).
While it may be relatively easy for managers to devise work plans for
large-scale dairy farms, ensuring that they are carried out might be
much more difficult and costly. Compared to a compact factory with
standardized tasks, monitoring labour activity is more difficult as farming activities are spread over large areas and most tasks are non-
standardized. Without adequate supervision, workers will have the
opportunity to shirk.

The problem of monitoring the factors controlling output quality is
extremely complex in dairying. Unlike industrial processes, the individual biological processes which make up dairying are each subject
to considerable potential variability in performance. Further, dairying
requires the melding of a number of biological activities. Changes in
one biological process can interact with a number of other biological
processes which ultimately affect herd performance or output quality.
When different people are responsible for the individual processes, it
becomes difficult to isolate the root cause of a problem. Furthermore, as
there is often a lengthy time lag between a failure in the system and an
observable effect on the quality of output, identifying the source of the
problem and the individual responsible may not be possible. Close
supervision of all activities in large-scale operations is likely to be prohibitively expensive.

If adequate supervision on the job is not possible, one alternative
might be to provide incentives to induce the desired performance from
workers. Incentives only work well, however, when output can be easily
monitored and processes are independent (Jensen and Mecklin, 1992).
These are not the attributes of a dairy system. Most outputs in the dairy
industry have a large number of quality characteristics which complicate the monitoring process.

Workers on dairy farms are unlikely to accept incentive schemes
because their activity will depend on a number of other activities which
are the responsibility of different workers. The delay between the point
in time at which decisions are taken and when the outputs of a dairy
system can be assessed make it impossible to separate out the contribution of individual components of the system in order to establish appropriate incentives. Hence, using either managerial orders and/or incentives to coordinate a dairy system employing large numbers of people has
inherent inefficiencies when compared to managerially compact family
dairy farms.

Of course, considerable improvement can be made within Agrokombinate Red Wind's existing structures. Major gains in productivity can come from altering the hierarchical management structure. A single manager responsible for all dairy operations on an individual former collective/state farm could be appointed. Forage production and storage, breeding decisions, nutrition, herd health, milking, waste disposal, day-to-day equipment maintenance, sanitation, on-farm holding of milk, heifer rearing and the monitoring of milk and forage quality should all be the responsibility of the dairy manager.

The dairy manager should adopt a team approach to coordination which involves all the managers of individual processes in decision making. Managers of individual processes should be required to spend considerable time learning all the other processes so that an appreciation of the broader system and its problems can be gained. All levels of management and the work force need to be trained to approach problems from the point of view of the system. Managers should be provided with training in business and human resource management. A Japanese management model may, however, be more appropriate than the North American/European model. Japanese success in manufacturing can, in part, be attributed to their alternative approach to western style hierarchical managerial orders for the coordination of labour activity. In the Japanese management model, workers are encouraged both to collectively take initiative and to be involved in problem solving. The emphasis in the Japanese model is on the system rather than on worker specialization.

It was not the intent of this example to provide a model for the transformation of the dairy industries in the former command economies. An attempt has been made to point out the difficulties which must be overcome when privatization and the creation of markets are considered within the context of complex biologically-based systems such as dairying. Operationalizing the transition from vertical coordination based on the command system, to one where vertical coordination is based on market forces, is a complicated problem. Conditions need to be created which encourage new entrants to dairying, not just family farmers but also input suppliers and processors. To the extent that Agrokombinate Red Wind typifies the organization of dairying, considerable economic power will remain within the existing industrial structure. In its own interest, Agrokombinates such as Red Wind are likely to attempt to stifle competition. As a result, the development of an industrial structure which can support family farm dairying is likely to be inhibited.

## 12.3. Other Perishable Commodities

Modern meat processing and distribution systems require elaborate cold chain capabilities. Any break in the chain means that meat may spoil, or more catastrophically, become a breeding ground for life-threatening organisms. This cold chain capability is largely absent in the former command economies. As a result, most livestock products are cured by turning them into sausage or other smoked products. A smaller proportion is canned. The processing and distribution of fresh fish is similarly constrained.

Fresh meat and fish are sold through very short distribution systems. The meat is cooked shortly after being purchased. Sanitary standards are low in butcher shops and, as yet, an effective inspection system does not exist. The longer the fresh meat chain, the greater the losses due to spoilage.

Consumers often have little choice in the cuts they receive and no chance to inspect the meat before purchase. Fresh meat is often offered to customers on a first come, first served basis in the order that the carcass is butchered. Chunks of meat are often still simply handed out over the counter with the customer having no chance of refusal in favour of a preferred cut. As meat is generally in short supply, customer refusals are unlikely to have any significant effect on quality. Without consumer pressure, there will be little incentive to improve cold chain capability along the meat and livestock chain.

It seems unlikely that sufficient funds will be available for any large-scale investment into establishing cold chains. The investment required for hygienic slaughtering and processing facilities is extremely high. Cold storage facilities and refrigerated trucking fleets are very costly. Refrigerated counters in supermarkets or specialized meat shops are expensive both to purchase and to operate and maintain.

The constraints on the provision of cold chain capability suggests the development of a more traditional butcher-based meat retailing sector in the intermediate run. Animals will be shipped live from farms to the place of slaughter. Wholesale cattle/meat markets will develop which will play a role similar to that which Smithfields traditionally played in London or the Rungis market plays outside Paris. Butchers will purchase live animals which they will either pay to have slaughtered at a facility at or near the cattle market, or rent such facilities and slaughter the animals themselves, or take the animals to their retail shop location for slaughter.

While this system has low cold chain requirements, the evidence from modern market economies suggests it is relatively inefficient. Transporting animals rather than meat is expensive and leads to death losses and shrinkage. There also appear to be considerable economies

of scale in meat processing. Further, centralized slaughter reduces the
collection costs associated with by-products. The collection of hides,
bones, offals, blood, inputs to pet foods, etc. is expensive when slaughter takes place at a large number of butcher shops. Many potentially
valuable products will simply be treated as waste. This waste reduces
the value of the entire animal. In many modern market economies, traditional butcher shops have virtually disappeared. They simply cannot
compete with the supermarket system and the cold chain which underlies it.

As a cold chain is a system, each stage along the chain has to be
undertaken in a timely manner and diligence exercised in the enforcement of food safety procedures. As with dairy products, the complexity
of the system and the delays before spoilage problems emerge make it
very difficult to identify the stage at which a problem is created. Incentives do not work well in such complex systems, suggesting that
improvements to efficiency can only come from better monitoring capability. Better monitoring will improve the manager's ability to effectively direct production. Considerable vertical integration is likely to
arise in any processing and distribution system for fresh meat and fish.

While the domestic market can, to a large extent, continue to be
supplied with sausage, smoked products and canned meat and fish, the
lack of cold chain facilities for meat, poultry and fish will mean that
considerable export potential will be foregone. Consumers in modern
market economies require meat to be fresh and handled in a sanitary
fashion. As yet, there is no evidence that this capability exists in the
former command economies.

Perishable fruits and vegetables also require a holistic system if
these products are to be transported over intermediate to long distances
and arrive fresh. In the CEECs and the NISs, spoilage rates tend to be
very high. In the command system, emphasis was placed on transforming perishable products into non-perishables rather than
attempting to develop delivery capabilities for fresh produce. One result
of liberalization, at least for areas that have reasonable transportation
links with western Europe, is that fresh fruit and vegetables tend to be
supplied from the distribution points in western Europe. Wholesalers
and some retailers have been quick to invest in the facilities required
to prevent the spoilage of imported products. This has inhibited the
development of the domestic fresh produce industry.

As with meat and dairy products, the complexity of the factors
affecting the quality of fruit and vegetable products purchased by consumers makes it difficult to design incentives to achieve the desired
quality. Further, even simple techniques such as washing, careful handling and the visual sorting and removal of deteriorating products are not

well accepted. Fruit and vegetables are often sold out of the back of a farm truck on an 'as is' basis.

Of course, many fruits and vegetables are seasonal. Little cold storage capability exists. As a result, it is not possible to store seasonal products for distribution throughout the year. Foreign supplies become the only fresh products available out of season. Investment in storage facilities is unlikely to be profitable unless other aspects of a system for processing and distributing fresh produce are put in place. Putting sophisticated systems for handling fresh fruits and vegetables in place for relatively short seasonal supplies is hard to justify. The slow development of local markets and small-scale jobbers taking advantage of arbitrage opportunities seems more likely than investment in large-scale processing and distribution systems. A niche market which serves those sufficiently well off to afford high prices will continue to develop for imported fruits and vegetables.

## 12.4. Conclusions

An efficient processing and distribution system for perishable products was one of the last developments among agribusinesses in modern market economies. In part, this late development arose because processing and distribution must be approached as an integrated system. These systems are characterized by high levels of investment, considerable technical sophistication and an exceptional degree of monitoring effort. In the transitional economies, the former is not likely to be forthcoming, while the latter two are likely to be beyond the capability of the existing management and quite possibly resisted. As a result, considerable potential, particularly in terms of exports, may be foregone. While alternative systems which convert perishables into storable commodities such as cured meat and canned vegetables will continue to be used, high levels of waste are likely to continue.

Agribusinesses in the CEECs and the NISs have to be particularly careful when purchasing foreign technology and equipment. Prior to purchasing imported equipment, an appreciation of how the individual piece of equipment fits within the system is required. Sophisticated milk processing equipment being supplied from buckets and milk cartons sitting in the sun on a loading dock waiting for transport will mean that the imported technology is largely wasted. Unfortunately, the hierarchical management system and large-scale facilities in the command economy led to labour specialization, lack of cooperation and a failure to think of food production, processing and distribution as a holistic system. As a result, not only do the technical problems of a systems approach have to be addressed but also the psychological aspects of the perception which managers and workers have of tasks and how those tasks fit within a system. Transition in industries which deal in highly perishable foods is likely to be particularly slow.

# Notes

1. A cold chain is a system whereby a perishable product is never exposed to temperatures (or other environmental forces) above that necessary to minimize the deterioration of the product. The cold chain usually encompasses all the steps between primary producer and final consumer.
2. This assumes that production quotas, or other restrictions on farm output sanctioned by governments, are not in place. Of course, technology-based economies of scale will also affect the size of firms – the size of the herd will increase. This simply means that it is possible to coordinate a larger herd with the same amount of effort.
3. While the facts are somewhat stylized, the example is based on an actual dairy operation in the Moscow area.
4. It is assumed that few potential family farmers will have sufficient capital when establishing their farm to build and equip a milking parlour.
5. There appears to be no evidence that technical diseconomies of scale exist in dairying.

# References

Anderson, C.L., Hobbs, J.E. and Kerr, W.A. (1992) Transactions, costs and the benefits of trade: liberalizing the Japanese importing system for beef. *Asian Economic Journal* 63, 289–301.

Eggertssen, T. (1990) *Economic Behaviour and Institutions*. Cambridge University Press, Cambridge, 385 pp.

Gady, R.L. and Peyton, R.H. (1992) A food processor's perspective on trade and investment opportunities in Eastern Europe and the former Soviet Union. *American Journal of Agricultural Economics* 74, 1179–1183.

Hobbs, J.E., Kerr, W.A. and Gaisford, J.D. (1993) Transforming command economy distribution systems. *Scottish Agricultural Economics Review* 7, 135–140.

Jensen, M.C. and Meckling, W.H. (1992) Specific and general knowledge and organizational structure. In: Werin, L. and Wykonden, H. (eds) *Contract Economics*. Blackwell, Oxford, pp. 305–360.

McNeal, A.O. and Kerr, W.A. (1995) Extension for Russian agricultural industrial complexes: lessons from a dairy project. *European Journal of Agricultural Education and Extension* 2, 49–57.

Weleschuk, I.T. and Kerr, W.A. (1995) The sharing of risks and returns in prairie special crops – a transaction cost approach. *Canadian Journal of Agricultural Economics* 43, 237–258.

# 13 The Horn of Plenty: Competition from Imports

The tremendous growth of international trade since World War II is an important fact of modern economic life. During the period from 1945 to 1989 an 11-fold increase in the real value of world trade flows helped fuel a fivefold increase in the world's real GDP (Husted and Melvin, 1995). During the same period, however, international trade posed severe difficulties for the command systems of the Soviet Union and the Central and Eastern European Countries (CEECs). Further, many of these international trade difficulties have remained and new problems have arisen in the course of the transition to a market-based system.

International trade provides mutual gains for participating countries. Comparative advantage is the cornerstone of so-called inter-industry trade where a country exports one good, say wheat, in return for the import of a different type of good, say lamb. Comparative advantages and disadvantages exist because countries differ in terms of natural resource endowments and climates, accumulated capital and technology, production experience, tastes, average income-levels and income distributions. Much of the trade between developed and less developed countries is based on comparative advantage because of the large differences in tastes, endowments, technologies and experience. Further, comparative advantage is always a two-way street. If Japan has an advantage in producing electronic equipment relative to textiles in comparison with China, then China has an advantage in producing textiles relative to electronics in comparison to Japan. Thus, both countries can gain from exporting in accordance with their comparative advantages.

Intra-industry trade occurs when a country exports and imports the same good or at least different varieties of the same good. For example, Germany may export white wine to France, while France exports red wine to Germany. Indeed, much of the phenomenal increase in international trade among modern market economies has been trade of this intra-industry type. To a certain extent, intra-industry trade arises from comparative advantage. For example, many countries both import and export the same food products to take advantage of the opposite patterns of seasons in the northern and southern hemispheres. There are, however, several other increasingly important sources of gains from intra-industry trade that go well beyond comparative advantage. Countries that have small markets are able to exploit economies of scale and scope when they produce for the world market.[1] At the same time that producers are able to rationalize production, consumers have access to more variety because of the availability of imports. So-called pro-competitive effects also arise as domestic producers come into competition with foreigners. Thus, international trade also makes markets more competitive and lessens the distortionary effects of market power.

Despite the rapid growth in world trade, and in spite of the pervasive view in most economics textbooks, international trade is not frictionless. The transaction costs and even the transport costs associated with international trade are typically higher than those associated with intra-national trade. The higher are transaction and transport costs, the lower are trade volumes and the smaller are the gains from trade. Transport costs tend to be higher for international transactions simply because the distances involved are typically greater. Further, firms face additional risks in international transactions. Commercial risk is greater and the threat of opportunism is more serious because multiple legal jurisdictions are involved. Additional risk arises from fluctuating currency values. Market economies have been able to develop institutions for reducing risk. Just as there are costs associated with risk, there are costs associated with risk avoidance. Commercial risk can be reduced by involving trading houses or financial institutions[2] and disputes can be resolved via international arbitration. Exchange-rate risk can be reduced by using forward or futures markets to hedge or close risky positions arising from the provisions for future payments and receipts in contracts.

Governments, both intentionally and unintentionally, take measures that affect the cost of international transactions. By maintaining fully convertible currencies, governments greatly reduce the costs of international transactions. Although firms do face exchange-rate risk when they engage in international as opposed to intra-national trade, they need not resort to counter-trade or barter-type goods-for-goods transactions. The limitations of counter-trade are severe because a country is forced to balance its trade on a bilateral basis with each of its individual trading partners.

Governments also intervene in trade flows. Tariffs and non-tariff barriers (NTBs) restrict trade, while export and other subsidies augment trade. Even when explicit barriers are absent, customs and inspection lead to extra brokerage charges on international transactions. Rational governments have every incentive to intervene directly and indirectly in international trade as they pursue domestic political goals such as re-election and related economic goals such as supporting the incomes of various sectors or groups, stimulating regional development and promoting overall growth. Of course, interventions in international trade flows in the agricultural sector have been widespread. Since it is natural for countries to intervene, free trade is definitely not the equilibrium state for the international economy even though it would be an efficient state in the absence of other forms of market failure. While each country's own interventions may further its interests, or at least those of its government, the actions of other countries typically work in the opposite direction.[3] Consequently, multi-lateral trade liberalization is typically in the interest of most, if not all, countries.

Market-oriented countries have tried to deal with this self-defeating or 'prisoners' dilemma' aspect of international trade through multilateral efforts at the global and regional levels. At the global level, efforts at trade liberalization have been made through the General Agreement on Tariffs and Trade (GATT) and now the World Trade Organization (WTO). At the regional level there have been many attempts at liberalization. The European Union (EU) and the North American Free Trade Agreement (NAFTA) are among the largest and most far-reaching regional attempts at liberalization and international economic integration. We will return to this multi-lateral dimension of international trade relations in Chapter 22.

## 13.1. International Trade Under the Command System

The command system made it extremely difficult for the countries to realize the gains available from trade for two key related reasons. First, prices did not serve a resource allocation role by signalling the relative scarcity of commodities. Thus, high domestic prices were not an appropriate signal for imports nor were low prices an appropriate signal for exports. Second, international trade was an excessively cumbersome process involving extremely high transaction costs.

The command system was intrinsically contradictory to the forces of international trade. Since the overall economic plan emphasized capital goods over consumer goods to promote domestic industrialization, the former tended to have artificially low administrative prices. Similarly, the final prices of consumer necessities, including many food items, were kept artificially low and quantities were rationed by queuing rather than by price. In broad terms, price signals would have suggested that command economies export capital goods and consumer necessities in return for consumer luxuries. The export of capital goods and consumer necessities would have decreased the quantities available for domestic use and raised prices, while the import of luxury goods would have increased domestic consumption and reduced prices. Thus, an open international trade regime would have served to partially negate the goals of the central plan and the highly distorted prices that resulted. While keeping international trade flows to a minimum was possible and often deemed desirable from a pure planning standpoint, eliminating trade entirely proved to be impossible. Thus, a centrally-managed system of international trade was necessary in order to protect the planning process.

In order to prevent spontaneous trade, the Soviet Union and the CEECs had non-convertible currencies throughout the communist era. Official trade was then conducted by State Trading Agencies. While unofficial trade was not entirely eliminated, it was typically relegated to a black-market fringe, often involving casual transactions with foreign visitors. In the first instance, official international trade was import driven under the command system. Where industries had extreme difficulties in reaching targets it was expedient to allow imports to make up the materials balance deficit. In turn, where planning targets were easier to reach, surpluses could be exported. In a sense, the difficulty or ease of meeting targets acted as an indicator of comparative advantage, albeit one that was less transparent, less objective and less precise than market prices.

The planning process necessitated inflexibilities in international as well as national trading arrangements. Consequently, short-term swings in world prices could wreak havoc on planned trade. In order to preserve a balanced budget with respect to foreign exchange, State Trading Agencies often sought long-term contracts or counter-trade (barter) contracts. For the agrifood sector, the difficulty in adjusting short-term trade flows in the event of weather-related variations in harvests was a particular problem. International trade was conducted so that the value of trade flows balanced on a bilateral or pair-wise basis rather than a multilateral basis. State Trading Agencies also tended to favour the import of final goods rather than intermediate goods so that planning with regard to downstream industries would not be compromised in the event that sudden price changes required the reduction or cancellation of imports. While trading houses have arisen spontaneously to reduce transaction costs and facilitate trade between market economies, the State Trading Agencies of command economies are rightly associated with very high transaction costs because of the complexities of identifying exchanges in a non-market context coupled with the problem of budgeting foreign exchange.

Ironically, the economic difficulties posed by trading with modern market economies were typically smaller than those associated with trading with other centrally planned economies (Ethier, 1995). Since there was no supra-national planning authority to govern production and distribution across international boundaries between centrally planned economies, both countries had to identify a trade as beneficial and attend to the bilateral balance of trade just as they would do in the case of East–West trade. In addition, there was a significant pricing problem for trade between two command economies. In trade with modern market economies, planners had world prices as a natural and transparent benchmark for consummating trades. When one command economy traded with another, however, there was no such point of reference. The administrative prices of each good differed across countries and were clearly inappropriate for the actual conduct of international trade because they reflected accounting requirements rather than scarcity values. Consequently, and as a further irony, international exchanges within the Communist bloc were typically consummated at prices arrived at by adapting the average world prices that were observed in trades among Western countries. Nonetheless, certain key natural resources, especially oil, were made available to the CEECs on terms that were much more favourable than prevailing world prices.

Although transaction costs were particularly high for trade between the countries of the Communist bloc, this type of trade grew in importance. In part, this was due to lower transport costs arising from closer geographic proximity. As is often the case, however, political matters were decisive in the economic realm. The Cold War led to political restrictions and risks that worked against East–West trade. Meanwhile, the Council for Mutual Economic Assistance (CMEA) was formed in 1949 to facilitate trade between Communist countries. By the late 1950s, trade among the CMEA countries had risen from less than one-third to almost three-quarters of the members' total trade, but later East–West trade recovered to about 45% of total trade (Ethier, 1995). Most of the intra-CMEA trade involved the Soviet Union; there was, for example, relatively little trade among the CEECs (OECD, 1995).

The CEECs, particularly Poland and Hungary, accumulated large debts on Eurodollar markets in the late 1970s and 1980s in an attempt to finance their imports despite the philosophy of preserving a balanced budget by barter-based administrative trade. Like many less developed countries (LDCs), the CEECs began to borrow significantly in the period between the first and second major OPEC oil-price shocks (i.e. from 1975 to 1979) when inflation rates in the developed market economies were high and real interest rates were extremely low (Köves, 1992). Both the CEECs and LDCs were caught by the abrupt change in US monetary policy in 1979. A debt crisis was precipitated as real interest rates rose dramatically, the US dollar appreciated and the terms of trade deteriorated for many debtor countries as the developed market economies slipped into recession in the early 1980s. Countries such as Mexico and Poland, that were diverse in most other respects, found themselves on the verge of default. Throughout the 1980s, the trade of the CEECs was conducted against the backdrop of the enormous burden of foreign indebtedness.[4] The required austerity measures in the CEECs diminished both consumption and investment opportunities and culminated in the rising tide of political discontent. For Russia, the dismantling of the Soviet Union has also contributed to a significant debt problem because it agreed to assume the entire foreign debt of the Union in order to retain all former Soviet assets located abroad (OECD, 1995).[5] Throughout the region, the legacy of foreign debt remains a constraining factor in the transition to market-based economic systems.

In the remainder of this chapter, we focus on the import side of trade flows and trade policy for the CEECs and the NISs during the transition to market-based economies. Exports are examined in Chapter 14. While it is clear that imports and exports are intrinsically linked elements in a general equilibrium system, the overall picture is complicated by external borrowing and repayment. A nation's external budget consists of its current and capital accounts. Within the current account, credits in the form of exports of goods and services plus any net receipt of unilateral transfers such as grant-type foreign aid are weighed against debits in the form of imports of goods and services plus interest payments on accumulated foreign debt. Any deficit in the current account must be reflected in new international borrowing at the commercial or official (i.e. central bank or governmental) levels. Thus, the interest payments on foreign debt act as a constraining factor on living standards and domestic investment in the CEECs and the NISs. This has led some authors such as Köves (1992) to exaggerate the importance of foreign indebtedness as a source of the problems in the transition. The degree to which external indebtedness actually acts as a financial constraint on the economic transition depends largely on the stance towards foreign investment in general and direct foreign investment by transnational firms in particular. In the case of the CEECs and the NISs where domestic financial markets are underdeveloped (see Chapter 17), transnationals would typically act as a source of finance either from their own retained earnings or by borrowing on the financial markets of the modern market economies.[6] Thus, foreign direct investment in the CEECs and the NISs typically amounts to new market-driven borrowing with the transnational itself assuming the risk.[7] The issue of foreign direct investment is taken up in Chapter 15.

## 13.2. Competitiveness: Problem or Symptom?

Many firms in the CEECs and the NISs – be they smaller new firms or large ex-command firms – have felt overwhelmed by foreign competition. This has been particularly true for agribusinesses which were ill-equipped to deal with the short run consumer desertion to new, higher quality and fresher imported foods. Indeed, in many cases the legacy of the command system has worked against domestic firms and in favour of foreign exporters. There have been widespread problems on the marketing, finance and organizational fronts. Under the command system each command firm was tied to very few downstream buyers and upstream sellers. Wider business connections and marketing skills were unnecessary. Further, command firms did not have any incentive to improve product quality nor to maintain a good reputation among customers.

Financing has also been a problem in facing up to the foreign competition. On the one hand, the underdeveloped domestic financial system has made it difficult for new firms to obtain credit. On the other hand, ex-command firms have often been privatized with large debt overhangs. As we saw in Chapter 2, the accumulation of debt under the command system was largely arbitrary. Budget constraints were soft and firms were not typically required or even expected to pay back their debts. Nevertheless, in order to protect the integrity of the privatized financial institutions themselves, governments have often elected to carry forward the debts that accrued under the command system. As part of the process of transition, budget constraints are being hardened although the pace for such reform varies widely across countries. Thus, to an increasing extent, ex-command firms must struggle with their debts. Finance-related issues will be discussed further in Chapter 17.

Organizational difficulties have also hampered the efforts of many ex-command firms to compete with foreigners. Many privatized state enterprises have been inclined to minimize layoffs rather than pursue the efficiencies that would result in the maximization of profits. Such problems have arisen both where there has been a separation of ownership and control and where there has not. On the one hand, where privatization has created a highly dispersed pattern of ownership, as with voucher-style privatization, control has often remained with old-style management. On the other hand, there are situations such as the former state and collective farms in the NISs where official privatization has led to *de facto* collective ownership. In both types of situation, there has been a strong incentive for managers not to deal with the widespread problem of underemployment by laying off workers.

Lack of competitiveness, however, is not itself a problem. Rather, it is a symptom of the underlying problems relating to financing, marketing, quality of output and underemployment that we have just discussed. Over the long haul, foreign competition provides the impetus for domestic firms to deal with many of these underlying problems. Further, in Chapter 5 we saw that the problem of market power is a widespread legacy of the command system in the CEECs and the NISs. Consequently, the pro-competitive effects of a liberal international trade regime are likely to be extremely important in reducing the market power of sellers and buyers in the former command economies.

The economy-wide implications of import competition should not be overstated either. Undoubtedly, numerous firms in many sectors will fail in the course of the transition. Though foreign competition may be blamed, this is part of the natural process of adjusting to a market economy. Further, given the uneven debt loads that have been inherited, some well-managed firms will go bankrupt while other more poorly managed firms will remain solvent, at least temporarily. Nevertheless, the very success of exporters entering markets in the CEECs and the NISs produces general equilibrium effects through the balance of payments that ease the pressure on domestic firms. Import growth creates a shortage of foreign exchange that pushes towards the depreciation of the domestic currency. This in turn makes imports themselves less attractive and domestic goods more attractive.

## 13.3. The Import Side of a Post-transition Trade Policy

The CEECs and NISs, therefore, should be moving towards a new trade policy system that is characterized by a high degree of openness towards imports.[8] This does not necessarily mean that the goal should be one of completely unrestricted access for imports, however. Since the modern market economies do not pursue free trade, especially in agriculture, producer interests in the CEECs and NISs are likely to be able to either resist or reverse unilateral moves towards pure free trade. Further, it may be politically wise to maintain some vestiges of protection, say tariffs in the 5–10% range. Such tariffs can be removed as concessions in future global or regional trade liberalization.

The case against a high degree of protectionism is very strong. The CEECs and, with the exception of Russia, the NISs are small countries that are destined to have little pricing power on most world markets. For small countries with no market power, free trade is the optimal policy. Russia, however, is sufficiently large that it could influence the world market for some products through its trade policy. Indeed, the Soviet Union is known to have had a significant impact on world grain prices when it entered the world market following poor domestic harvests. The standard argument against a policy of national-welfare-maximizing tariffs is that a country will face retaliation that renders it worse off overall. This would clearly be the case for Russia. Since Russia is currently not a member of the GATT and WTO, it is not committed to lowering or at least maintaining trade barriers. Members of the WTO, however, would not be obliged to extend Most-Favoured-Nation (MFN) status to Russia.[9] Thus, Russia would be likely to face stiff retaliation if it became highly protectionist.[10] In the case of Russia, the pro-competitive benefits of a liberal trade regime give a second extremely compelling argument against protectionism. In Russia and the other NISs, even more than in the CEECs, there are pervasive bilateral mono-poly problems where both buyers and sellers have the potential to exercise significant degrees of market power. Imports can act as an important competitive force in reducing market power. Wisely, Russia has applied to join the WTO, but its initial tariff commitments remain in doubt.

The infant-industry argument could also be used by decision makers in the CEECs and the NISs to try to justify import-substituting policies. The infant-industry argument for protection claims that certain industries would develop a comparative advantage if only they were given assistance to get off the ground. In developing countries, imperfections in capital markets and technological and marketing externalities (so-called problems of appropriability) have, in the past, been used to argue for protectionist policies of import substitution. Similar arguments, however, can be used to mount a more cogent case for export promotion. On the practical side, the success of apparently export-promoting countries, such as Korea and Taiwan, has further eroded support among economists and policy makers for import substitution. Export promotion and import substitution are discussed more fully in the next chapter.

While low barriers to imports are crucial, there are three other important features of the liberal post-transition trade policy regime. First, tariffs, rather than non-tariff barriers (NTBs), should be the exclusive instrument of trade policy. The primary problem with quantitative measures such as import quotas or voluntary export restraints (VERs) agreed to by foreigners is that they are anti-competitive (see Helpman and Krugman, 1989). Given the pervasive problems of market power, quantitative restrictions on imports are particularly deleterious in the CEECs and NISs. To illustrate this point, suppose that there is a single domestic firm, but that the world market is competitive because there are many foreign firms. First consider a tariff. The domestic firm would trigger a flood of imports if the price is raised above the world price plus tariff. Thus, the domestic firm is effectively a price taker, albeit at a price higher than the world price. By contrast, with quantitative restrictions the domestic firm can raise prices with impunity. Since the volume of imports remains fixed no matter how high the price is raised, quantitative measures help preserve market power and prevent competition. Tariffs also have other secondary advantages over NTBs. In the first place, tariffs generate revenue for cash-strapped governments. Further, as economic growth occurs and demand rises, tariffs allow an increase in trade volumes while quantitative measures do not. Finally, quantitative measures almost inevitably lead to discrimination among exporting countries whereas tariffs do not.

The non-discrimination issue is important in its own right. Indeed, the second important additional feature of the post-transition trade policy is that tariffs should not discriminate across countries. This means that each CEEC or NIS should extend MFN status to all of the other CEECs and NISs at least on the basis of reciprocal concessions. Although it has been shown that the command system provided obstacles to the development of trade among the countries of the Communist bloc, it would be natural for intra-regional trade to become significant as the CEECs and the NISs become more market oriented. As a country such as Hungary joins the GATT and WTO, it is obliged to grant MFN status to all other members.[11] It need not, however, grant MFN status to other CEECs and NISs that are not yet GATT members. Indeed, Hungary intends to maintain higher tariffs against non-GATT members such as Russia and the Ukraine (OECD, 1995).[12] No doubt, there is pressure from domestic producers to maintain higher tariffs against non-GATT countries and, especially in the case of imports from Russia, political pressure as well. This pressure should be resisted since it results in trade erosion and/or trade diversion. Trade volumes with the other CEECs and NISs will be smaller and more distorted, and, if the discrimination is sufficiently pronounced, trade will be displaced to high-cost producers in the rest of the world.[13] For the CEECs and NISs whose economies are already in a precarious position, it is important to avoid the efficiency losses from a discriminatory trade regime.

The third important additional feature of the post-transition trade policy regime is that all commodities be subject to a uniform *ad valorem* tariff; there should not be discrimination across goods. Otherwise, the import-competing sub-sectors that are favoured with higher than average tariff rates will have an artificial incentive to become too large relative to those with lower than average tariff rates.[14] Although a low uniform tariff rate allows for limited protection to the import-competing sector as a whole, no sub-sectors are favoured or penalized. The fact that world price signals are still manifest under a uniform tariff is particularly important for the CEECs and the NISs as they grope for comparative advantages and disadvantages from an initial state of highly distorted prices and institutionalized state trading.

While small deviations from uniformity in nominal tariff rates may seem inconsequential, the resulting differences in effective protection can be much more dramatic (Corden, 1966). For example, suppose that both processed meats and live animals are tradable goods and that initially both are subject to the same nominal tariff rate of 7%.[15] In this case, the effective rate of protection on processed meats would be 7% since the tariff structure allows for an additional 7% in value added in the domestic industry compared to the world market. Now, by contrast, suppose that live animals enter tariff free, while processed meats are still subject to a 7% tariff. There is room for additional value added in the processing industry and the effective rate of protection is in excess of the apparent 7% nominal rate because the input is subject to a lower tariff. Or, consider the case where live animals are subject to a 20% nominal tariff rate. Since the input is subject to a higher tariff, it is more difficult to add value in the domestic processing industry. Thus, the effective rate of protection is less than the apparent 7% nominal rate and could even be negative. In order to utilize resources rationally at the various stages of production in the agrifood sector and throughout the economy, it is important to have uniform rates of effective protection that arise with a uniform nominal tariff structure.

Legislative commitments to a uniform tariff rate also have the significant advantage of limiting the possibilities for rent seeking. This is very important in the CEECs and the NISs given the history of non-arms-length connections between firm managers and government officials under the command system. Of course, temporary deviations from uniformity may be warranted in the event of terms of trade shocks. Where congestion arises in the infrastructure relating to training and relocation, there is a strong case for temporary support to declining industries (Leger and Gaisford, 1996).[16] For example, suppose that a sudden sharp decline in grain prices ultimately requires large-scale rural–urban migration and significant retraining of the migrants. Temporary support to grain production which is gradually phased out would avoid excessive congestion of vocational schools and other training facilities and allow an orderly move to lower grain outputs. Such temporary anti-surge or safeguard measures differ from countervailing duties and anti-dumping duties in that there is no presumption that foreign trade practices are unfair.

## 13.4. The Transition to a New Trade Policy: Theory

Of course, major microeconomic adjustments are not costless. Further, at the macroeconomic level the equilibration processes that would eventually eliminate balance of payments deficits and, especially, unemployment are likely to be long and painful. Thus, there does remain the question of how much foreign competition should be permitted and how soon. At what speed should the CEECs and the NISs move towards a new liberal trade regime on the import side such as the one described in the previous section? How should this trade policy transition be coordinated across sectors? As with many questions concerning the pace of transition, expert opinion in the trade area tends to be divided between those proposing a big bang or shock therapy, and those advising gradualism. On the one hand, shock therapists stress the simultaneity of many of the required reforms including trade liberalization and fear political and economic back-sliding if the transition is not rapid. On the other hand, gradualists point to the likelihood of massive economic disruption undercutting political support for the reform process.

With respect to reform in general and trade policy reform in particular, Harberger (1992) has argued for balance. He has suggested that one year is too rapid, ten years is too long, and five years is about right. Given concerns over food security, such a balanced approach is particularly important in the agrifood sector. On the other hand, where domestic agribusinesses cannot provide food security, open access to foreign food supplies may be the only sensible policy option. Once trade restrictions have been removed, even to rectify problems with food security, they should not be re-imposed. Fernandez and Rodrik (1991) have argued that there is considerable inertia with respect to trade policy reform. Even though the winners are able to compensate the losers, risk-averse individuals will tend to oppose reform when there is uncertainty over who the winners and losers will be. For this reason, resolving the uncertainty as quickly as possible will help consolidate political support for reform. On the other hand, if trade policy reform proceeds too rapidly, the microeconomic and macroeconomic costs of adjustment will rise dramatically and erode support for reform. In other words, shock therapy can produce additional losers by moving too fast.

In gauging the appropriate duration of time for phasing in trade policy reforms, it would, perhaps, be wise for the CEECs and the NISs to pay attention to the deeds done in the developed market economies as well as the words proffered by their experts. There are several important and related lessons in this regard. On the trade front, at least, shock therapy has certainly not been the chosen route when the developed market economies have altered their trade policies and entered into new institutional arrangements and agreements pertaining to the conduct of trade. Whether global or regional, virtually all trade liberalization efforts have been implemented gradually to ease the adjustment process. This is true of the various rounds of the GATT, it is true of the Canada–US

Free Trade Agreement and the North American Free Trade Agreement, and it is also true of each of the steps towards greater integration in western Europe. Where it has been anticipated that adjustment would be easy, scheduled trade liberalization has been rapid, but where adjustment difficulties were expected, the schedule for liberalization has been more protracted. Some of the more contentious elements of the Uruguay Round GATT agreement pertain to export subsidies on agricultural goods and trade in textiles and clothing which are to be phased in over ten years. Nonetheless, the timetables for liberalization for all goods are pre-set to facilitate orderly adjustment. Where unexpected difficulties have arisen, pre-arranged escape or safeguard clauses have provided temporary breathing space and where adjustment has proved easier than expected, countries have sometimes agreed to the accelerated implementation of cuts to trade barriers.

Given the large magnitude of the required adjustments to achieve a liberal trade policy regime in the CEECs and the NISs, it may be reasonable to argue for a phase-in period that is up to ten years in duration. While the credibility of the trade reform package would be called into question by a phase-in period that is too protracted (e.g. over ten years), an even more serious credibility problem would arise from moving too swiftly towards a liberalized trade regime and then being forced to temporarily but systematically retreat towards higher tariffs. Nevertheless, the move from direct quantity controls on imports inherited from the command system to equivalent tariffs should be accomplished as rapidly as possible. The substitution of new quantitative measures such as quotas, tariff-rate quotas and VERs should be strongly resisted for the reasons discussed in the previous section.

The transition in trade policy should begin with tariffication and then the tariffs themselves should be cut in accordance with a pre-announced and legislated schedule. This would provide a credible reduction in uncertainty and provide a more stable environment for investment. In Chapter 22, it is argued that membership in the GATT and WTO can also help make credible a country's commitment to trade liberalization. As in the case of agreements to liberalize trade among market economies, safeguard or escape clauses should also be legislated in order to deal with documented import surges that could not have been initially expected. While it must be acknowledged that the abuse of such safeguards could undermine credibility, the absence of any safeguards at all would make government commitments to liberalization even more dubious.

After the process of tariffication is complete, tariff cuts should proceed most slowly on commodities where initial tariffs are high and/or resources are relatively immobile since larger adjustment costs are to be expected in such instances. For example, tariffs on grains and vegetables might be eliminated rapidly, those on live animals and meats more slowly and those on fruits and wines more slowly still. Nevertheless, tariff cuts should proceed in all sectors simultaneously rather than sequentially. Harberger (1992) argues compellingly that sequential tariff cuts add substantially to adjustment costs. When one import-competing sub-sector experiences tariff cuts, its output will decline. As resources move out of the first import-competing sub-sector, outputs will expand elsewhere in the economy. In particular, there will be temporary expansions in the outputs of other import-competing sub-sectors that will be subject to tariff reduction and overall output contraction in the future.

## 13.5. The Transition to a New Trade Policy: Practice

The conduct of trade policy by the CEECs and the NISs on the import side has unfortunately, but understandably, diverged considerably from the principles laid out above. As in many other areas, the transition in the trade policy of the CEECs is more advanced than in the NISs. For countries such as Kazakhstan and the Ukraine, much of the trade with other NISs remains on a barter basis (OECD, 1995). Russia's ability to import is limited by the international debt load it assumed at the time of the dissolution of the Soviet Union. The overall import picture for the Ukraine and Belarus is clouded by mounting debts to Russia stemming from a chronic inability to pay for oil and gas imports at world prices (OECD, 1995).

At least in the CEECs, there now seem to be several discernible trends as well as a plethora of differences in detail. With the onset of liberalization, the CEECs and the Baltic countries moved towards convertible currencies and away from trade monopolized by State Trading Agencies. The latter were an integral part of the materials balance approach of central planning. Most of these countries began with very liberal import policies that contributed to a surge in imports. Food imports were very popular in the early days of liberalization. Having started with import policies that were, perhaps, too liberal, these countries have tended to drift towards much more protectionism. In a manner that is all too familiar in modern market economies, producer interest has come to be the key driving force setting import restrictions, particularly when domestic producers have come under pressure from strong import competition. Further, the timing of modifications in import policy have been erratic and, given the major economic adjustments, significant modifications to trade regimes have been alarmingly frequent. In late 1994, Hungary raised its average import duty on food products from 24% to 46% by adjusting tariffs not bound[17] by the GATT

(OECD, 1995), and, after an initial liberalization of agricultural trade in 1990, Polish protection has escalated markedly in subsequent years (OECD, 1995). Similarly, the Czech Republic has moved towards high tariffs on those imported agricultural commodities that compete with domestic producers (OECD, 1995). The three Baltic countries set their initial tariffs at very low rates but Latvia and especially Lithuania have recently imposed significant increases in tariffs on agricultural products and Estonia is under pressure to follow suit (OECD, 1995). Russia has also begun to introduce tariffs on food products in response to producer pressure (OECD, 1995).

As is frequently the case when producer interests dominate, import policy has often been extremely haphazard. While tariffs have been used extensively, NTBs have also been used regularly. The Ukraine, for example, uses import quotas as well as tariffs (OECD, 1995). On agricultural goods, some of the CEECs have instituted variable (or, sometimes, fixed) levies modelled after those of the European Union (EU).[18] On a temporary basis, Poland has applied variable levies to a wide variety of agricultural products, often in addition to tariffs (OECD, 1995).[19] The Czech Republic has also subjected many imported agricultural products to additional but non-variable levies (OECD, 1995).

Individual goods have been treated on a piecemeal basis with regard to the degree of protection as well as the types of trade restricting policies used. Although Bulgaria's average tariff rate was 18%, the average tariff on agricultural products was over 25% and 63% of agricultural imports were subject to a tariff of over 40% (OECD, 1994). Romania has increased its effective protection of many food products not only by increasing the tariffs and other border measures applicable to the products themselves, but also by reducing the tariffs on agricultural inputs (OECD, 1995). While Albanian tariffs on food items range from 5 to 30%, the tariffs on inputs such as fertilizer fall into a similar range (OECD, 1995). Thus, the rates of effective protection on farming activities may not be systematically higher than nominal rates even though they are certainly not uniform across goods.

On an overall basis, it would be difficult to avoid the conclusion that the chaotic conduct of trade policy in the CEECs and the NISs to date has exacerbated the underlying uncertainty, diminished the incentive to invest and contributed to the depth of the slump in economic activity. The one bright spot for the future is that all of the CEECs and almost all of the NISs have either joined the GATT and WTO or are in the process of doing so. By mid-1995, the Czech Republic, Hungary, Poland, Romania, the Slovak Republic and Slovenia had become GATT members. A year later, working groups were considering Albania, Bulgaria, Croatia, Macedonia and all of the NISs except Turkmenistan and Azerbaijan (WTO web-site, September 1996).

On the import side, as the Uruguay Round disciplines are phased in, the CEECs and the NISs will be forced to convert NTBs into equivalent tariffs. For example, Hungary completed its required tariffication by early 1995 (OECD, 1995). The degree to which the Uruguay Round will force the CEECs and eventually the NISs to gradually reduce tariffs is, however, necessarily limited:

> CEEC GATT members, like developing countries, were allowed to offer ceiling bindings essentially unrelated to base period conditions, rather than tariff bindings based on estimates of tariff equivalents of trade restrictions in the base period. Tariffs bound by CEECs were therefore bound at very high levels compared to previous actual tariff levels. These ceilings set limits to tariffs, but governments may decide not to take advantage of them to increase actual protection. However, it is likely that producer interests will use this 'water in the tariff' to press for policies that will take fuller advantage of the scope offered to increase protection.
>
> (OECD, 1995, p. 16)

Thus, in spite of the GATT, the door is open to the type of creeping producer-driven protectionism that is well established in the developed market economies themselves.

In addition to tariffication and gradual liberalization, a gradual move towards a uniform tariff structure was identified as an important desirable feature of trade policy towards imports. Unfortunately, there is virtually no evidence of a move towards uniform tariffs in practice. As a result, there will continue to be arbitrary differences in the degree of effective protection from world prices for different goods, and an important opportunity to use world prices to signal the efficient allocation of resources will be missed. The importance of GATT/WTO commitments and the limitations thereof is discussed further in Chapter 22.

## 13.6. Conclusions

In the first flush of liberalization, imports of long banned or severely restricted foodstuffs abounded. Packaged brand name products, exotic fruit and high quality items were most in evidence. Consumers, long denied any choice, were suddenly presented with a huge range of novel products. There was a strong movement away from low quality and boring domestically produced foodstuffs. Local agribusinesses and farms were unprepared for this demand shock and the process of transformation in the agrifood sector did not get off to a good start. The ready availability of appealing imported food products was the most obvious manifestation of the new economic order. Of course, reality set in quickly as inflation reduced real incomes and currency adjustments increased the prices of imported food. Consumers, for the most part, were forced to return to domestic produce.

Foreign foods, however, have not disappeared from the markets and retail shelves of the CEECs and the NISs. They may only be affordable to the emerging elite or on special occasions but they may still provide a positive externality in the transition process. One of the most powerful images of transition in Poland is neat rows of uniform shining oranges imported all the way from the US displayed next to bruised and dirty apples being sold out of the back of a truck from a former collective farm ten kilometers away. While the oranges were more costly, they were not excessively so, hence, questions were inevitably raised as to why oranges were so cheap in comparison to apples. For those with an entrepreneurial spirit, the oranges provided the spur to finding out how things could be done better. Individual consumers berated the stolid apple merchant on the shoddiness of his produce. It was a low cost (although maybe not in self-esteem) means of acquiring the information on consumer tastes and the quality of competitors' products. The transmission of this type of information is essential if the production orientation of the command era is to be replaced by the consumer orientation required in a market economy.

While the natural reaction of those running agribusinesses which must compete with superior quality imported food products may be to ask for protection, they will simultaneously begin to attempt to make their products more competitive with imports. There is a model product to emulate and an agrifood chain which can be studied. As suggested above, governments need to follow a balanced approach to trade policy. In particular, markets for imported foodstuffs should remain open – embargoes on imports should not be used – so that the transaction cost reducing demonstration effect of high quality imports is available to domestic agribusiness firms and potential entrepreneurs.

# Notes

1. With the exception of Russia, small market size will be an increasingly important issue for the CEECs and the NISs.
2. For example, some banks specialize in document collection procedures that require a foreign buyer to make payment before gaining access to the goods. Documentary letters of credit provide for a further reduction in commercial risk for the seller, since payment is insured even if the buyer refuses to take possession (Kerr and Perdikis, 1995).
3. Johnson (1953) and Dixit (1987) have examined the formal Nash equilibrium in tariffs for a stylized two-country, two-good world in which each country pursues a simple national-welfare-maximizing goal. Clearly, in the real world of international trade there are many countries trading many goods, pursuing many goals and using many policy instruments. Nevertheless, the central point still remains; in the real world with all its complexity, free trade is not an equilibrium.

4. Only Romania was successful in paying down its debt during the 1980s and Köves (1992) claims that this contributed to '... economic chaos and increasing political oppression' (p. 6).

5. In November 1994, Russia's debt was estimated to be US$112.7 billion.

6. Transnational firms do not always act as sources of finance. When a transnational firm from one modern market economy begins operations in another modern market economy, it often raises capital on the host country's financial markets.

7. In principle, the CEECs and NISs could resort to new official or governmental borrowing to finance investment. This would usually be unwise. Since the transition to a market-based economy naturally involves a narrower role for the state, such official borrowing should be seen as an extremely undesirable continuation of the old planning role. We argue further against state targeting of industries in Chapter 14.

8. A World Bank country study (1993) has expressed similar views on the key elements of post-transition and transitional trade policy to those expressed in the following two sections.

9. A country's MFN tariff on a particular good is the lowest tariff that country offers to any other country on that good. Under the GATT, a country is obliged to extend its MFN tariff to all GATT members. There are two caveats to this simple rule. First, when a country is a member of a preferential trading arrangement such as a free trade area, the goods of partner countries enter tariff free and the MFN tariff is the lowest tariff offered to any outside country. Second, under the rules of the GATT, developed countries are permitted to give a preference to goods from less developed countries (LDCs). For a developed country that grants such preferences, the MFN tariff is the lowest tariff offered to any country other than LDCs.

10. While Johnson (1953) and Kennan and Riezman (1988) have shown that a country with sufficient market power might be better off even after retaliation, it is extremely unlikely that Russia would wield that much control over world markets.

11. As of early 1995, the Czech Republic, Hungary, Poland, Romania and Slovakia had already become GATT members while applications for membership were at an advanced stage for Bulgaria, Estonia, Latvia and Lithuania (OECD, 1995).

12. Russia, in turn, has not routinely granted MFN status to the Baltic countries and the CEECs. Since Russia and Ukraine along with most of the other NISs have recently applied for GATT/WTO membership, these opportunities to deny MFN status may not be permanent.

13. Implementing higher tariffs against non-GATT countries is, in effect, the reverse of forming a preferential trading arrangement such as a customs union or free trade area. Since the formation of a preferential trade arrangement leads to trade creation and/or trade diversion, a move towards discriminatory tariffs results in the erosion of trade and/or trade diversion.

14. It should be acknowledged that uniform tariffs will not typically be second-best, let alone efficient, if the average tariff is to remain positive rather than equal to zero (i.e. one distortion is to remain). Nevertheless, the information requirements preclude solving the full second-best problem and appropriately targeting each import-competing sub-sector. It can also be noted that, if a large country used tariffs to maximize national welfare, it would set the tariff rate on each good equal to the inverse of the foreign export elasticity. Again, different goods would have different tariff rates. It has been shown that even in the case of Russia, which does possess the potential for some pricing power on world markets, the anti-competitive effects of national-welfare-maximizing tariffs on the domestic market in combination with foreign retaliation would be extremely detrimental.

15. If there are any other tradable inputs into meat processing, we assume that they are also subject to tariffs at the 7% rate.

16. Leger and Gaisford (1996) argue for temporary production subsidies rather than tariffs to avoid consumption-side distortions. At least during the initial phases of the transition in the CEECs and the NISs, however, government budget considerations are likely to dictate the use of tariffs.

17. Under the GATT, a tariff which is bound has a fixed rate which can never be raised in future. A bound tariff can, however, be unilaterally lowered. Once lowered it is

bound at the new rate. Bound tariffs are the norm in the GATT and a different tariff arrangement must be specially negotiated.

18. The variable levies of the EU's Common Agricultural Policy are adjusted in the wake of movements in the world prices so as to maintain stable domestic prices. These variable levies will be converted to tariffs under the Uruguay Round GATT Agreement.

19. These levies, however, will eventually be converted to tariff equivalents as part of Poland's GATT commitments.

# References

Corden, W.M. (1966) The structure of a tariff system and the effective protection rate. *Journal of Political Economy* 74, 221–237.

Dixit, A.K. (1987) Strategic aspects of trade policy. In: Bewley, T. (ed.) *Advances in Economic Theory, Proceedings of the 5th World Congress of the Econometrics Society*. Cambridge University Press, Cambridge UK, pp. 329–362.

Ethier, W.J. (1995) *Modern International Economics*. Norton, New York, 623 pp.

Fernandez, R. and Rodrik, D. (1991) Resistance to reform: status quo bias in the presence of individual-specific uncertainty. *American Economic Review* 81, 1146–1155.

Harberger, A. (1992) Strategies for the transition. In: Clague, C. and Rausser, G.C. (eds) *The Emergence of Market Economies in Eastern Europe*. Blackwell, Cambridge, Massachusetts, pp. 1–22.

Helpman E. and Krugman, P.R. (1989) *Trade Policy and Market Structure*. MIT Press, Cambridge, Massachusetts, 191 pp.

Husted, S. and Melvin, M. (1995) *International Economics*. 3rd edn, Harper Collins, New York, 544 pp.

Johnson, H.G. (1953) Optimum tariffs and retaliation. *Review of Economic Studies* 21, 142–153.

Kennan, J. and Riezman, R. (1988) Do big countries win tariff wars? *International Economic Review* 29, 81–85.

Kerr, W.A. and Perdikis, N. (1995) *The Economics of International Business*. Chapman and Hall, London, 274 pp.

Köves, A. (1992) *Central and East European Economies in Transition*. Westview Press, Boulder, Colorado, 150 pp.

Leger, L.A. and Gaisford, J.D. (1996) Costs of employment, terms of trade shocks and policy responses: efficiency versus equilibrium. The University of Calgary Economics Department Discussion Paper No. 96-02.

OECD (1994) *Agricultural Policies, Markets and Trade in the Central and Eastern European Countries, the New Independent States, Mongolia and China: Monitoring and Outlook 1994*. Centre for Co-operation with the Economies in Transition, Organization for Economic Co-operation and Development, Paris, 249 pp.

OECD (1995) *Agricultural Policies, Markets and Trade in the Central and Eastern European Countries, Selected New Independent States, Mongolia and China: Monitoring and Outlook 1995*. Centre for Co-operation with the Economies in Transition, Organization for Economic Co-operation and Development, Paris, 232 pp.

World Bank (1993) *Kazakhstan; the Transition to a Market Economy.* A World
    Bank Country Study. International Bank for Reconstruction and Develop-
    ment, Washington, 234 pp.

# 14 Entering the International Market: Realizing Export Potential

In the previous chapter, lack of competitiveness relative to foreign exporters was identified as a major issue for both new firms and ex-command firms in the Central and Eastern European Countries (CEECs) and New Independent States (NISs) of the former Soviet Union. Lack of competitiveness was, in turn, symptomatic of underlying problems on the marketing, product-quality, finance and organizational fronts. Precisely because of the lack of competitiveness, it was argued the CEECs and the NISs should move gradually towards a post-transition trade policy that allows a high degree of openness to imports. In order to take maximum advantages of the efficiencies and especially the pro-competitive effects of trade, there should be no non-tariff barriers (NTBs) imposed against imports, and tariff rates should be low (i.e. in the 5–10% range), uniform across goods and non-discriminatory across countries. The transition to such a liberal import policy should include tariffication and gradual tariff elimination. Further, commitments to scheduled liberalization both in domestic legislation and through membership in the General Agreement on Tariffs and Trade (GATT) and World Trade Organization (WTO) could help provide a more certain investment environment during the transition. Finally, during the transition and thereafter, safeguard or escape clauses should allow for additional temporary protection in the event of unanticipated terms of trade shocks.

The purpose of this chapter is to examine the export side of international trade. The difficulties that agribusinesses face as they attempt to export to foreign markets tend to be even more daunting than those that they face in their home markets. Particularly in regard to potential exports to the modern market economies, lack of marketing experience and poor product quality are major impediments to export success. Further, it is often argued that the modern market economies compete unfairly by erecting high barriers to food imports and/or promoting their own exports. Thus, the question arises as to whether the CEECs and the NISs should adopt policy measures in support of exports.

## 14.1. Unfair Foreign Competition?

It is certainly the case that the agricultural support policies of the modern market economies have resulted in highly distorted world markets. Support policies have affected input utilization and the product mix as well as the overall level of agricultural output within individual modern market economies. In many, but not all instances, the aggregate effect of this support has been to generate additional production and reduce world prices.[1] At the economy-wide level, the reduction in world prices generated by foreign agricultural support programmes has been unfortunate or even unfair for the CEECs and the NISs. Given the higher world prices that would prevail in the absence of foreign support, the CEECs and the NISs would have more domestic production and higher net exports (i.e. they would import less, switch from importer to exporter status or export more). In extreme cases, it is even possible for foreigners to support particular goods to such an extent that no domestic production would take place in the CEECs or the NISs.

Davies (1995) has shown that in the long run foreign support policies do not reduce the competitiveness of individual farms in the CEECs and the NISs. Rather, the lower world price leads to lower values for agricultural inputs such as land that are highly sector specific. This means that newly privatized land in the CEECs and NISs would have been worth more in the absence of foreign support policies, but it does not mean that domestic producers are at a competitive disadvantage. Lower implicit or explicit land rentals exactly compensate for lower world prices so that a producer on the margin of the industry just earns zero profit in long run equilibrium.

On the other side, the value of the support policies themselves becomes capitalized into higher land values in the modern market economy countries. For example, the high domestic prices that result from an export subsidy or the high producer price that follows from a production subsidy leads to a bidding up of land prices. Thus, there is a windfall gain that accrues to initial land owners. These higher land values imply higher explicit or implicit rental costs on land. Thus, in long run equilibrium, a prospective entrant to the industry in modern market economies would still earn zero profits.

Interestingly enough, foreign support policies for agricultural products are probably beneficial on an overall basis to some of the CEECs and NISs. Suppose that a country would import a particular good both at the undistorted world price and the lower world price that result from foreign support policies. In such a case, the country in question would benefit from the foreign support programme because the gain in consumer surplus would outweigh the loss in producer surplus. By contrast, consider another country that would export the good both at the undistorted world price and the lower world price that results from foreign support policies. This second country must be worse off because the loss in producer surplus would outweigh the gain in consumer surplus. Since Albania, Slovakia, Poland, Romania and Slovenia all had negative agricultural and food trade balances for at least two of the three years in the 1992–1994 period (OECD, 1995), there is circumstantial evidence that they may have gained from the agricultural support policies in the modern market economies at least in the short term. Of course, this evidence is far from overwhelming since foreign support is not uniform across all food products and since the CEECs' agricultural trade was not in equilibrium during this period. Further, complications arise because of the trade-distorting policies of the CEECs themselves.

In summary, the agricultural support policies of modern market economies do not place individual agricultural producers in the CEECs and NISs at a permanent competitive disadvantage. Such foreign support policies, however, do reduce the degree of comparative advantage of the CEECs and NISs in agricultural activities relative to non-agricultural activities. Nonetheless, the degree of comparative advantage in non-agricultural products relative to agricultural products increases. While the aggregate outputs of food products that receive foreign support will be smaller in and, in extreme cases, could be driven to zero, aggregate output will expand in other sectors. This analysis of export competitiveness suggests two important and related questions: (i) is it possible to make inferences concerning a country's export advantages; and (ii) is it possible to identify specific sectors or sub-sectors that should be the target of export-promoting policies? In the next two sections, answers are provided for each of these questions.

## 14.2. Identifying Comparative Advantage

It is useful to pursue the question of comparative advantage. There is considerable disagreement as to whether each of the CEECs and the NISs have a broad comparative advantage in agricultural or non-agricultural production. On the one hand, Liefert (1994) used a revealed comparative advantage analysis to claim that the former Soviet Union as a whole had a comparative disadvantage in grains and meats relative to heavy industry. On the other hand, Johnson (1992) has argued that the former Soviet Union has the potential to become a significant net exporter of grains, and Tangermann (1992) has suggested that the CEECs have the potential to become significant net exporters of many temperate crops. Both authors cite the reductions in consumption that are likely to accompany the removal of food subsidies as well as large productivity gains as underlying factors contributing to the availability of exports. Munk (1992) concedes that the CEECs might have a comparative advantage over the modern market economies in the absence of support policies in those countries but not in the presence of such policies.[2]

A country's degree of comparative advantage over the rest of the world in a particular good is given by the ratio of the autarky prices.[3] Since autarky is only a counterfactual state and does not actually exist even in the case of the CEECs and the NISs, empirical inferences are highly problematic at best. Even if it was possible to identify the degree of comparative advantage on a disaggregated basis, this information would be of limited usefulness in drawing conclusions concerning export patterns. Exports are only positively correlated with degrees of comparative advantage when there are three or more goods and two or more factors (Dixit and Norman, 1980). Suppose that a country's degree of comparative advantage over the rest of the world is highest in manufactures, lower in grains and lowest in meats. Such a country could well import manufactures as well as meats in return for grain.[4] While a country, on average, exports the goods for which it has the greater degrees of comparative advantage, this is not true in every instance.

The empirical problem of determining comparative advantage and drawing inferences for the trade pattern are compounded by the fact that comparative advantage is not static. There are dynamic aspects of comparative advantage that arise from technological innovation and technology transfer as well as factor accumulation and changes in relative demands that arise as incomes and income distributions change.[5] Differences in structural adjustments in the CEECs and the NISs clearly have the potential to affect the trade pattern in profound and hard-to-predict ways. The removal of subsidies and price distortions in the food sector will generate higher domestic food prices relative to other goods and reduced food consumption. As noted by Johnson (1992), this will tend to reduce food imports (including imports of raw materials for food processing due to the decline in derived demand) or enhance exports depending on whether the country was an importer or exporter to begin with. The economic collapse in the communist countries was related to severe incentive problems, particularly with regard to the development, adoption and diffusion of new technologies in non-military applications. Thus, acquiring and applying up-to-date technologies is also likely to have a profound impact on the pattern of trade in the CEECs and the NISs (see Johnson, 1992; Liefert, 1994).[6] Export patterns are complicated further by intra-industry trade considerations. For example, economies of scale and different product characteristics may be important in industries such as agricultural inputs and highly processed food products.

At least for the immediate future, unusually high transactions costs are likely to impair a country's export potential in general and, thereby, indirectly limit imports as well:

> Contract disputes have become a major problem in trade. Traders often have no experience and contracts often omit some important items such as specification of product quality. Disagreements over quality of items delivered may lead to breaches of contract. In addition contractors may not be able to fulfil agreements because production falls short of their expectations. In other cases, contractors may have found a better deal after the original commitment was made. This type of opportunistic behaviour is not unexpected in a society in the early phase of transition from a command to a market economy. Opportunism may be more common in international trade than in domestic trade, as foreign trading partners know . . . each other less well than domestic partners and adverse repercussions are smaller. The fear of breach of contract has become so strong that some farm managers are only prepared to deliver their produce if they have already received their trading partner's products. Transactions costs are high because of insufficient information on international prices, high uncertainty about trade policy, and macroeconomic developments, and delays by farms in meeting their financial obligations (since they often receive payment for their output many months delayed).
>
> (World Bank, 1994, p. 147)

While these comments were made in specific reference to the Ukraine, they have much broader applicability.

It would appear, therefore, that it is not possible to draw very firm inferences on a country's natural or free-trade export pattern on the basis of underlying fundamentals such as factor endowments. Of course, market forces are perfectly capable of determining a country's trade pattern. Nevertheless, the inability to determine a country's natural export pattern does suggest that there may also be difficulties in targeting sectors where export-promoting policies would be desirable.

## 14.3. Export Promotion: Theory

Individual firms in the agrifood sector will be competitive regardless of whether agrifood products receive support. Nevertheless, if a CEEC or

NIS directly or indirectly subsidizes its agricultural exports (or protects import-competing activities) the aggregate outputs of food products would rise. For this reason, export promotion serves producer interests. We can now turn to several possible rationales for export promotion by the CEECs and the NISs.

The infant industry argument is a core element of both import-substituting and export-promoting strategies of developing countries. The infant industry argument suggests that an industry would develop a comparative advantage if only it were given assistance to get off the ground. The infant industry could be promoted by a tariff, an export subsidy plus a tariff, or by a domestic subsidy on production, capital investment, etc. Since an export subsidy raises the (non-subsidized) domestic price above the world price, a tariff or non-tariff barrier is needed to prevent the re-entry of exported goods. A production subsidy is superior to a tariff as a means of promoting an infant industry because it does not introduce a consumption distortion (i.e. tax consumption). Imposing an export subsidy plus tariff may have an advantage over a tariff alone because it does not introduce an export barrier. That is to say, when exports are subsidized the domestic price that is earned by the firm does not suddenly fall to the world price as the industry goes through the transition from importing to exporting. For example, if wine is subject to a tariff, then it will be sold at the world price plus the tariff if there are no exports but it will be sold at the world price if there are any exports. Thus, to make the leap into export markets, the price obtained by producers has to fall by the full amount of the tariff.[7]

A practical problem with the infant industry argument is that it leaves an open avenue for rent-seeking activity where firms lobby for support from government. Thus, it is perhaps not surprising that many so-called infant industries never grow up. This problem seems to be especially pronounced with the import substitution strategy because of the export barriers that arise with import substitution. Whether because of the absence of export barriers or not, countries that have relied more on export promotion throughout the 1970s and 1980s, such as Korea, appear to have out-performed countries that have relied on import substitution, such as India.

There is a crucial theoretical dimension of the infant industry argument that is not obvious. Since the industry does not start or expand on its own, the expected present value of profits or net private benefits must be negative. In order for it to be socially desirable to have the industry start or expand, the expected present value of net social benefits must be non-negative. Thus, for a true infant industry, there must be some externality that causes the net social benefits to exceed the net private benefits.[8]

Recent applications of the infant industry argument in developing countries have centred on externalities arising from imperfections in capital markets and so-called appropriability problems involving the adaptation of technology and marketing (Krugman and Obsfeld, 1994). These types of market failure may also be plausible in the case of the CEECs and the NISs. If capital markets do not function properly and costly investments are required early on but revenues will materialize later, an industry may grow at less than the efficient rate, or not even start at all. Certainly, new smaller-scale farms and indeed new agribusinesses throughout the agrifood sector may face severe capital constraints. On the other hand, larger state and collective farms, whether restructured or not, as well as privatized state enterprises in the agrifood chain, often face soft budget constraints which make capital available on a preferential basis. Thus, imperfections in financial markets do not provide grounds for systematically subsidizing production or exports in agriculture. Further, the subsidization of individual firms' exports or production on the basis of accessibility to financing would entail very high information collection costs and open the door to frightening rent-seeking possibilities.

A particular sector of the agrifood industry may also grow too slowly if it is costly for the initial entrant(s) to adapt technology or marketing. Given the former isolation of the Communist bloc from the modern market economies, difficulties (appropriability externalities) of these types are also likely to run across sectors. Consequently, export or production subsidies to agriculture or any other sector cannot be justified or even rationalized on the basis of this argument.

Since the early 1980s a new strategic trade policy argument has been formulated for export subsidies (see Brander and Spencer, 1985 and Helpman and Krugman, 1989). An export subsidy can shift profits from foreign firm(s) to domestic firm(s) in a world oligopoly. Essentially, the export subsidy reduces the effective marginal cost of the domestic firm(s) allowing a credible commitment to higher output levels. There are a number of general caveats to this argument (see Grossman, 1986). The argument is weakened when firms are not wholly domestically owned, when there is a larger number of domestic firms, or when there are domestic consumers of the product in question. The argument is reversed or, in other words, becomes an argument for export taxes if the world oligopoly competes in prices rather than quantities.[9] In view of the need to target strategic industries, thereby opening the door for rent seeking activities, many would question the general usefulness of the strategic trade policy. In any event, the applicability of the oligopoly argument to agriculture at the level of farm products, is extremely dubious at best since farm production tends to be highly competitive, or would be competitive in the absence of government intervention. The strategic trade policy argument is even more precarious in the case of the CEECs and NISs. Especially in the NISs, agricultural production is still dominated by large former state and collective farms which may or may not have undergone some reorganization. In such instances, we have seen that a crucial source of gains from trade arises from increasing competition for such large firms. Subsidies on exports would weaken these competitive pressures just as would be the case with restrictions on imports (recall Chapter 13). The same argument is even more important for some of the large privatized processing firms which could have some international market power. Domestic gains from reducing their market power through competition are likely to outweigh any advantages garnered in international markets.

In the modern market economies, overall economic growth in the presence of low income elasticities for most food products coupled with technological progress in agriculture has produced a long term decline in the prices of agricultural versus non-agricultural goods. In response to the resulting rural–urban adjustments, governments have implemented a great many policies including export subsidies to try to support the incomes of their domestic farm sectors. There is no doubt that a great many of these policies have been misguided both in their design and duration. Clearly, US and EU export subsidies had very deleterious beggar-thy-neighbour effects on world grain markets in the recent past. Moreover, agricultural supports in general, and export subsidies in particular, have developed an extraordinary degree of inertia because the rents they generate have been capitalized into land values and other fixed assets such as storage facilities.

The current situation in the CEECs and the NISs is fundamentally different from that which has prevailed for decades in modern market economies since massive economy-wide adjustments are required in the former command economies. Thus, the case for favouring agricultural as opposed to non-agricultural exports is very weak. Further, the higher domestic prices that would result from subsidizing agricultural exports could have very unfortunate effects given the precarious nature of domestic food supply chains (see Chapter 5). It is also important for the CEECs and the NISs to avoid entrenching the rents from export subsidies for trade policy as well as domestic reasons. The GATT prohibits export subsidies on non-primary goods, and the provisions of the Uruguay Round will increasingly limit the use of export subsidies in agriculture.

## 14.4. Export Promotion Versus Export Restriction: Practice

In practice, there is considerable variation in export policies across the CEECs and the NISs. It is useful to consider the conduct of export activity under the command system and its immediate aftermath in order to understand the wide spectrum of export practices that now exists. In Chapter 13, we saw that trade activity was orchestrated and conducted by state trading agencies under the command system. Particularly in some of the NISs, state trading agencies or their successors have retained an active role at least on a temporary basis. Although much of the initiative for export activity quickly devolved to the enterprise level, the government still retained overall control by requiring export licences. In some of the CEECs and most of the NISs, export licences are still required for many goods including large numbers of agricultural products.

Where export restrictions remain, they should be phased out rapidly within a one to two year period. Export restrictions are especially counter-productive since, in addition to restricting trade, they reduce the amount of hard currency that countries obtain and make it more difficult to initiate and maintain the convertible currencies which are central to market-based reform. Further, as with any licensing controlled by bureaucrats, the allocation procedure for those exports that are allowed is susceptible to corruption. Within the gamut of export restrictions, export taxes are preferable to licences for many of the same reasons that tariffs were shown to be preferable to non-tariff barriers in the previous chapter. Export taxes do not discriminate across firms or countries and they reduce the opportunities for non-productive rent seeking. Further, export taxes eliminate bureaucratic vicissitudes concerning the granting of licences and, thereby, provide a more certain trading environment in which exporting firms can develop better reputations. Finally, export taxes are not at odds with economic growth since export volumes may rise as the economic expansion occurs. Thus, the conversion of remaining export licences to export taxes should be an immediate priority for the CEECs and the NISs (see World Bank, 1993).

In the agrifood sector, CEEC policy orientation ranges from export-promoting measures modelled after the EU to highly restrictive policies that are reminiscent of the command era. Even within a single country, there is often a wide range of export promoting and restricting policies. Hungary has introduced export subsidies on some grains and grain products and Slovakia has also made some use of export subsidies (OECD, 1995). The general direction of Czech trade policy with respect to exports remains ambiguous. Some commodities such as milk products have been granted export subsidies, while other products have required non-automatic export licences so that the supply of food for domestic consumption can be protected if the authorities deem it necessary (OECD, 1995). At the other end of the spectrum, key elements of the command system's food philosophy have resurfaced in new forms in Bulgaria. In spite of significant tariffs on many food imports, Bulgaria has attempted to maintain low food prices for consumers by taxing or prohibiting the export of numerous food products (OECD, 1995). In 1991 and 1992, Albania prohibited most food exports to deal with a domestic food crisis (OECD, 1995).[10]

Much of the old command-economy trade structure remains in place in many of the NISs. In Belarus, most food exports continue to be arranged on a bilateral basis. Further, the re-imposition of price controls on many food items has led to unintended black-market exports which have added to domestic shortages (OECD, 1995). In Kazakhstan, state trading agencies retain monopoly trading rights over important agricultural products such as grain and cotton (OECD, 1995). In the Ukraine, non-monetary barter transactions have been on the increase in recent years. Further, many exports remain under direct state control via quotas and licences that ensure that the government will capture the hard currency (World Bank, 1994; OECD, 1995). Licences for Ukranian grain exports remained in place until 1996.

## 14.5. Conclusions

It should be clear that trade policies for the agrifood sectors of the CEECs and the NISs are inconsistent, reactive to the pleadings of special interests and still largely controlled by the heavy hand of government. This is true both across countries and across sectors within countries. Hence, it resembles the agricultural trade policies of modern market economies. Given that market prices are seldom, it seems, allowed to guide the allocation of resources to and within the agrifood systems of modern market economies, it is often difficult to argue that the CEECs and the NISs should put their faith in markets. This is particularly true when international prices are badly distorted by the subsidy policies of modern market economies and the prime target market for the CEECs and the NISs, the European Union, is heavily protected.

The fundamental problem, however, for the agribusiness sector of the CEECs and the NISs is to improve their efficiency. Meeting foreign competition, both in the domestic market and in export markets, is extremely important to this process. Further, while international prices of agricultural products are significantly distorted, they will remain distorted for a long time. The discipline imposed on agricultural policy making by the Uruguay Round has long phase-in periods and, in any case, only rolls back most subsidies and is far from eliminating them. Hence, these distorted prices represent the international trade environment which will be faced by agribusiness firms in the CEECs and the NISs in the foreseeable future. The important point to emphasize is that, ultimately, the asset value of agricultural land will reflect the underlying efficiency of the agrifood system. Eventually, all policy benefits will be capitalized into this asset value. Since well-specified property rights do not yet exist and the agribusiness sector is still far from equilibrium, asset values in land remain unclear. Hence, there is an ideal opportunity to accept world prices without initiating trade policy distortions because a vested interest in land does not yet exist. As world prices are, for the most part, distorted in ways which might be considered detrimental to the CEECs and the NISs if markets and property rights were fully developed, by following the least distortionary trade policy which is politically acceptable, the CEECs and the NISs should be in an excellent position to take advantage of improving prices over the long run.

The resources which might have been used in implementing formal trade policies, such as export subsidies, will be much better spent in reducing the transaction costs faced by agribusiness firms attempting to deal with the international market. Lowering information costs, including price information, product information and the costs of education related to learning how to conduct international business should have a high priority.

# Notes

1. For example, domestic subsidies on agricultural production or input utilization and export subsidies increase output levels and drive down prices on world markets. In the case of domestic subsidies, domestic consumers also pay lower prices, but the reverse is true in the case of export subsidies where domestic consumption is not subsidized. By contrast, supports based on production quotas lower production and raise domestic and world prices.
2. Oher studies relating to future trade patterns include Hayes *et al.* (1995) and Duchene and Sanik-Leygonie (1993).
3. Autarky is when no trade takes place.
4. At autarky prices, which are the basis for determining comparative advantage, manufactures would be cheaper relative to grains at home than in the rest of the world and grains would be cheaper relative to meats. Under free trade, however, the factor costs would be different. In fact, if there are no transport costs and all three goods are produced at home and in the rest of the world, the costs of each good must be the same in both locations under free trade. Suppose that the move to free trade causes rental prices on factors used intensively in manufactures as well as meats to rise relative to factors used intensively in grains. As a result, the production of manufactures could be reduced sufficiently to cause some importation.
5. There is a growing realization that endogenous growth, propelled by endogenous research and development and endogenous technology transfer, is very important in a trade context (see Grossman and Helpman, 1991).
6. The conundrums associated with the acquisition of modern but foreign-developed technology through various forms of foreign direct investment are discussed further in Chapter 15.
7. It is noteworthy that if both export subsidies and tariffs were set at a common *ad valorem* rate, then all domestic prices would be pushed above world prices by the

same percentage. Thus, all domestic price ratios would be equal to the corresponding world price ratios and there would be no distortions arising from trade policy. With balanced trade, no revenue would be raised overall by the system of uniform tariffs and export subsidies. Thus, free trade has the advantage of generating no administrative costs.

8. For example, orchards provide a positive externality for apiaries since blossoms provide pollen for bees. (As it happens, apiaries also provide a positive externality for orchards since bees pollinate orchards.) Thus, the net social benefit of fruit production exceeds the net private benefit or profit because of the external benefit provided to honey production.

9. An oligopoly that competes in prices (e.g. the firms must make commitments to prices) is called a Bertrand oligopoly, while an oligopoly that competes in quantities (e.g. the firms must make production commitments) is called a Cournot oligopoly (see Tirole, 1988; Helpman and Krugman, 1989; Gravelle and Rees, 1992).

10. In 1992, Albania also received large shipments of food aid to help weather the crisis.

# References

Brander, J.A. and Spencer, B.J. (1985) Export subsidies and international market share rivalry. *Journal of International Economics* 18, 83–100.

Davies, A.S. (1995) Integrating the agricultural sectors of the European Union and the transforming economies. Paper presented at a conference entitled: Agricultural Harmonization Between the European Union and the Central and Eastern European Countries, University of Nitra, Nitra, Slovakia, May.

Dixit, A.K. and Norman, V. (1980) *Theory of International Trade.* Cambridge University Press, Digswell, UK, 339 pp.

Duchene, G. and Sanik-Leygonie, C. (1993) Foreign trade liberalization and redeployment in the former USSR. *European Economy* 49, 135–152.

Gravelle, H. and Rees, R. (1992) *Microeconomics.* 2nd edn. Longman, London, 752 pp.

Grossman, G.M. (1986) Strategic export promotion: a critique. In: Krugman, P.R. (ed.) *Strategic Trade Policy and the New International Economics.* MIT Press, Cambridge, Massachusetts, pp. 47–68.

Grossman, G.M. and Helpman, E. (1991) *Innovation and Growth in the Global Economy.* MIT Press, Cambridge, Massachusetts, 359 pp.

Hayes, D.J., Kumi, A. and Johnson, S.R. (1995) Trade impacts of Soviet reform: a Heckscher-Ohlin-Vanek approach. *Review of Agricultural Economics* 17, 131–145.

Helpman E. and Krugman, P.R. (1989) *Trade Policy and Market Structure.* MIT Press, Cambridge, Massachusetts, 191 pp.

Johnson, D.G. (1992) World trade effects of the dismantling of socialized agriculture in the former Soviet Union. Paper presented to the International Agricultural Trade Research Consortium, St Petersburg, Florida, December.

Krugman P.R. and Obsfeld, M. (1994) *International Economics.* 3rd edn. Harper Collins, New York, 795 pp.

Liefert, W.M. (1994) Economic reform and comparative advantage in the newly independent states. *American Journal of Agricultural Economics* 76, 636–640.

Munk, K.J. (1992) The development of agricultural policies and trade relations in response to the transformation in Central and Eastern Europe. Paper presented to the International Agricultural Trade Research Consortium, St Petersburg, Florida, December.

OECD (1995) *Agricultural Policies, Markets and Trade in the Central and Eastern European Countries, Selected New Independent States, Mongolia and China: Monitoring and Outlook 1995.* Centre for Co-operation with the Economies in Transition, Organization for Economic Co-operation and Development, Paris, 232 pp.

Tangermann, S. (1992) Central and Eastern European reform, issues and implications for world agricultural markets. Paper presented to the International Agricultural Trade Research Consortium, St Petersburg, Florida, December.

Tirole, J. (1988) *The Theory of Industrial Organization.* MIT Press, Cambridge, Massachusetts, 479 pp.

World Bank (1993) *Kazakhstan; the Transition to a Market Economy.* A World Bank Country Study. International Bank for Reconstruction and Development, Washington DC, 234 pp.

World Bank (1994) *Ukraine; the Agricultural Sector in Transition.* A World Bank Country Study. International Bank for Reconstruction and Development, Washington DC, 148 pp.

# 15 Direct Foreign Investment: Handling Agribusiness Transnationals

This chapter examines how the Central and Eastern European Countries (CEECs) and the New Independent States (NISs) have dealt with the difficult problem of agribusiness transnationals. Well financed and highly skilled transnational agribusiness firms have tended to regard the markets emerging in the former command economies as having huge potential. While these firms realize there are difficulties in doing business in these markets during the transition, many have taken a long term view and have attempted to position themselves to take advantage of future growth. Many local entrepreneurs and managers in the former command economies, however, perceive agribusiness transnationals as a threat, a threat that enjoys considerable competitive advantages. Governments have also been wary of allowing transnational agribusiness firms to gain a strong position in such a sensitive sector of the economy. Hence, how to define the role of agribusiness transnationals has proved to be a conundrum for the economies in transition.

## 15.1. The Conundrum

The question of the role which foreign agribusiness firms should play in the agrifood sectors of the former command economies has proved to be one of the most vexing problems for policy makers in those countries. Policy makers are faced with attempting to satisfy what appear to be fundamentally conflicting industrial policy goals. On the one hand, there is the goal of creating a modern, domestically owned and operated commercial industrial sector (of which agribusiness is a part) to ensure economic independence. On the other hand, there is the goal of acquiring foreign technology and expertise to increase the rate of economic growth. The need to acquire up-to-date technology and expertise is particularly important in the agricultural and food sectors due to concerns over food security. In particular, expertise from modern market economies is needed to establish an effective system of vertical coordination based on markets to replace the visible hand of the state planner.

The conflict between the goal of establishing a successful domestically owned industry and the need to acquire foreign technology and expertise arises because the most efficient, and sometimes the only, means of ensuring the acquisition of up-to-date technology and managerial expertise is through direct investment by transnational agribusiness (Offerdal, 1992). While there are other means of acquiring up-to-date technology such as the purchase of rights to the use of patents and licensing arrangements, in the absence of strong legal protection for industrial processes and intellectual property, these alternatives are unlikely to be available. While the former command economies have enacted legislation to protect intellectual property, enforcement is at best weak and more often nonexistent (Camdessus, 1992).

Strong, well enforced patent and intellectual property laws allow the firm to sell the rights and thus capture the gains from research and development. When intellectual property protection is weak, once the process moves outside the direct control of the firm, the potential for diffusion without compensation arises. Hence, individual foreign agribusiness may rationally refuse to sell or license their products and processes. In short, the licence value of a process in the CEECs and the NISs will be less than the profits obtainable from self-operation and self-diffusion of the process.

Developing a modern agrifood sector is an ongoing process. Often, officials in the former command economies appear to act as if they believe that modernization can be accomplished with a one-time transfer of technology. In reality, there are continuous search costs for firms not engaged directly in research and development in the identification and evaluation of technology developed by others. A once and for all technology transfer is much less important than the relatively continuous acquisition of improvements in existing products and, in particular, processes (Archibald and Rosenbluth, 1978).

In addition, a significant proportion of new technology originates in private corporations. As it is unlikely that agribusiness transnationals will undertake a significant amount of research within the former command economies, research and development activities can be expected to remain concentrated in modern market economies. Of course, the situation in the former command economies differs from that in developing countries because there is a large cadre of well educated scientists

and engineers. It is, however, very difficult for foreign agribusiness firms
to tap into this group on a large scale due to the poor enforcement of
intellectual property rights. Transnational agribusiness firms are more
likely to hire the best and the brightest of the existing research cohort
and move them to their research facilities in modern market economies.
Officials in the former command economies have expressed concern
regarding this brain drain. In the absence of the resources to offer competitive compensation for scientists and engineers, the emigration of
highly qualified individuals will continue. In a few cases, scientists
have been able to form private research groups and successfully bid on
research sub-contracts let by foreign firms, but these remain the exception. Agricultural research establishments dating from the command era
have also attempted to bid for research contracts but their rapidly ageing
equipment and poorly maintained facilities have, for the most part,
made them uncompetitive. In particular, the almost total breakdown in
the supply of spare parts for sophisticated equipment has meant that
they have often not been able to fulfil the research commitments of the
contracts they are able to secure.

As agricultural and agribusiness research will continue to take place
in the modern market economies, a conduit for ongoing transfers of
technology is required. Given the reluctance of foreign firms to transfer
technology directly to firms in the former command economies, it may
only be feasible to allow the transfer to occur through intra-firm channels which terminate in foreign corporate outlets within the country
rather than to restrict technology transfers to firm-to-firm market transactions across international borders.

In theory, transfers of technology between the divisions of a transnational corporation have two positive impacts for the receiving country:
(i) the direct modernization of the affected industries which should improve
international competitiveness; and (ii) the externalities associated with
the demonstration effect and improvements to human capital. Models of
technological transfer between local subsidiaries of transnational firms
and local firms in the host country have been formally developed by
Findlay (1978) and Wong and Blomstrom (1992). Das (1987) suggests that
technology transfers by transnational firms create externalities which are
a benefit to the host country as a whole. According to Marton:

> the most important channel of technology transfer has been through
> foreign direct investment by multinational enterprises based in developed
> countries. The growth of corporate multinationalization has been
> accompanied by technological flows from parent companies to their
> foreign subsidiaries and affiliates.
>
> (Marton, 1986, p. 412)

Additional support for the view that transnational firms are a conduit for technology is provided by Vernon (1971, 1975, 1977), Hymer (1960) and Streeten (1977).

Such benefits are not costless. Transnational agribusiness firms must be allowed a chance at profitable operation in order to induce them to locate in a liberalizing command economy. This implies that, at any given time, the opportunities for private domestic enterprises are constrained by foreign competition in those sectors where transnational corporations choose to operate. In many cases, it means that domestic firms are forced to compete on an unequal basis with international agribusiness firms. This inequality is based on the international agribusiness firm's superior access to information and human capital and, possibly, on its ability to acquire market power. As a result, the development of the indigenous domestic industry is retarded (Barnett and Muller, 1974; Amin, 1976; Biersteker, 1976; Hymer, 1979). In addition, given their superior management, their access to capital markets and their ability to follow relatively unrestricted employment policies, agribusiness transnationals should be more than a match for existing large-scale enterprises controlled by agroindustrial complexes in the transitional economies.

Market opportunities are not static, but rather ever changing and evolving. Given opportunities for profit and growth, international agribusiness firms may well expand and eventually dominate the privatized agribusiness sectors of an economy in transition. As a result, the goal of ensuring economic independence through the creation of a domestically owned and operated agribusiness sector will be frustrated. On the other hand, large privatized, domestic agroindustrial complexes may have sufficient market power (and/or political influence) to compete successfully with international agribusiness firms (Gady and Peyton, 1992). If the competition is too strong, and particularly if it is perceived as being unfair, transnationals will be eliminated or forced to withdraw from the market. Their absence, however, reduces the flow of technology and, as a result, beneficial externalities and international competitiveness will be reduced. The goal of modernizing the agrifood sector will be frustrated.

The relative competitiveness of domestic and transnational agribusiness firms can be influenced by the judicial and policy regimes which define the business parameters in the agrifood sector as well as in the economy as a whole. Those responsible for shepherding the liberalization process must attempt to develop and implement policies that provide a stable balance between profitable transnational agribusiness firms and a viable domestically owned industry. Such a balance may prove illusive.

In some instances, the agrifood sector may gravitate towards either: (i) total domination of an industry by transnational agribusiness firms; or (ii) the total disappearance of foreign agribusiness firms. If either of these cases, with their politically undesirable side effects (i.e. unacceptable levels of foreign control or absence of modern technology) are being approached, policy makers may be forced to introduce radical policy measures to prevent the evolution to an extreme situation. There may even be radical changes in government itself and, subsequently, dramatic changes in policy (Kerr, 1993).

Frequent or radical changes in policy are likely to slow the process of modernizing the agrifood sector in the former command economies. Only in a stable environment will agribusiness firms (both foreign and domestic) make productive (as opposed to speculative) and longer term investments. As the process of liberalization progresses, obtaining an acceptable balance between international agribusiness firms and domestic firms may become the central problem for policy makers. As yet, other problems have had a greater priority. The high transaction costs and considerable risks which have characterized the agrifood sector, as well as the wider economy, have made it difficult for foreign agribusiness to operate successfully. This has tended to keep profits low for both domestic and foreign firms and, as a result, foreign investment has not yet taken place on a large scale. Still, there are examples of international agribusiness firms that have been acting strategically to position themselves to be able to dominate industries, or at least to act as regional monopolies, as long term profitable opportunities become available. They are willing to forego current profits for the prospect of long term gain.

Of course, the discussion above is very simplistic because there exist alternative arrangements to wholly owned subsidiaries of foreign agribusiness firms and domestically owned and operated firms. As pointed out in Chapter 9, these arrangements include partnerships between foreign and domestic firms, joint ventures, franchise agreements, licensing agreements and strategic alliances. These arrangements, however, have often led to franchiser–franchisee, licenser–licensee or intra-firm jockeying for strategic advantage which is a microcosm of the larger foreign firm/domestic firm competition which takes place in the marketplace. Failures among these intermediate forms of organization is high, partly because of the absence of an effective legal system and partly because business ethics have yet to develop fully in the transition economies. Foreign firms complain of opportunistic behaviour by entrepreneurs in the former command economies, while domestic businesspeople complain about excessive rigidity, lack of trust and pervasive secrecy in their dealings with foreign agribusiness firms. Foreign agribusiness firms have also been forced to make difficult decisions regarding whether to acquire existing agribusiness enterprises or to make greenfield investments. Both have pitfalls associated with them.

## 15.2. Government Policies and International Agribusiness Strategies

For the most part, if a balance between sufficient foreign investment to produce a modern agribusiness sector and a modern domestically owned sector is desired, then the balance has not yet been achieved. In most former command economies there is too little foreign investment in the agrifood sector to provide the impetus for modernization. On the other hand, privatized local industry has probably not yet gained an unassailable competitive position. The same conditions which deter privatization may also deter direct foreign investment in the agrifood sector. Hence, the primary problem is one of creating conditions whereby any form of private agribusiness can prosper. There does seem to be considerable evidence that, if the conditions for the success of privatized domestic firms are created, then international agribusiness will be willing to invest.

Problems with privatization, and hence lack of foreign investment, have been greater in the agrifood sector than in many other sectors due to the reluctance of governments to liberalize food prices. While there may be many good reasons for this reluctance – to prevent the increase of monopoly power, to keep food prices low as part of broader social welfare goals and to provide food security – the effect has been to make many existing agricultural input enterprises and food processing facilities unprofitable.

Further, as existing food enterprises have been lumbered with debts incurred (and which were never expected to be repaid) in the command era, they are unappealing for both domestic and foreign investors. In an attempt to foster strong domestically owned industries, the governments of the CEECs and the NISs have intervened to limit foreign ownership of existing privatized enterprises or to limit foreign decision-making discretion in other ways. As a result, foreign investors have not been forthcoming. For example, in its study of agricultural reform in Armenia, the World Bank commented:

> Experience in Central and East European countries indicates that foreign participation in privatization of food processing and distribution could substantially speed up the creation of a competitive sector. However, foreign investors in general require majority ownership or at least a decisive role in the management of operations and much more freedom in marketing and currency transactions then those which Armenia can offer at present.
>
> (World Bank, 1995, p. 53)

Transnational corporations have identified a large number of barriers to investing in the former command economies. These include slowly developing business infrastructure, difficulties in recruiting well trained local staff and enforcement of commercial law. The major constraints identified in an OECD survey of existing foreign investors as well as firms who had considered investing, but had decided not to pursue it, suggested that:

> Central bureaucratic, administrative and legislative issues were the main external constraints, although their existence impeded the investment process rather than prevented it outright. Protracted and complex negotiations on approval procedures are now the standard irritant expected by seasoned investors. Frequent changes of government officials and difficulties in finding decision makers who would accept responsibility are a common complaint. Moreover, inconsistent views and policy changes further impede the investment process. Corporations mentioned that they received conflicting information from different ministries. Government strategies were often ill-defined and ambiguous and lacked a determined, long term strategy.
>
> (OECD, 1994a, p. 11)

As governments in the former command economies have not found a means to solve the dual problem of acquiring foreign technology and expertise to facilitate the modernization process and yet foster a strong domestic industry, bureaucrats at all levels are faced with inconsistent or opaque policies regarding foreign investment. As a result, they tend either to procrastinate or act on their own initiative. In the latter case, the current resolution of the investment question may be entirely different to that provided by the next bureaucrat whom potential foreign investors encounter. Often, after being wined, dined and apparently courted by officials in the former command economies, foreign investors find themselves confronted by a vast amount of red tape when they attempt to actually initiate their investment. According to Fischer:

> The potential role of foreigners has been a matter of concern in all the formerly socialist economies. Countries want the benefits of foreign expertise and foreign finance, but they are concerned that, in the absence of domestic sources of finance, foreigners will acquire a larger part of industry at fire sale prices. Accordingly, at the same time as countries seek foreign expertise in the form of technical assistance or management contracts, they make provisions to control the share of foreign ownership.
>
> (Fischer, 1992, p. 237)

For example, almost all of the former command economies have attempted to provide tax incentives or tax holidays for foreign firms as a part of an official policy to encourage foreign investors. According to the OECD:

Incentives offered in the region are often tied to foreign investment. These can be in the form of special tax holidays under the income tax or special reliefs from customs, duties or turnover taxes. The incentives are sometimes directed at firms that are 100 percent owned by foreigners and at other times offered to joint ventures, often with as little as 30 percent foreign ownership.

(OECD, 1995a, p. 47)

Hence, while incentives for foreign investment appear to exist:

The concern raised most frequently by the private sector was its ability to predict the tax consequences of its investment and other decisions. The issue is particularly important for the long term, capital intensive investments that most governments in the regions are attempting to attract. In most countries, there were common concerns expressed about a number of problems which made it difficult for firms to predict the tax outcomes of their actions.

(OECD, 1995a, p. 22)

These difficulties included: (i) vague and imprecise tax laws; and (ii) frequent changes to laws and regulations.

The general result of these mixed signals, however, has been that the levels of foreign direct investment have remained small in most countries and far less than the levels desired (United Nations, 1992). Investment by foreign agribusiness remains small and limited in the scope of activities (Henry and Voltaire, 1995). As a result, modernization and technical transfer are at rates which, as yet, have had little impact on the agrifood sector.

Hungary, however, provides an example of the dangers which may arise when few or no controls are put on foreign direct investment. According to the OECD:

There are no restrictions applicable specifically to foreign investment. Foreign investors are permitted to own 100 percent of the shares of a Hungarian company and, since 1991, no special permission is needed for majority foreign ownership. Companies with foreign participation are subject to the same general controls as apply to all Hungarian businesses.

(OECD, 1995a, p. 90)

Relative to other command economies, this openness has led to a considerable inflow of foreign direct investment. Hungary was the recipient of more than a quarter of all investment in the former command economies including the former Soviet Union. Foreign investment in the Hungarian economy amounted to over 11% of GDP, compared to 6.7% in Czechoslovakia and 1.3% for Poland (OECD, 1995a). Investments in Hungary also appears to be on a larger scale than in other former command economies.

In agribusiness, this has led to the strategic purchase of key sectors of the Hungarian industry. The sectors targeted were those where a considerable degree of market concentration was possible and included vegetable oils, sugar, distilling, brewing, confectionery and tobacco and comprised approximately 30% of food industry output (OECD, 1994b). On the other hand, little foreign investment has taken place in the widely diffused meat, dairy and grain milling sectors.

Targeted foreign direct investment, for example, led to the purchase of the entire Hungarian vegetable oil industry by a single transnational. A private monopoly was created which also has monopsony power over local oilseed supplies (OECD, 1994b). All three state owned confectionery companies were privatized with a large degree of foreign ownership, one firm was purchased outright. The four existing tobacco manufacturing plants were purchased by foreign concerns. Two of the five major breweries were acquired with majority foreign interest. The only two (of nine) refrigeration firms to be privatized were purchased by foreign interests. Three of the five distilling companies privatized are joint ventures with majority foreign ownership.

As a result, the OECD concluded that:

> Foreign interest has been greatest in the less traditional industries. The vegetable oil industry is totally foreign owned, as is tobacco. While the need for foreign capital and expertise and to retain some scale economies is not in doubt, the high degree of privatization achieved may have been at the expense of a competitive industry.
>
> (OECD, 1994b, p. 14)

The control over key sectors of the Hungarian agrifood industry gained through strategic purchases by transnational corporations has led to concerns regarding future privatizations in agribusiness. While there has not been a major policy shift regarding foreign direct investment:

> The high share of ownership by foreign multinationals in the privatized food industry has created pressure to favour greater Hungarian ownership in the firms still to be privatized.
>
> (OECD, 1994b, p. 14)

Strategic investment by transnational corporations has not, however, been confined to Hungary. In the Czech Republic:

> There was some direct foreign capital investment when privatizing some of the larger Czech enterprises with monopolistic market positions. The three most important examples of this were the tobacco industry (Phillip Morris), Ceske Cokoladovney, the largest confectionery enterprise (privatized with the participation of BSN and Nestle), and one of the four largest vegetable oil factories, which was sold outright to Unilever.
>
> (OECD, 1995b)

A similar pattern of foreign agribusiness investment, but with smaller overall industry shares, is observed in Poland. The inability of individual firms to obtain a monopoly position in segments of the Polish industry may simply reflect the large size of the Polish agrifood complex. As in Hungary and the Czech Republic, however, well known multinational food corporations hold an important share of the confectionery, brewing and tobacco industries as well as in baby food and pasta (OECD, 1995c).

Hence, it would appear that the fears of those charged with overseeing the process of transition regarding the ability of transnationals to dominate the national market may have been justified. If given the opportunity, they will act strategically to gain a market advantage. Unfortunately, neither those who continue to regulate the activities of transnationals nor those who have implemented more open regimes have been able to find the means of securing the desired balance. As a result, one can expect changing policy regimes towards direct foreign investment as politicians and bureaucrats tinker with their policies in search of the illusive balance (Kerr, 1993).

## 15.3. Forms of Direct Foreign Investment

Direct foreign investment in the former command economies has been, for the most part, restricted to a few of the wide range of governance arrangements which are possible. While almost all of the forms of governance were tried, particularly in the early period of transition, three forms of foreign direct investment are still common – joint ventures, acquisitions and greenfield investments.

Given the transaction costs involved, other forms of governance have largely been rejected by foreign firms wishing to operate in the former command economies, although some examples of each can probably be found. In many cases, when the alternative forms of governance continue to exist, it is only because the foreign investor is constrained by political factors.

### 15.3.1. Licensing

For example, licensing arrangements are no longer common. This is because the poorly enforced legal systems in the former command economies have meant that either: (i) licence fees are not paid; or (ii) production of the product or the use of industrial processes do not remain within the control of the licensor. Licensors are not able to put in place cost-effective monitoring systems. As a result, if foreign firms wish to pursue a market for their products in transitional economies, they are more likely to opt for a joint venture or greenfield investment where they can exercise more direct control over their products or processes.

### 15.3.2. Franchising

For similar reasons, franchises have been largely withdrawn. The successful operation of franchises as a form of governance requires two pre-conditions: (i) local franchisees who understand the importance of reputation and, hence, the need to maintain quality and to follow the procedures laid down by the franchisor; and (ii) the ability to induce the franchisee to adhere to the agreement. The former is absolutely necessary if the franchise is to be successful over the long run. Those who have sold franchises in the transition economies have often found that they simply cannot rely on local firms to maintain standards. Even if the actual holder of the franchise is conscientious in attempting to maintain standards, poor employee training and attitudes can lead to slippage. The local franchisee often finds it very difficult to convince employees of the need for service, quality control, etc., because these concerns simply did not exist in the former command economies. Employees often do not listen to the local managers. When the quality requirements are directly conveyed to employees by foreigners, however, they tend to carry more weight. As a result, the direct supervision possible from a joint venture has become the preferred governance structure.

To work well, franchises also need a well functioning market economy. In the former command economies, breaches of quality or service often result from the inability to acquire proper inputs, services or permits. These problems are, for the most part, beyond the control of the local franchisee. They might, however, also be the result of shirking or opportunism by the franchisee. The foreign firm attempting to monitor its franchise may have difficulty ascertaining the truth of the matter. In a well functioning market economy, persistent difficulties in obtaining high quality supplies simply do not arise. As a result, it is fairly easy to identify breaches of an agreement which are caused by mismanagement of the franchise. Put simply, the monitoring costs associated with the franchise arrangement are too high in the former command economies. While the direct managerial supervision which joint ventures allow cannot solve the problems arising from a poorly developed market economy, it can eliminate (or at least reduce) shirking and/or opportunistic behaviour on the part of a franchisee.

When the franchisee does not understand the need for strict quality control, then a franchise arrangement is bound to fail. There are often large opportunities for short term profits from substituting low quality inputs, counterfeit products or inferior production methods for franchised products. This is particularly true in the case of experience goods whose quality can only be discerned through consumption or credence goods which cannot even be judged after consumption because the consumer has insufficient knowledge or information (see discussion in chapter 11). Faced with opportunism by franchisees which could ruin the reputation of the franchisor, joint ventures with possibilities for direct monitoring are preferred by most firms which would normally use a franchise in modern market economies.

### 15.3.3. Partnerships

Partnerships have also proved unpopular. As in traditional partnerships, each partner is legally responsible for the actions of the other partners (hence, risking their assets), the absence of civil legal remedies and the volatility of the transition economies have made partnerships appear very risky to foreign investors. Limited partnerships (which are more akin to joint ventures) have been more common. In particular, limited partnerships have been common between emigrants, or their children, living in modern market economies and partners in the transition economies. Small retail food outlets, restaurants or local food processing such as sausage making have been the types of investments which have involved partnerships with foreigners. Typically the foreign partner provides some capital and possibly a process common in the west, but hitherto unknown in the transition economy. The foreign investor is, in reality, a silent partner with the local partner(s) responsible for running the endeavour. In many cases, these partnerships are family based and the capital transfer provides the foreign partner with the opportunity to help those left behind in the old country.

### 15.3.4. Stock purchases

Foreign firms have not typically purchased controlling interests in firms listed on the nascent stock markets of transformation economies. For the most part, stock markets do not function with the transparency of those in modern market economies. Insider trading is common and large-scale market manipulation is possible. These developing stock markets tend to be extremely volatile and speculative. A solid base of blue chip companies has not yet developed. As a result, shares listed on the stock market are considered very risky by foreign firms seeking long run agrifood investments. The more direct control which acquisitions or greenfield investments allow have been preferred forms of governance.

### 15.3.5. Joint ventures

According to the OECD:

> The most popular form of investment in the CEEC and the NIS is the joint venture, either with the establishment of a new entity or, more frequently, incorporating certain operations of a local partner.
>
> (OECD, 1994a, p. 29)

Joint ventures seem to be a governance structure well suited for foreign direct investment in agribusiness in the former command economies. They provide for the limited risk of capital. They allow either direct control of the assets invested or at least the ability to directly monitor the activities of management in the host country. They allow the development of profitable enterprises without a commitment to those segments of a business which do not appear to have a profitable future. Local expertise can be tapped to help deal with the bureaucracy and to explain local conditions and customs. Joint ventures can be easily dissolved if either the economic/political environment deteriorates or the competitiveness of the venture is not as anticipated. Joint ventures, with their direct local involvement, are more politically acceptable than greenfield investments or acquisitions. They would appear to represent the type of opportunities for technology transfer which are envisioned by the host country when it encourages foreign investment.

The advantages of joint ventures must be weighed against the costs. As pointed out in Chapter 9, one of the most obvious disadvantages is the danger that a foreign investor creates its own future competition. A second disadvantage of joint ventures is that the objectives of the parties to the joint venture may be incompatible. This can lead to disagreement between the two managements and conflicts of interest between the joint venture participants. In-depth discussions are required prior to joint venture agreements to ensure that both parties fully understand one another's objectives. Local entrepreneurs and managers in the former command economies often have so little familiarity with standard western business practice that they have no real understanding of their foreign partner's concerns. Further, in the former command economies, where investment capital is extremely scarce and, in many cases, incomes are sufficient to provide little more than the necessities of life, prospective joint venture partners have been willing to agree to anything simply to obtain access to investment funds. Once the funds are obtained, they have been invested differently than specified in the agreement or simply siphoned off for personal consumption.

Trust can be developed over time, but it cannot be assumed. As a result, perhaps the major disadvantage of a joint venture is the managerial costs of coordinating, controlling and monitoring the actions of joint venture partners (Middleton *et al.*, 1993). As pointed out in Chapter 9, joint ventures often fail. While exact figures are not available for agribusiness joint ventures in the former command economies, the divorce rate appears to be very high. In some cases, the transnational partner has found it necessary to buy out or take over its foreign partner to secure its investment. According to the OECD:

> Interestingly, we noted a numbers of instances where the initial investment had taken the form of a joint venture, because there was no other method at the time, with the western partner subsequently buying out the local share once 100 percent ownership was possible. One hundred percent ownership was desired because it gave the investor, in their opinion, full control over operations, technology, location, management style and decision making, ensuring a greater chance of success.
>
> (OECD, 1994a, p. 31)

In many cases, however, it has simply meant abandonment of the venture and writing-off the losses. Many transnationals have simply withdrawn from the market in the former command economies.

As discussed in Chapter 9, while joint ventures which require foreign businesses to invest large amounts of capital are perceived to be extremely risky; technical joint ventures and marketing joint ventures have increased in popularity (Davis, 1990). Marketing and technical joint ventures appear to satisfy government goals pertaining to the transfer of the skills necessary to modernize the agrifood sector. Marketing joint ventures serve the dual purpose of providing a low-cost mechanism for agribusiness transnationals to enter the market and for firms in host countries to acquire on-the-job training in marketing which can subsequently be utilized in the promotion of the products they produce which do not fall under the joint venture.

Marketing and technical joint ventures also serve as very useful signalling devices regarding joint venture partners. By not offering investment funds up front, potential joint venture partners looking only to act opportunistically are deterred. As a result, the partners who do enter into these types of joint ventures are more likely to be genuinely interested in business development. Further, over time and without risking capital, transnationals can gain valuable insights into the abilities and character of their joint venture partners. This can be crucial information when investment joint ventures are considered in the future (Middleton *et al.*, 1993). While the reluctance of transnationals to commit large amounts of investment funds is frustrating to genuine potential joint venture partners, given the high information costs associated with establishing who is a genuine partner, such risk averse behaviour by transnationals should not be unexpected.

### 15.3.6. Acquisitions

Acquisition of existing agribusiness firms in the former command economies has not been widespread. This is because acquisitions are often encumbered with a large number of restrictions which constrain the transnationals' freedom of action. In many cases, the acquisition can only be made if the transnational is willing to assume the firm's existing debts. These are often significant, given the book values of the capital transfers which were made under the command system.

If a firm is acquired, the transnational may be lumbered with a labour force which is too large because overmanning was so common in the command era. Further, the skills and motivation of the existing workforce may be at levels which are not acceptable to the transnational. Dismissing the existing labour force may, however, be difficult. A similar problem may arise with the existing management. They may have specialized knowledge which is essential for the operation of the firm. As a result, they cannot be dismissed even if they are otherwise incompetent.

The poor quality of the capital goods and infrastructure owned by existing firms may not be apparent at the time of purchase, leaving the international agribusiness with a relatively worthless asset. These risks tend to work against direct acquisitions.

In a number of countries, foreign acquisitions have been discouraged during the privatization process. Given that agricultural products often require a systems approach, being prevented from acquiring a segment of a system may make other acquisitions which are part of the system unattractive. Systems are particularly vulnerable to the problems associated with bilateral monopolies along the food production and distribution chain.

As discussed above, acquisitions by international agribusinesses appear to be most common when they will lead to the ability to exercise market power. Strategic acquisitions can be used very effectively. As Henry and Voltaire suggest:

Significantly, foreign investment has focused in countries with fairly
advanced and structured distribution, and strong nodes in these systems
... In Hungary, foreign direct investment has targeted the dominant
players in that part of the distribution system that benefits from a strong
presence in the higher income urban market.

(Henry and Voltaire, 1995, p. 277)

In some cases, it is the ability to exercise market power or to erect barriers to entry which is being acquired. The productive assets of the
acquisition may only be incidental to the acquisition decision. In a
survey of firms from a variety of modern market economies who had
made acquisitions in the former command economies, the OECD
(1994a) found that the main reason cited for acquiring a majority
shareholding was to gain market share.

### 15.3.7. Greenfield investment

Greenfield sites have been a less popular vehicle for investment than
acquisitions and joint ventures (OECD, 1994a). Without some local participation, operating in the local economy may be extremely difficult.
Local managers may be hired, but their knowledge of the technology,
marketing and business support requirements of the type of operation
which foreign firms wish to establish may be extremely limited. Greenfield investments may also be viewed with suspicion by the host country government because, while they are likely to be extremely modern,
they are not an efficient mechanism for technological transfer. The
absence of local involvement in management limits the technology
transfer to that which arises from having labour working with modern
technology.

Greenfield investments may have a compatibility problem when
interfacing with the rest of the food chain. Modern facilities require high
quality inputs, sophisticated distribution systems and efficient business
and technical support services. Hence, greenfield investments have
been restricted primarily to a few areas of the economy. There have
been a number of greenfield investments in the agricultural inputs
sector. These distribution centres, and in some instances production
facilities, tend to be extensions of the transnationals' operations in
modern market economies. They tend simply to distribute products
manufactured in multinationals' facilities outside the host country. In
the case of agricultural machinery, foreign firms may also carry out
repairs using imported spare parts and equipment purchased in modern
market economies. Reliance on local suppliers is kept to a minimum.

In the case of agroinput manufacturing facilities, essential raw materials are imported from modern market economies. This removes the difficulties associated with local disruptions in input supplies and the problems associated with the inconsistent quality of local inputs. Of course, the benefits received by the local economy are reduced commensurately.

The other major area of greenfield investment has been in food distribution and retail. These greenfield investments tend to act as conduits for the distribution of food products imported from modern market economies. They are particularly evident in markets which are in close proximity to modern market economies and rely on good transportation links. As distances increase, and the possibility of transportation difficulties rise, their numbers decrease significantly. Greenfield distribution warehouses and retail outlets often will not carry local products, they simply do not have to rely on local products to be successful.

In some notable cases, greenfield investments have represented, in all but name, vertically integrated systems. For example, McDonald's in Moscow created virtually an entire vertically integrated supply system within Russia to ensure that its quality specifications were met consistently.

Greenfield investments have also tended to be restricted to new product lines where only poor local substitutes exist. High technology agroinputs, niche market food products and business services have been set up via greenfield investments. Agribusiness transnationals have been reluctant to make greenfield investments in direct competition with existing agribusiness enterprises in the transitional economies. As governments in the former command economies have not been willing to allow large-scale bankruptcies in the agroinputs and food processing sectors, competing directly with state and former state enterprises simply means that those enterprises will receive larger subsidies. As a result, the potential profitability of greenfield investments is reduced, and the modernization of the sector is slowed.

The problems associated with the refusal of governments to allow former state enterprises to be wound up and liquidated are particularly difficult in the agrifood sector as a result of political concerns over food security. In many cases, without government intervention, the enterprises would simply be abandoned with workers forced onto the unemployment rolls. Some governments have refused to accept the fact that many of their agrifood enterprises may never be profitable. According to Paliwoda:

> The more Western-oriented economies now show signs of willingness to accept very low liquidation prices to entice buyers. In situations where the political economy is uncertain, capital stock is obsolete, and workers

show signs of intransigent behaviour, liquidation prices are the only prices on offer.

(Paliwoda, 1994, p. 30)

When greenfield investments are proposed in direct competition with failing state enterprises, governments need to realize that there is a direct trade-off between supporting inefficient existing enterprises with subsidies and modernization via foreign investment. If a greenfield investment is being proposed, the agribusiness transnational is suggesting that the creation of a wholly new enterprise is a more efficient alternative than the acquisition of an existing enterprise. Hence, the modernization goal is not likely to be realized unless subsidization for firms competing with greenfield investment ceases.

## 15.4. Conclusions

On balance, transnational agrifood companies still represent too small a presence in the economies in transition to provide a conduit for widespread modernization of the industry (Henry and Voltaire, 1995). The need for modernization is certainly acute, and policy makers need to re-think their strategies for attracting foreign investment in the agrifood sector. In the Russian agrifood industry for example: 'One-fourth of the production apparatus is older than the standard service life. Over 40 percent of the equipment base is obsolete' (Nefedov, 1994, p. 51). While foreign investment cannot be the only means of modernization, it seems clear that considerably more investment than has taken place to date is possible.

There are two major reasons why foreign investment in the agrifood sector has been curtailed. First, a business environment conducive to sustained profitable operations has not been created. Second, governments concerned about possible foreign domination of agrifood industries have put in place, or more correctly not removed, barriers to the operation of transnational agribusiness firms. In the case of the former, international agribusiness, as with domestic businesses, must have a business environment within which they have a reasonable opportunity to operate profitably. As suggested throughout this volume, the high transaction costs associated with doing business in the former command economies, combined with the risks associated with poor macroeconomic management and inconsistent and conflicting government policies, have made many potentially profitable enterprises non-viable. This has inhibited domestic as well as foreign investment in new agribusiness ventures. Unprofitable existing agribusiness enterprises which were set up in the command era, in most cases, have been able to obtain subsidies keeping them in business. As a result, firms with obsolete technology remain in place and inhibit the modernization process. Ongoing subsidization of domestic agribusiness firms further reduces the incentives for transnational agribusiness firms to invest in the former command economies.

Transnational agribusiness firms are further discouraged from investing in the former command economies by discriminatory regulations aimed at limiting their ability to compete with domestic firms and to expand. These regulations are attempts by domestic politicians to prevent the monopolization of domestic industries by transnationals. Given the low level of foreign investment by agribusiness firms, government concerns are premature in most cases. The concerns themselves, however, are not unfounded. Where business conditions do allow for at least a reasonable chance at success, transnational agribusiness firms are willing to invest. The strategic investments which have been made to acquire market power in key individual sectors in Hungary, the Czech Republic and Poland suggest that governments concerned with developing a viable domestically owned and operated agribusiness sector should be willing to act to curb the power of transnationals.

Achieving a desired balance between transnational and domestic firms is likely to prove very difficult – as it has in developing countries (Kerr, 1993). Just because achieving an objective is difficult, however, does not mean that it should be abandoned. It appears that for most of the former command economies, the opportunity remains to develop policies which will strike the correct balance. Most fundamentally, the opportunity remains because governments have not yet created a business environment where international agribusinesses wish to risk their investments. When, or if, they create a business environment conducive to foreign investment, it is important that the set of policies in place affecting agribusiness transnationals have been devised in such a way that the desired balance can be achieved.

# References

Amin, S. (1976) *Unequal Development*. Monthly Review Press, New York, 540 pp.

Archibald, G.C. and Rosenbluth, R. (1978) *Production Theory in Terms of Characteristics – Some Preliminary Considerations*. Discussion paper no. 78-19, Department of Economics, University of British Columbia.

Barnett, R. and Muller, R.E. (1974) *Global Reach, The Power of the Multinational Corporations*. Simon and Schuster, New York, 360 pp.

Biersteker, T.J. (1976) *Distortion or Development? Contending Perspectives on the Multinational Corporation*. MIT Press, Cambridge, Massachusetts, 199 pp.

Camdessus, M. (1992) *Economic Transformation in the Fifteen Republics of the Former USSR: A Challenge and Opportunity for the World*. Address to Georgetown University, School of Foreign Service, International Monetary Fund, Washington DC, April 15.

Das, S. (1987) Externalities and technological transfer through multinational corporations. *Journal of International Economics* 22, 171–182.

Davis, R. (1990) Doing business in Japan: joint ventures. In: Elton, D.K., Kerr, W.A., Klein, K.K. and Penner, E.T. (eds) *Selling Beef to Japan*. Canada West Foundation, Calgary, 197–200.

Findlay, R. (1978) Relative backwardness, direct foreign investment and the transfer of technology: a simple dynamic model. *Quarterly Journal of Economics* 92, 1–16.

Fischer, S. (1992) Privatization in Eastern European transformation. In: Clague, C. and Rauser, G.C. (eds) *The Emergence of Market Economies in Eastern Europe*. Blackwell, Oxford, pp. 227–243.

Gady, R.L. and Payton, R.H. (1992) A food processor's perspective on trade and investment opportunities in Eastern Europe and the former Soviet Union. *American Journal of Agricultural Economics* 74, 1179–1183.

Henry, R. and Voltaire, K. (1995) Food marketing in transition in Eastern and Central Europe. *Agribusiness: An International Journal* 11, 273–280.

Hymer, S. (1960) The international operations of national firms: a study of direct investment. PhD dissertation, Massachusetts Institute of Technology.

Hymer, S. (1979) *The Multinational Corporation: A Radical Approach*. Cambridge University Press, Cambridge, 323 pp.

Kerr, W.A. (1993) Domestic firms and transnational corporations in liberalizing command economies – a dynamic approach. *Economic Systems* 17, 195–211.

Marton, K. (1986) Technology transfer to developing countries via multinationals. *The World Economy* 7, 409–426.

Middleton, J., Hobbs, J.E. and Kerr, W.A. (1993) Poland's evolving food distribution system: joint venture opportunities for British agribusiness. *Journal of European Business Education* 3, 36–45.

Nefedov, V. (1994) The agro-industrial complex of the CIS. *Problems of Economic Transition* 37, 50–62.

OECD (1994a) *Assessing Investment Opportunities in Economies in Transition*. Centre for Co-operation with the Economies in Transition, Paris, 183 pp.

OECD (1994b) *Review of Agricultural Policies: Hungary*. Centre for Co-operation with the Economies in Transition, Paris, 222 pp.

OECD (1995a) *Taxation and Foreign Direct Investment: the Experience of the Economies in Transition*. Centre for Co-operation with the Economies in Transition, Paris, 165 pp.

OECD (1995b) *Agricultural Policies, Markets and Trade in Central and Eastern European Countries, Selected New Independent States, Mongolia and China: Monitoring and Outlook 1995*. Centre for Co-operation with the Economies in Transition, Paris, 232 pp.

OECD (1995c) *Review of Agricultural Policies: Poland*. Centre for Co-operation with the Economies in Transition, Paris, 285 pp.

Offerdal, E.C.F. (1992) Taxation of foreign direct investment. In: Tanzi, V. (ed.)

*Fiscal Policies in the Economies in Transition*. International Monetary Fund, Washington DC, pp. 232–253.

Paliwoda, S.J. (1994) *Investing in Eastern Europe: Capitalizing on Emerging Markets*. Addison-Wesley Publishing Co., Wokingham, 160 pp.

Streeten, P. (1977) Self-reliant industrialization. In: Wilber, C.K. (ed.) *The Political Economy of Development and Underdevelopment*. Random House, New York.

United Nations (1992) *World Investment Report 1992: Transnational Corporations as Engines of Growth*. Transnational Corporations and Management Division, Department of Economic and Social Development, New York, 358 pp.

Vernon, R. (1971) *Sovereignty at Bay: The Multinational Spread of U.S. Enterprises*. Basic Books, New York, 326 pp.

Vernon, R. (1975) *Multinational Enterprises in Developing Countries: an Analysis of National Policies*. United Nations Industrial Development Organization, New York.

Vernon, R. (1977) *Storm Over the Multinationals: the Real Issues*. Harvard University Press, Cambridge, Massachusetts, 260 pp.

Wong, J. and Blomstrom, M. (1992) Foreign investment and technological transfers. *European Economic Review* 36, 137–155.

World Bank (1995) *Armenia: the Challenge of Reform in the Agricultural Sector*. Washington DC, 198 pp.

# 16 Reducing Transaction Costs Through Market Institutions

The creation of a system of commercial law where no system exists is easy to envision. Legal systems can be created by governments, and there are a number of models to follow. Conceptually, establishing institutions to facilitate the financing of transition is not as straightforward as putting commercial legal systems in place. However, successful banks have been created by governments in modern market economies, stock markets have been fostered, and there are regulatory systems for the financial sector which can be copied. Envisioning the creation of market institutions is much more difficult. In part, this is because market institutions exist in almost infinite variety. Most of the difficulty, however, arises because the vast majority of market institutions have come into being as a result of private actions in response to demands for the services that such institutions provide. Seldom are market institutions created by government. Hence, while an array of existing market institutions can be identified, isolating the factors which make them the relevant institution in a particular environment is, at best, a poorly developed science.

The market institutions which exist in modern market economies have evolved over long periods of time. This evolution has little do with direct policy actions taken by governments. This absence of government involvement might suggest that in the wake of privatization and the freeing of prices, market institutions will arise spontaneously in the Central and Eastern European Countries (CEECs) and the New Independent States (NISs). Of course, to some extent they already have been developing. Farmers' markets are the most obvious manifestation of this evolutionary process. As discussed in Chapter 8, however, modern market economies have long since abandoned farmers' markets, except as a niche marketing technique. Farmers' markets are simply transaction cost inefficient. Consumers in the former command economies are sufficiently sophisticated that they are unlikely to settle for them for very long.

One major difficulty with waiting for market institutions to develop naturally is that the agrifood system in the former command economies must supply large urban areas which are often far from the centres of agricultural production. The existing enterprises in the food supply chain are large scale and designed to supply large numbers of customers. While small businesses can operate relatively efficiently using transactions based on the development of personal trust, large businesses cannot. This is one of the reasons for the success of small businesses in the later stages of distribution and retail as well as small scale food processing. It also explains why most successful small businesses have failed to grow.

Developing trust based on personal relationships is costly in terms of a businessperson's time. Large firms have many suppliers and customers and, hence, must organize a considerable number of transactions. It is too costly to develop personal relationships with all transaction partners. Large firms require institutional arrangements which allow them to undertake transactions with suppliers and customers who are unknown. This means that institutions must exist which ensure, for example, that payment can be expected more or less automatically. In other words, the entrepreneur/manager will not have to expend time and energy organizing this aspect of the transaction. Large agribusiness firms in the former command economies cannot expect that even this most basic of business requirements exists.

Complex transactions have a time dimension. This means that circumstances may change and chances to act opportunistically may arise. The former requires that contingencies need to be specified when the transaction is being negotiated while the latter requires *ex post* penalties in the case of non-performance. As discussed in Chapter 4, in modern market economies, legal contracts are used to organize complex transactions.[1] Enforcement of contracts in the former command economies is weak. As a result, opportunism abounds. Agribusiness firms are forced into undertaking complex transactions only with those whom they trust. This acts to limit the number of potential transaction partners and reduce competition (see Chapter 5).

Compared to their counterparts in modern market economies, business people in the former command economies spend a great deal of their time organizing transactions. In short, this means that they have less time for other activities such as marketing, product development, improving productive processes and staff training. In a well functioning market economy, business people spend almost no time organizing transactions.

The key to the establishment of market based agribusiness sectors in the former command economies is the fostering of market institutions to reduce transaction costs. No economy can be modern without these institutions. As yet, little attention has been given to the problem of fostering these institutions. While they may evolve over time, the evolutionary process is unlikely to be sufficiently rapid to allow the survival of most of those enterprises from the command era which are capable of transition or to prevent the stifling of new enterprises of sufficient size to take advantage of economies of scale. Failure to foster the development of market institutions will either lead to the re-imposition of elements of the command system in the name of food security or the development of a small scale, inefficient and, therefore, expensive food supply system which slows the entire process of economic development.

## 16.1. Fostering Market Institutions

The primary function of market institutions is to facilitate the organization of transactions. Improving the ways in which transactions are organized is the process of reducing transaction costs. If there is no other message from Section III of this book, it is that the transaction costs faced by firms in the agrifood sectors of the former command economies are extremely high. Transaction cost theory suggests that over time the forces of competition will lead to agricultural supply chains being organized by the set of governance structures which have the lowest cost. Second guessing the most transaction cost efficient set of market institutions is akin to attempting to pick future winner industries to support in government industrial policies. While some obvious losers may be easily discarded, actually selecting the most transaction cost efficient mechanism is far more difficult. Hence, it is probably more important to provide an environment where firms, groups of firms, industries and even whole sectors can establish the market institutions which they best think will service their needs. Governments also need to be ready to allow market institutions to fail if they are not successful. Forcing farmers to use an auction market simply because firms or governments have expended funds for its establishment is not prudent. Farmers may have simply found that informal one-on-one arrangements with processors are less costly (Hobbs, 1995).

Transaction costs are often divided into three categories:[2] (i) information costs; (ii) negotiation costs; and (iii) monitoring costs. How to foster an environment to lower each of these costs will be discussed in turn.

### 16.1.1. Lowering information costs

Information costs relate to identifying other firms which have products or services the firm might wish to procure or who could be markets for products produced by the firm. They also relate to the ability to assess the reputation of firms with which transactions may be undertaken. Acquiring information regarding quality specifications causes firms to incur costs as does the acquisition of price information.

Improved communication systems is one key to lowering information costs. All former command economies have been making efforts to improve their telephone systems. However, waiting lists for phone services remain long in many countries. State run telephone systems (often now in partnership with a foreign telecommunications firm) remain the norm. This is a hold over from the previous regime where individuals' communications were of interest to the state. Given the current state of technology, phone systems based on land line technology no longer need to have monopoly rights for the provision of phone services. Telephone monopolies are even less justified when the telephone system can provide services which are no longer based on land lines. Telephone communications should be opened up as widely as possible including allowing competing firms access to state operated land line systems.

In times of excess demand for state phone services, individuals providing phone connection services or those who allocate lines are able to extract bribes for the provision of those services. If prices were allowed to ration phone services, the rents would at least accrue to the phone company and could be used for re-investment. As a result, over time the cost of service should fall. Further, a bribe based system creates an incentive for those who benefit from the system to resist changes which would reduce their incomes, slowing the reform process in state monopolies. Competition among phone service providers will decrease the ability to extract rents through corruption.

Phone systems require the provision of associated services such as directories, yellow pages, long distance assistance, etc. There is no need for these associated services to be provided by state telephone systems. They can be provided privately. Private firms can collect, print and sell directories. They can be contracted to provide long distance information. The state phone system, however, should not be allowed to stand in the way of private telephone companies obtaining information relating to numbers, the locations of phones and their affiliation whether personal, business or government. State phone services should give priority to improving and streamlining these information systems. They may wish to distribute the information themselves, but they should not be given a monopoly on its distribution. Competition, or the threat of competition, should act to reduce the information costs for agribusiness firms and lead to improvements in the dissemination of information.

It may appear quaint or outdated to focus attention on the telephone system in the era of information super-highways, computer networks, etc. Certainly, these newer technologies are playing an important role in reducing information costs for many firms in modern market economies, but few of those firms would be willing to trade in their telephones and rely on these new systems alone. As yet, the telephone remains the most important, and lowest cost, means of acquiring information in modern market economies. In many cases, the newer technologies piggy-back on the telephone system.

An efficient, reliable and secure postal system should also be a priority. While businesses in modern market economies have replaced the mail service with fax or other electronic systems for a considerable portion of their communications, the mail – as evidenced by junk mail – still provides a low cost means to distribute information directly to consumers or business customers. The ability of the postal system to lower information costs should not be underestimated.

In this vein, tariffs and other import restrictions as well as domestic taxes should be removed on all printing and duplicating equipment. Import restrictions and taxes should be removed from paper. Relative to the command system, market economies require very large quantities of paper to distribute information. Firms in the former command economies often face shortages of paper. Making available even simple information such as price lists, much less the sophisticated print advertising that characterizes both firm-to-firm sales efforts and consumer marketing, is either not possible or prohibitively expensive for agribusiness firms in the CEECs and the NISs.

Competition in electronic media should also be encouraged, particularly in radio. Radio provides a relatively low technology, low cost means of making information available through purchased advertising time. In modern market economies, television remains a high cost medium, but local radio can be a very effective business information medium.

Beyond doing everything possible to broaden access to low cost information channels for firms, there are probably few direct initiatives that governments can take to lower information costs. Governments do, however, have a reactive role to play in providing information for businesses. Governments need to be willing to provide certain forms of business information which they have a comparative advantage in collecting. It should be remembered that communist governments were very secretive. Hence, providing information may not be a role bureaucrats are, as yet, comfortable with. What there needs to be, along with the willingness to provide information, is an effective means by which the information needs of businesses can be translated to the government.

As pointed out in Chapter 9, this conduit is provided by business associations in modern market economies. Agribusiness firms often belong to a number of specialized business organizations whose roles vary from establishing standards to informal exchanges of information. Employees of agribusiness firms often belong to professional organizations – for purchasing agents, engineers, accountants, etc.

These organizations tend to provide a forum whereby business needs are synthesized and organized for presentation to government. These needs are broadly defined and include lobbying for trade barriers, subsidies and regulations. An extremely important function of these organizations, which is much less prominent, is that they provide the means to relay the industries' information needs to governments. Agribusiness organizations and agribusiness professional associations have developed good working relationships with government departments. Prices, costs, volumes and a host of other information is collected, processed and distributed by government agriculture and business departments. They often also provide analysis of the data collected. Firms may not make direct use of the information, but this information provides a considerable proportion of the information and the most low cost information which is used by agribusiness consultants in the preparation of studies commissioned by firms.

While most of this information could be collected privately and sold, governments typically have some cost advantages in collecting information. Governments can compel firms to provide information which they would not give voluntarily. Governments can also process information so as to ensure firms' anonymity which increases the likelihood that they will cooperate in providing information. As governments maintain a widespread network of offices, particularly in rural areas, this existing network provides a low cost means for collecting information from widely dispersed agricultural enterprises.

As yet in the former command economies, organizations of agribusinesses and associations of agribusiness professionals do not play a prominent role. Part of the reason for this is that such organizations may appear as the antithesis of competition to newly privatized agribusiness firms which expect to compete. The benefits of cooperating with competitors to achieve common goals may not be immediately apparent to agribusiness firms attempting to determine how to compete.

When organizations representing agribusinesses have been formed in the former command economies, they tend to concentrate their efforts on lobbying the government for subsidies, special concessions or protection. Their potential role in lowering information costs is not well understood.

Part of the reason why both agribusiness organizations and associations of agribusiness professionals have not expanded their role as a conduit for information needs is that they do not expect governments to provide information. Governments, hence, need to be more productive in advertising their willingness to provide information.

By promoting conferences or meetings with the topic of information requirements on the agenda, governments can force agribusiness organizations to think about their information requirements. Foreign experts can play a role by discussing the types of information which are typically made available by governments in modern market economies.

Once agribusiness organizations begin to assess their information needs, they may find that the organizations themselves may provide a lower cost source of information than the government for at least some information. As in modern market economies, agribusiness organizations themselves may collect and disseminate information.

These organizations can also act to lower the cost of information regarding a firm's reputation. This is particularly important in the former command economies where reputations have not had time to develop and many new firms exist. Agribusiness organizations can develop codes of conduct for their members, establish mechanisms for accepting and verifying complaints regarding their members' actions and for making information regarding complaints available. They may wish to develop certification procedures and disciplinary measures. These services can considerably reduce information costs relating to the reputation of firms with which one is contemplating entering into a business arrangement.

Agribusiness organizations, probably in conjunction with government departments, can also work to establish commercially useful grading systems for the heterogeneous products which result from biological production processes. As suggested in Chapter 8, grading systems can reduce the information costs associated with sorting by buyers. Further, standardized grading systems reduce the information costs of transactions which are carried out over considerable distances. Grading systems also reduce the information costs for producers by providing clear signals regarding the combination of product characteristics which will provide the greatest return.

Some aspects of reducing information costs are better developed than others in the former command economies. Agribusiness organizations, often in conjunction with government agriculture departments or local governments, organize a large number of trade shows. For the most part, these have provided showcases for foreign products or venues where domestic exporters can have their products examined by foreign buyers. Over time, more trade shows with a domestic orientation are being organized.

For the most part, however, information costs for firms remain high. Reducing information costs, particularly for geographically dispersed agribusinesses, has not received enough attention. Without addressing these high costs, much of the effort being put into educating agribusiness managers will be wasted. Modern managers require large quantities of high quality information to effectively use their skills.

### 16.1.2. Lowering negotiation costs

As outlined in Chapter 4, negotiation costs encompass a very wide range of business activities relating to consummating a transaction. Two crucial aspects of negotiation costs relate to financing transactions and the commercial legal environments. These aspects of negotiation will be dealt with in considerable detail in Chapters 17 and 18 respectively.

One aspect of the financial system will, however, be dealt with here. In the command era there was no need for commercial banking services as they are understood in modern market economies. Accounting transfers were simply made in the planning ministries. As a result, no mechanism for the large scale transfer of funds were required. In modern market economies, banks and financial institutions, in addition to their financing role, play a fundamental role in business to business transfers of funds. Relatively automatic transfer of funds lowers transaction costs for businesses.

Agribusinesses in the former command economies must still conduct much of their business in cash. Part of the difficulty lies in a low level of trust compounded by a legal system which does not effectively curb opportunism. Transactions are not carried out by cheque or other promissory notes because delayed payments may simply not materialize and the holders of promises to pay have no practical recourse to the courts to enforce payment. As a result, purchases are paid for in cash.

Cash payments are also required because banks, particularly when different geographic areas are involved, do not have the capability to clear cheques. Both for private firms, and banks themselves, the costs of verifying whether the promised funds are actually available are very high.

Payment in cash is costly and risky. Cash must be physically transported to the place of transaction. It must be transported by trustworthy employees. When large sums are involved, the entrepreneur or manager may have to carry the money personally. This clearly has a opportunity cost in terms of managerial time. It is a task which few business people in modern market economies would ever be faced with. Of course, moving large quantities of cash about leaves firms open to robbery. Security personnel have to be hired or protection from robbery purchased from organized crime.

As any small retail store owner in modern market economies realizes, matching the available stock of currency (the float) to the demands for currency – for example, simply being able to make correct change – is difficult. Employees being dispatched to the bank with handfuls of large notes are often observed. For large agribusiness firms in the former command economies who must conduct many daily transactions in cash, these problems are compounded. The solution is to keep large quantities of cash on hand at any given time. This creates opportunities for pilfering. As a result, the costs of monitoring the activities of employees increase. Security systems to prevent robbery are also required. Cash is not petty for agribusiness in the former command economies. Holding large quantities of cash also has a large opportunity cost in terms of foregone interest.

Conducting transactions in cash may impose severe cash flow constraints on certain segments of the food supply chain, inhibiting its development. Delayed payment is an accepted practice in modern market economies with 30, 60 and 90 days to pay being normal. While these delays impose costs somewhere along the supply chain, their persistence suggests that they represent an efficient method of organizing transactions. In particular, cash payment on delivery constrains the activities of small firms. The result is sub-optimal inventory levels, a planning process made inefficient by short run cash flow concerns and overly frequent and small orders. Privatized fertilizer dealers with ten bags of product in stock waiting for the cash from the sale of those ten bags so that more fertilizer can be purchased is a common occurrence. Given the critical timing of fertilizer application, the inevitable delays associated with the cash-based system will lead to lower yields.

While considerable attention has been given to the financing role of the new or reformed banks in the former command economies, too little attention has been given to reducing the dependency on cash-based transactions. Of course, in those countries with closer ties to modern market economies, much more progress towards the creation of a transaction cost reducing banking system has been made. In rural areas, however, the commercial banking system remains underdeveloped, partially because of the poor performance of primary agriculture. Special agricultural banks exist in most of the countries of the CEECs and the NISs but they act more as channels for financial transfers to rural enterprises than as facilitators of transactions. As a result, agribusiness situated in rural areas remains cash-based. Rural banks clearly need to expand their role as facilitators of transactions.

The ability of agribusiness firms to negotiate complex transactions is greatly inhibited by the absence of a functioning commercial legal system. The courts cannot be relied upon to produce consistent judgements – partially because they have little in the way of precedent to guide them. As a result, the court system lacks the transparency required to write contract provisions which can be relied upon to provide legal protection. Further, and more important, the state often cannot be relied upon to enforce the judgement of the courts. In some countries this problem stems from corruption, in others to the absence of a clear mandate for the police and a lack of resources for enforcement. The previous heavy (and political) role of security services is not yet forgotten in former communist countries and, as a result, governments have tended to tread cautiously when giving the reformed police powers and resources.

When it is difficult to write clear contracts and/or when they cannot be effectively enforced, complex transactions must be accomplished using alternative mechanisms. The mechanism most often used is the building of long-term, trust-based business relationships (Kerr, 1996). Trust-based relationships, however, are inefficient. They require very costly personal investments by entrepreneurs or managers and take a long time to mature. This has two effects. First, the large personal investment in time means that the number of possible transaction partners for agribusiness firms is reduced. Further, once the heavy investment in establishing trust is made, it reduces the likelihood of switching transaction partners. This adversely affects competition and, hence, increases the probability that bilateral monopoly situations will be perpetuated. Second, it slows down the negotiation process meaning that profitable business opportunities may be foregone.

The inability to use formal legal contracts also makes for difficult negotiations with foreign agribusiness firms who are used to using contracts to facilitate complex transactions (Kerr, 1996). Many foreign agribusinesses signed contracts with eager enterprises in the former command economies in the early days of liberalization only to find out that the contracts provided no guarantee of performance and that no effective recourse was available.

For foreign agribusinesses, the obvious alternative to contracts is to vertically integrate. Hence, an incentive is provided for agribusiness transnationals to take a direct role in the agribusiness sector of the former command economies. Their direct involvement may adversely affect the development of domestic agribusiness enterprises and even thwart the goal of establishing a competitive domestically owned and operated agribusiness sector (Kerr, 1993).

As will be discussed in Chapter 18, developing an effective commercial legal system is likely to involve a long process of institution building. While the fostering of a commercial legal system is a key component in the development of a modern market economy, other forms of institution building may provide a transaction cost reducing interim step. One obvious candidate is the increased use of private commercial arbitration. The procedures for arbitrating complex agribusiness contracts are well established internationally. Arbitration can overcome the problems associated with the absence of transparency in the legal system by allowing the parties to a contract to choose the legal system whose principles will be used in deciding contract disputes (Kerr and Perdikis, 1995). This facilitates the use of contracts.

Private commercial arbitration also has well established rules for appointing arbitrators to hear any dispute, for selecting those nominated as arbitrators and for ensuring the transparency of arbitrators decisions – all in aid of removing any bias in the arbitration. Arbitration also allows for decisions to be made on the basis of sound business practice rather than strict legal interpretations of contracts. This may be a considerable benefit in the agribusiness sectors of the former command economies where the future business environment is difficult to predict and, hence, fully accounting for all contingencies in a contract is often impossible.

Agribusiness firms in modern market economies are familiar with arbitration and are likely to be more willing to accept it as a dispute mechanism than a foreign and less than transparent legal system. In fact, there is a long tradition of using arbitration to settle disputes with foreign agribusiness firms dating from the communist era. As communist courts were agents of party policy and, hence, openly biased, an alternative had to be found to induce foreign firms to enter into business transactions with state enterprises. Arbitration was the mechanism chosen, albeit state fostered.

Private commercial arbitration cannot prevent overt opportunism because arbitration either relies on voluntary compliance with the arbitrator's judgements or must rely on the courts to enforce arbitration. As the courts in the former command economies cannot yet be relied upon for enforcement, agribusiness firms remain vulnerable. Promoting the use of arbitration, however, is still likely to be transaction cost reducing. Many disputes are simply honest disagreements regarding contract clauses or what should be done when contracts are incomplete and do not involve overt opportunism. Arbitration can be used to expeditiously solve such disputes. In addition, and probably more important, firms can use their compliance with arbitration awards as a means to establish a reputation. If the arbitration institutions which are established have a mandate to publicize their verdicts and the compliance of the parties, agribusinesses have a transparent mechanism by which to establish their reputation. This reduces information costs and widens the set of firms with which one would consider entering into complex transactions. The investment in building trust through personal business relationships is reduced. This will enhance competition.

While it may be possible for agribusiness firms to use general private commercial arbitration institutions, the fostering of arbitration institutions which specialize in agribusiness disputes may be a worthwhile activity for governments. The nature of agricultural production which is characterized by heterogeneous outputs, perishability, uninterruptable production systems, long lags between the production decision and marketing, seasonality, timing constraints, etc., often adds a new level of complexity to contracts. Arbitrators, unlike judges, can have specialized knowledge of production processes and normal business practices in agribusiness. Specialization also makes it easier for firms to learn the characteristics of individual arbitrators. This facilitates the ability of arbitrators to establish reputations regarding their knowledge and fairness. Poor arbitrators are more easily weeded out. Agribusinesses' trust in the arbitration institutions will increase.

Fostering the establishment of arbitration institutions may also be a role which foreign agribusiness organizations can undertake. Such initiatives may be particularly important where there is little trust of the government by the domestic agribusiness sector. Cooperation between domestic and foreign agribusiness organizations can lead to private agribusiness arbitration institutions where the government plays a minimum role. Of course, private arbitration is no substitute for a well functioning commercial legal system, but it is likely to provide a transaction cost reducing intermediate step. Arbitration is likely to continue to operate in a parallel fashion to the judicial system as is the case in modern market economies.

Beyond these broad based institution building initiatives, there is clearly a very wide range of specialized institutions which facilitate the accomplishment of transactions. While not exclusively agricultural, auctions and futures markets are firmly rooted in the agribusiness sectors of modern market economies. As pointed out in Chapter 8, auctions provide a competitive and transparent pricing mechanism for heterogeneous products. They also provide a mechanism for forcing a sale of perishable products. Futures markets provide a means by which buyers and sellers can reduce the risks associated with the long lags between the point in time when production decisions must be made and when products come to market. They also provide a market where those willing to assume risk for the possibility of high return can transact with those willing to trade lower potential returns for lower risk.

Auctions probably represent the type of institution which can currently be fostered to assist in reducing negotiation costs while futures markets should probably await the further development of markets. Fostering auctions in the agrifood sector can probably be accomplished at relatively low cost. The essential element is the training of auctioneers. In modern market economies, this is largely accomplished through relatively inexpensive short courses (usually less than a month) provided by the private sector. Hence, the training of auctioneers in the former command economies could be done inexpensively. Transferring the training skills to local private or state training institutions can be accomplished with relative ease. Having the courses run on a commercial basis should be possible and, in any case, state subsidies would not need to be large.

Trained auctioneers are simply entrepreneurs who have a service to sell. While many auction markets in modern market economies have substantial facilities, for the most part they are not necessary. The roving auctioneer with a bull horn (megaphone), a couple of side men to take bids and a clerk to record bids and accept payment is still common. Livestock can be sold by auctioneers who travel from farm to farm along with the buyers. This method of selling livestock would seem particularly applicable where large former collective and state farms remain in being. The number of animals available for sale could justify the attendance of both the auctioneer and a sufficient number of buyers. Of course, having a sufficient number of buyers is required for a successful auction. It is the role of the auctioneer as an entrepreneur to ensure that the sale he/she is organizing is sufficiently interesting to attract buyers.

Auctions can take place in all sorts of venues – selling truck lots in parking lots, in rented warehouse facilities, at wineries, etc. Auctioneers can broaden their business to items such as used machinery and, when permitted, land.

Auctioneering is primarily a business based on reputation. Hence, the auctioneer has an incentive to ensure that payments are received and made. Auctioneers have the right to exclude either buyers or sellers from their sales. Excluding those who have poor reputations enhances the reputation of the auctioneer and provides a transparent confirmation of the reputation of those allowed to participate.

As in modern market economies, many auctioneers will fail. As their training is inexpensive, however, this does not represent a large waste of resources. The auctioneer entrepreneurs who survive will lower the negotiation costs for many market participants. The most successful auctioneers will accumulate sufficient capital, or reputation to borrow capital, to invest in permanent facilities. Hence, investing in the training of large numbers of auctioneers is probably a wiser institution building strategy then constructing expensive auction facilities.

It may be appropriate for direct government investments in auction markets to be made when very perishable commodities are to be traded – flowers, fresh fruit, etc. Modern facilities can reduce wastage and attract foreign buyers for high value commodities.

While the fostering of auctions appears to be a relatively low cost exercise in institution building to reduce negotiation costs, fostering futures markets may not be. This does not mean, however, that futures markets should be impeded from developing naturally if interest in such markets warrants their establishment.

At best, futures markets are likely to serve only a small segment of the firms which constitute the agribusiness sector. Even in modern market economies with well developed low cost information systems and high levels of business education, only a small proportion of agri-businesses and even a smaller proportion of farmers actively participate in futures markets.

To effectively utilize futures markets, participants require large amounts of high quality information. As these information sources do not exist as yet in most former command economies, market participation is likely to be thin. Futures markets also depend upon the existence of well functioning and transparent spot markets. As these are not well developed, hedgers in particular, will have little upon which to base their expectations. The net results will be thin markets populated by risk loving speculators. This combination can lead to market manipulation and futures market fluctuations which are not based on the market fundamentals of the underlying commodity's business and economic environment. Poor systems for regulating the activities of futures exchanges exacerbate the problem. The evidence suggests that not only futures markets but other forward markets and stock markets in the former command economies can be characterized in this way.

As a result, the fostering of futures markets on the basis that they provide a positive externality in the form of information on future prices for those who do not actively participate in futures markets cannot be justified. One has only to look at the history of the formative years of the major US exchanges to find evidence of the consequences of thin, speculative futures markets.

There are many other specialized institutional arrangements which exist in modern market economies which act to reduce negotiation costs. Over time, one can expect those which are appropriate to conditions in the former command economies to develop naturally. The real question is which institutions should be actively fostered by governments or those providing foreign assistance and advice to the agribusiness sector.

As the discussion of auctions and futures markets illustrates, each institution will have to be evaluated on its individual merit given the current state of the country's agribusiness environment. One suspects that, for the time being, attempting to foster leading edge institutions modelled on those in modern market economies may not be successful given their heavy information requirements and scale economies. Older, more basic, and in some cases former, western market institutions may be more appropriate.

### 16.1.3. Lowering monitoring costs

Monitoring costs are associated with activities which happen after a transaction has been agreed. They include the enforcement costs associated with ensuring that the terms of a contract are lived up to or in recovering compensation when the terms of a transaction are breached. Hence, while the existence of a functioning legal system or arbitration mechanism may reduce negotiation costs by allowing contracts to be used as a method of organizing complex transactions, the costs associated with legal cases or arbitrations are enforcement costs.

Enforcement costs are high in the former command economies because alternatives to courts or arbitration must be found. One method is to have all products sold on the basis of *caveat emptor* (buyer beware). This provides for no legal or other recourse once the buyer takes possession of the commodity – as a result bulls may be sterile, poultry products may have salmonella, fertilizer may be cut with sand, seed may not have been cleaned, flour may be inhabited by pests. With no legal recourse under *caveat emptor*, buyers who are risk averse will only be willing to pay less to offset the risk. One ends up with the classic market for 'lemons' problem with a dynamic element (see Chapter 4). As buyers will only pay for the lowest expected quality, firms have no incentive to improve their quality. A system of *caveat emptor* for products whose quality can be variable, which is almost always the case with biologically produced and perishable agricultural products, will lead to low quality products because no premium can be obtained for higher quality products.

The right of inspection and refusal to take delivery can reduce the risks associated with *caveat emptor*, but inspection is a costly and time consuming activity and the right of refusal puts the seller at risk of opportunistic behaviour by the buyer. As a result, sellers must add a premium to the price to offset this risk. The seller who delivers milk to a retail outlet, only to have it refused on some excuse (because the buyer has found a cheaper supplier) is faced with disposing of a perishable product at short notice. The price of milk will be adjusted to reflect the probability of such actions.

Actions which foster standardization in agricultural products and reputations among agribusinesses will reduce monitoring costs. The establishment of national grading standards will reduce sorting costs. By providing common quality information to all market participants, firms can then sort products according to the grade characteristics and buyers at least know that they are talking about the same product with sellers. Private grading of products by either the buyer or seller can take place, but disputes may arise as to the impartiality of the grading. An effective dispute settlement mechanism will then be required. Simple quality arbitrations by single arbitrators are common in agricultural markets (Kerr and Perdikis, 1995). However, these require that information on how a product actually grades be made quickly available to market participants so that they can ask for arbitration. In many cases, given the distances involved, combined with poor transportation technology or the perishability of the product (graded meat on a slaughter line), this is impractical in the former command economies. If sufficient competitive outlets and/or suppliers of products exist then competition will eliminate firms which practice biased grading.

As suggested in Chapter 8, grading can also be done by the government directly. Trained government graders can provide an independent assessment of heterogeneous products. This requires that graders not be corrupt. Given the high degree of corruption in most former command economies, to expect relatively low level government employees isolated in rural enterprises to remain uncorrupt is probably unrealistic at this stage. Monitoring their activities would be very expensive and, in the end, unlikely to be effective.

Privatized independent grading organizations jointly hired by buyers and sellers might provide a viable alternative. This would require seller organizations when large numbers of suppliers are involved. The grading firm would have a stake in building a reputation for being unbiased because either party could refuse to use their services in the future. The better the reputation of the firm, the more buyers and sellers will opt to use the service. Such firms could be fostered by providing government subsidized training for graders and tying other types of support such as loans or subsidies to an agreement by agribusiness firms to move to a system of grading.

Farm level quality assurance schemes can be used to reduce monitoring costs for buyers. These schemes often relate to aspects of production which cannot be detected by grading – the use of high quality animal feeds, close monitoring of pesticide use, controlled use of drugs, etc. Quality assurance schemes require clear performance standards. These could be developed by the government and applied nationally allowing for the economies of scale which arise from large markets, but they can be established locally by negotiations between buyers and a marketing groups. Members of the scheme guarantee that the practices specified in the scheme's code of practice have been undertaken and members monitor each other to prevent shirking or cheating. The marketing group's reputation will increase with effective self monitoring. Groups with better reputations should be able to command higher prices for their products and will be able to find additional buyers. Government promotion of group marketing schemes must be done very carefully, however, because (as pointed out in Chapter 8) they may appear to be a variant of communist collectives and resisted. It may be wiser to encourage buyers to promote quality assurance group marketing schemes through price premiums. Until the groups are well established, subsidies might be paid to assist buyers with the payment of premiums for quality assured products. After the schemes are established, the subsidies could be removed. If the premium was then not commercially warranted, the quality assurance scheme would collapse. Marketing groups do, however, appear to represent a low cost method for reducing monitoring costs.

Food safety is another area where monitoring costs can be reduced. Rigorous and well enforced food safety systems are required along any food supply chain. To ensure that the food they are handling is safe for customers, agribusinesses in a food supply chain need to have: (i) adequate internal food safety procedures; and (ii) monitoring procedures to ensure that the products they receive from suppliers are safe. Both of these requirements can be assisted by the government establishing standards for food handling procedures. Further, government can assist in the training of employees who handle food.

An effective food safety monitoring regime should also be put in place by government. If it does not exist, agribusinesses receiving food products from upstream suppliers must undertake expensive testing procedures to ensure the safety of their products. Testing at each stage of the food chain may be more expensive than an effective government food inspection system.

The food safety inspectorate is particularly important for food products which have export potential. Firms in modern market economies face stiff legal penalties and loss of reputation if they sell tainted food products (Kerr and Hobbs, 1992). If prospective importers believe that the inspection system for food safety is inadequate, the only alternative may be elaborate and costly systems to monitor their suppliers. This will reduce export sales. In some cases, the cost of such a monitoring system may prove to be prohibitively expensive.

Variable and poor quality agricultural products were an endemic feature of agricultural production and distribution in the command era. In the absence of competition among buyers or suppliers, effective monitoring of food quality would have been required to raise quality standards. This proved to be prohibitively expensive and, as a result, the system delivered only a very low quality product. Changing the outlook of employees and managers regarding quality has often proven to be difficult. Without a commitment to high quality, monitoring costs along the food supply chain will continue to be high. The first step, however, is simply to define what quality, including food safety, means for a wide range of food products. The issue of food safety is discussed in more detail in Chapter 19.

## 16.2. Conclusions

Fostering the development of marketing institutions has not been a priority for those charged with shepherding the process of transforming the agribusiness sectors of the former command economies. The process of privatization has consumed most of their energies and has received the lion's share of their attention. This is probably because it is a tractable problem. Privatization has a clearly definable end point, even if the question of who will own the privatized assets has no clear answer. The problem of prices has also received considerable attention. While freeing prices may have been the goal, for most countries achieving this objective in the food sector has not been easy. Partial price controls in agriculture remain the norm.

The need to foster market institutions in agriculture was less well recognized. The absence of market institutions, however, is one of the major reasons why there have been difficulties in the privatization process, both for agricultural land and agribusiness enterprises. It is also one of the major reasons why agricultural prices have not been deregulated. Without market institutions, transaction costs remain high, competition is inhibited and undesired consequences may arise during the transition processes – multinational enterprises expanding their influence, underdeveloped export markets, etc. Unfortunately, it has taken considerable time for the importance of market institutions to become evident, or at least to determine that they are unlikely to arise spontaneously at a rate which would not constrain the transition process in the agribusiness sector.

While the importance of market institutions may now have been recognized, policies which can foster their development are not as apparent as those associated with privatization or the freeing of prices. While it is possible to copy the market institutions which currently exist in modern market economies, e.g. formal written contracts for firms to copy, consultants to assist in negotiations, strategies for locating auctions, there is little assurance that they represent the right institution for that point in time or that marketplace.

The ideas suggested in this chapter – improving communications, finding ways for firms to improve their reputations, fostering agribusiness organizations, grading, etc. – are not the types of policy initiatives which decision makers can easily take on board. They lack immediate and predictable impact. In most cases, such policies will also fail to provide the potential to create a vested interest which can act as a continuing advocate for the policy. They do, however, provide the building blocks for the market institutions which underpin agribusiness sectors in all modern market economies.

In the absence of a strong political commitment, education remains the key. Agribusiness education can inform entrepreneurs and managers of alternative institutional arrangements. Visits to the modern market economies can provide examples for agribusiness executives from the former command economies to emulate. In their dealings with their counterparts in the CEECs and the NISs, foreign agribusiness firms should take pains to explain their requirements and how these requirements will facilitate the completion of the transaction. Lowering the information costs associated with learning about market institutions for those involved in new or newly privatized agribusinesses may be the most important first step to the establishment of market institutions.

## Notes

1. Contracts are, of course, not the only way to organize complex transactions. See Kerr (1996).
2. See Chapter 4.

## References

Hobbs, J.E. (1995) A transaction cost analysis of finished beef marketing in the United Kingdom. PhD Dissertation, University of Aberdeen.

Kerr, W.A. (1993) Domestic firms and transnational corporations in liberalizing command economies – a dynamic approach. *Economic Systems* 17, 195–211.

Kerr, W.A. (1996) Managing risk and the organization of transactions: a perspective from the Pacific Rim. *Journal Manajemen Prasetiya Mulya* 3, 1–7.

Kerr, W.A. and Hobbs, J.E. (1992) Consumer protection and non-tariff barriers to trade – an example from the new British food safety legislation. *Journal of Agricultural Taxation and Law* 14, 158–165.

Kerr, W.A. and Perdikis, W. (1995) *The Economics of International Business*. Chapman and Hall, London, 274 pp.

# 17 Financing Transformation

The financial systems that have been inherited from the command system by the Central and Eastern European Countries (CEECs) and New Independent States (NISs) bear little resemblance to those which are required by a vigorous market economy. Under the command system, the financial sector was underdeveloped both in terms of its functions and in the variety of institutions. The financial system was effectively a monobank system throughout most of the command era. For example, until the late 1980s, the entire financial system of the Soviet Union consisted of a very narrow range of banks that were specialized in: enterprise credit and central banking as required by the planning process (Gosbank), savings deposits (Sberbank) and foreign trade (Vneshekonombank). In 1987 the central banking and commercial banking functions were separated. An agricultural bank (Agroprombank) was one of three state-owned commercial banks that were created and charged with extending credit on a sectoral basis. Although other commercial banks were permitted, they tended to be unimportant prior to the dissolution of the Soviet Union. Consequently, each of the NISs initially inherited their regional branches from this rudimentary financial system. The CEECs were bequeathed similar highly concentrated banking systems structured on broad sectoral lines.

Under the command system, banks served primarily as an accounting adjunct to the planning process. Payment flows between firms were registered and the implementation of the plan was monitored. Credit was extended to overcome timing problems and any balances left outstanding merely allowed an *ex post* reconciliation of the plan with reality. It is extremely important to stress that when credit was extended under the command system, there was virtually no attention to risk assessment. Interest rates were very low and whether enterprise debts rose or fell was largely inconsequential. Both on the part of the borrower and the lender, the distinction between a loan versus a grant or subsidy was, to say the least, obscure. Since state-owned enterprises were in debt to the state, this inattention to creditworthiness was natural. Unfortunately, this means that the debts or credits amassed under the command system are largely unrelated to the quality of the past, let alone the future, management of state enterprises. On the one hand, if the debts of privatized firms are left intact, firms which are well-managed and would have been viable in the long run may go out of business unnecessarily during the turbulent adjustments of the transition. On the other hand, if enterprise loans are forgiven, a signal is sent that inattention to financial discipline may be acceptable in the future as well as the past.

## 17.1. The Financial Sector in Market Economies

Before we investigate the financial issues posed by the transition, it is helpful to consider the role of the financial system in the modern market economies. In such economies, the vast array of financial markets, institutions and instruments is testimony to a multiplicity of very important functions. It is helpful to begin by distinguishing between the macroeconomic and microeconomic functions of the financial system. In virtually all of the modern market economies, there is a two-tier structure where the central bank handles what are broadly macroeconomic functions and the commercial financial system handles more microeconomic matters. The central bank must effectively regulate the money supply and facilitate any government borrowing. On the monetary management side, the central bank must balance goals such as pursuing a low and non-accelerating inflation rate, achieving high employment and maintaining an orderly exchange rate regime. With fiscal matters in the hands of the government, it is essential that central banks operate at arms length from governments in order to maintain credible monetary goals. As described above, the NISs and many of the CEECs had established two-tier financial systems prior to the onset of the transition. Most have gone on to establish arms-length relationships between their central banks and governments, but, unfortunately, some such as Belarus have moved to sharply limit this independence (World Bank, 1996).

On the microeconomic side, a number of key functions relating to the commercial financial system of a modern market economy can be identified. To begin with, the commercial financial system provides

inter-temporal intermediation between savers on the one hand and borrowers or investors on the other hand. Financial resources are
assembled to allow large investments to take place. The commercial
financial system also provides intermediation relating to risk. Possible
projects are assessed with respect to risk and credit is allocated accordingly. Risk is pooled and shared and total risk is reduced through diver-
sification. The combination of intermediation with respect to time and
risk gives rise to monitoring and enforcement as a third broad microeconomic function. The management of firms must be effectively super-
vised in order to ensure that the capital with which they have been
entrusted is employed judiciously. Finally, as in the command system,
it is necessary that the financial sector fulfil a broad accounting role by
recording, tracking and reconciling financial transactions. Due to the
complexities of finance in a market based system, even this accounting
function is more onerous than under a command system. In a previous
chapter, we have already seen the difficulties that must be overcome to
accomplish the relatively simple requirement of clearing cheques.

In a modern market economy, it is sometimes but not always apparent which of the microeconomic functions are performed by which financial institution or market. The question of who, if anyone, monitors
the management of firms is a key issue. Information is asymmetric and
management effort and competence is not transparent. Even in cases
where outright fraud is not a problem, managers have different interests
than owners. For example, managers may desire perquisites and they
may have different goals such as maximizing the growth of the firm
rather than its stock market value. There is a principal-agent problem
where the owners or their representatives (the principal) must provide
appropriate incentives for management (the agent) to conform as closely
as possible to the owners' interests. Due to the divergence of interests
between owners and managers, some combination of costly monitoring
and incentive-based remuneration schemes would seem to be required.
While incentive devices such as stock options can link managerial pay
to firm performance, the extent to which managerial interests converge
on those of owners is incomplete.

Market forces and, in particular, the avoidance of bankruptcy also
place a degree of discipline on managers. In the CEECs and the NISs,
however, the prevalence of market power will seriously weaken the
degree of discipline provided by product-market competition. The
threat of hostile takeovers may also provide some discipline, but Frydman and Rapaczynski (1994) note that this is an extremely costly control device that requires well-established, liquid capital markets that are
not likely to be present in the CEECs and the NISs for many years.

At first glance, it is tempting to believe that shareholders would
assume an active role in monitoring management. Where ownership is

widely held, however, each shareholder has the incentive to free-ride on the monitoring of others. Even where ownership is more concentrated, it may be difficult to replace ineffective managers. Typically, active monitoring by banks and large institutional equity-holders are more important methods of monitoring management in the modern market economies. For example, the threat of bank interference, even if it does not involve liquidation, may help discipline managers.

The extent to which equity markets are effective in raising capital is another area that requires inspection. Here, appearances can be deceiving, as Stiglitz points out:

> Relatively little capital is raised in equity markets even in the United States and the United Kingdom. One cannot expect equity markets to play an important role in raising funds in the newly emerging democracies. Equity markets are also a sideshow in the allocation of capital.
>
> (Stiglitz, 1992, p. 183)

As it happens, in the CEECs and the NISs:

> ... both capitalization and share turnover on these formal [securities] markets have tended to be low by both developing and industrial country standards ... Accordingly, the new markets have raised only limited funding ... In very few countries has trading been active and had a disciplinary effect on managers.
>
> (World Bank, 1996, p. 107)

Nevertheless, equity markets are required by the transitional economies as a means to transfer newly allocated property rights and, over the longer term, to facilitate an efficient distribution of risk.

> Capital markets are especially needed after the initial distribution of vouchers and shareholdings in a mass privatization program, but also for the sale of state assets through direct share offerings.
>
> (World Bank, 1996, p. 106)

Governments tend to be heavily involved in financial markets in modern market economies. This involvement includes: minimum reserve requirements on banks, mandatory deposit insurance, rules that facilitate transparency and prevent insiders exercising advantage over other investors, etc. Further, central banks are required by statute to act as lender of last resort for commercial banks. While there is controversy among economists as to whether particular interventions by governments in financial markets are necessary, there is little doubt that some government actions are required.[1] Since there are pervasive problems of asymmetric information surrounding financial markets, a *laissez faire* approach would not be efficient. Additionally, as in other publicly-held firms, information on the management practices and ultimately the solvency of financial intermediaries, is a public good. Thus, there is an incentive for each individual saver to free-ride on the monitoring efforts of others. While banks and other financial institutions may assist in the monitoring of other firms, the requirement of monitoring the monitors remains.

The transition to market-based production requires a move from a very rudimentary financial system to a complex multi-function one. Whole new financial institutions and markets must arise and the rudimentary banking system inherited from the command era must be transformed through entry and/or adaptation. Further, the role of government with respect to both the macroeconomic and microeconomic functions of the financial system requires radical change. We now turn to these issues of transition.

## 17.2. Macroeconomic Climate in the Transition

Establishing effective monetary management by a central bank is a major task of the transition. Central banks must attend to the problem of price stabilization as prices are freed from command-style regulation, and the convertibility of currencies must be established to facilitate trade. There has been considerable controversy over the pace at which prices should be deregulated, monetary restraint pursued and the stabilization of the price level accomplished. On the one side, those advocating shock treatment emphasize the need for establishing credible inflationary control at the macroeconomic level as well as credible structural change at the microeconomic level. On the other side, gradualists emphasize the recessionary consequences and extreme microeconomic dislocation that would be likely to accompany shock treatment. Gradualist policies may in fact be more credible. As Adam (1995) points out, political strife may be the logical and, by extension, anticipated consequence of shock therapy. In such cases, policy changes or reversals become almost inevitable.

Theory aside, high inflation rates, or even bursts of hyperinflation, have been the norm. To date during the 1990s, Hungary and the Czech and Slovak Republics have managed to contain inflation in double digits, but the remainder of the CEECs have experienced at least one year of triple-digit inflation. Meanwhile, all of the NISs have experienced at least one year of hyperinflation where the price level rose by over 900%.[2] While the initial bursts of inflation that have accompanied the onset of transition have tended to be much higher in the NISs than the CEECs, the reverse is true of reported unemployment rates (see Table 1.1 in Chapter 1). Nevertheless, as noted in Chapter 1, the official unemployment statistics systematically understate actual unemployment in the NISs. Moreover, unemployment appears not to have peaked yet in the NISs. Of course, throughout the CEECs and NISs there is also significant underemployment of labour.

The extent of inflation or hyperinflation during transition has further microeconomic ramifications for the financial system. Inflation has a redistributive effect and it also serves to restrict the volume of credit extended by the financial system. Inflation, of course, acts as a tax on money balances. To the extent that the transition and, thus, the resulting inflation was unanticipated, wealth was transferred from creditors to debtors through the financial system. High expected rates of inflation are typically associated with a higher variance in the inflation rate as well. Thus, high inflation resulting from the economic transition is typically associated with greater risk to economic activity in general and to extending credit in particular. Further, since inflation erodes the real value of the principal of outstanding loans, the high interest rates associated with an inflationary environment entails an effective prepayment of the principal. That is to say, for a given real interest rate, the real value of the principal is effectively paid back faster in an inflationary as opposed to a non-inflationary setting.[3] The requirement to pay back the real value of the loan more quickly can lead to a reduction in the demand for credit. Finally, for firms in the agrifood sector there is a further danger from changes in relative prices in an inflationary environment. Since the spectre of inflation invites the reintroduction of price controls, revenue increases could fall well short of cost increases leading to widespread bankruptcy.[4]

## 17.3. Raising Capital for Firms: New Banks Versus Old Banks

There are wide variations across countries in the directions that have been taken by the commercial financial systems.

> The new entry approach involves the entry of a relatively large number of new banks, and in some cases the liquidation of old banks. Russia and Estonia have both taken this path ... The alternative rehabilitation approach, adopted by Hungary and Poland among others, stresses the recapitalization of existing banks, together with extensive programs to develop them institutionally and privatize them as soon as possible.
>
> (World Bank, 1996, p. 99)

The closer a country moves towards the entry approach, the more competitive its financial markets will be, but the greater the danger that depositor confidence will be undermined as some weaker, poorly-managed banks inevitably fail. For example, entry into the banking sector in Russia may have been so rapid that a general loss in confidence among depositors is a serious possibility (World Bank, 1996). As bank regulation is in its infancy in the CEECs and the NISs, it would be extremely naive to think that regulation would prevent all insolvencies. Even in the modern market economies, regulators have failed to act in time to prevent significant problems in the financial sector (e.g. the savings and loan crisis in the US). In a nascent competitive banking system, it would be particularly useful to implement some form of deposit insurance to prevent contagion effects among depositors from sweeping the entire banking system in the event of the failure of one or two significant banks.

It should be observed that the initial burst of high inflation or, in some cases, hyperinflation that seems to have been endemic to the initial stages of the transitions has had an important procompetitive benefit for the transformation of the financial system. Inflation and especially hyperinflation causes a dramatic reduction in the real value of both the assets and the liabilities on the balance sheets of the large existing banks. In countries such as Russia, the decapitalization of ex-command banks has drastically reduced their market share. This, in combination with the entry of many new commercial banks which was discussed above, has made the banking sector much more competitive. While a once-and-for-all decapitalization of old formerly monolithic banks does have its advantages, we will see later that at least for the farm sector, there are costs as well.

Not surprisingly, in countries such as Hungary, where there has been less inflation and less entry into the banking system, efforts have been focused mainly on rehabilitating ex-command banks. Where the focus is on rehabilitating old banks, there are likely to be fewer problems of depositors' confidence, but the efficiency gains from competition will be compromised. The problems of market power associated with the rehabilitation approach should not be underestimated.

> [S]ince the break-up of the old monobank was accomplished along the old territorial branch lines, there is relatively little competition among the existing financial institutions.
>
> (Frydman and Rapaczynski, 1994, p. 130)

The reorganized banks tend to be specialized along sectoral lines and they have regional monopoly power. Thus, sectoral specialization tends to accentuate the monopoly position of the banks. This is unfortunate because, in a more competitive environment, specialization would generate benefits from appropriately tailored financial services.

Of course, various blends of the entry and rehabilitation approaches are possible and probably desirable. Frydman and Rapaczynski suggest one plan that governments could follow would be to remove some:

... bad debts from the asset column of existing banks, and in exchange to relieve them of a corresponding amount of their own liabilities. This would mean that the state, in exchange for the banks' writing off some bad enterprise debts, would remove some deposit accounts from the banks and transfer them to a new institution, together with a sufficient amount of its own treasury bills to cover these liabilities. The owners of the new bank would have to put down a certain amount of equity and pay for the infrastructure of the new institution, but the equity contribution would be leveraged by its new deposits transferred from the old state banks.

A scheme of this kind would not eliminate the existing state banks altogether. Instead, it would 'shrink' them and allow for the creation of new large banks that would provide healthy competition for the old institutions. A move of this type would also make the state's pre-commitment to not repeating the bailout more credible: the new banks would not operate according to the old rules, and their existence might make future failures of the old banks conceivable.

(Frydman and Rapaczynski, 1994, p. 135)

In many of the CEECs and especially the NISs inflation has had a cleansing effect on the balance sheets of ex-command banks without requiring this type of micro-level reorganization by the authorities. Unanticipated inflation has reduced the real value of old non-performing loans. While it is true that the real value of old loans that were performing have been reduced by the same proportion, the real value of all old loans is now lower relative to newly generated business. Thus, old non-performing loans have become a smaller proportion of the total loan portfolio. A recent World Bank (1995) survey reports that bankers tend not to view the overhang of previous bad debts as a constraint in extending new credit to agriculture. While the report is correct in noting that banks have an incentive not to publicly reveal the extent of their problem loans, the erosion of the principal of those loans has also lessened the problem.

## 17.4. Financing Farms

Despite the possibility of steering a desirable middle course between the new and old bank extremes, the market power problem is likely to have a pronounced impact on major parts of the agrifood sector. Even where there has been significant entry of new banks, they tend to be concentrated in urban areas. Consequently, much of the agrifood sector will be left to deal with monopolistic banks.

Agricultural banks, like most specialized banks, are illiquid and often bankrupt and are likely to emerge from reform much smaller – if they

survive at all. New banks are usually reluctant to serve agriculture
because the risks are high, profitability is low, credit histories are short or
absent and land is poorly registered and difficult to collateralize.

(World Bank, 1996, p. 105)

Further, agricultural banks have tended to prefer reorganized state and
collective farms over small independent farms (World Bank, 1994). Continuing uncertainty over land ownership has also impeded the use of
land as collateral and, thereby, reduced the volume of agricultural
credit.

In response to the decline in food output and the lack of activity on
rural credit markets, some countries such as Bulgaria have, unfortunately, resorted to subsidized credit. While such credit subsidies seem
to respond to public food security concerns, the efficiency grounds for
such subsidies are lacking. What is worse, as Sturgess (1994) points out,
there are several adverse long-term effects. These include: stunting the
development of regular commercial financial relationships by using
explicit rationing rather than risk assessment to ration credit, reducing
the incentive of banks to monitor borrowers, and providing an incentive
for future rent-seeking by farmers.

## 17.5. Beyond Banks

Earlier, we saw that neither the competence nor effort of agribusiness
firms' management is visible to outside observers. Since agribusiness
managers are better informed than owners, asymmetric information
leads to situations where management could at least to some extent
pursue its own interest. While this is a generic problem common to both
command and market economies, it becomes particularly acute during
the transition.

> The combination of insider autonomy and extremely ill-defined property
> rights and expectations in the Eastern European regions yields a set of
> incentives that is seriously deleterious to the regions' economies ... This
> state of affairs leads to all kinds of attempts at wild appropriation, the
> diversion of enterprise resources to private uses (even if such uses are not
> efficient), attempts to maximize present wages and employment, a
> decapitalization of enterprises through lack of reinvestment and
> modernization, increase in future indebtedness to cover present expenses,
> and so forth.

(Frydman and Rapaczynski, 1994, p. 117)

In Chapter 7 we explored a particular farm-level manifestation of this
problem where the management of reorganized state and collective
farms had an incentive to maintain excessive employment in order to
obstruct the creation of independent family farms. Ultimately, as in the

modern market economies, banks and large institutional investors may fulfil the role of monitoring management, but this does not reduce the immediate problem for the transition.

Banks are not the only intermediaries that can potentially provide capital to firms in the CEECs and the NISs during the transition to a market economy.

> Many nonbank financial institutions such as portfolio capital funds (mutual funds), venture capital funds and leasing and factoring companies are well suited to the needs of transition economies.
>
> (World Bank, 1996, p. 106)

Venture capital funds would seem to be highly suitable at many points in the agrifood chain such as food processing and the manufacture of agroinputs. An advantage of venture capital funds is that they have strong incentives to assume a supervisory role over management (Stiglitz, 1992). Further, such funds can at least in part mobilize capital in the modern market economies. In view of the advantages of venture capital funds, it is not surprisingly that the assets managed by such funds in the CEECs had grown from nothing to over $4 billion by 1995 (World Bank, 1996).

Other financial institutions could potentially serve independent farms and the rural community in general.

> The emergence of new grassroots financial institutions such as post office banks, credit unions and possibly credit cooperatives should be encouraged by providing legal, regulatory and supervisory underpinnings as well as technical assistance.
>
> (World Bank, 1994, p. 92)

Such grassroots financial institutions would provide 'active peer monitoring of borrowers' and mobilize local savings (World Bank, 1996). In the agrifood sectors of modern market economies, it is common for suppliers to extend credit to farms. While this type of credit is not yet common in the CEECs and NISs, suppliers would have a clear incentive to assume the monitoring function. In an effort to encourage foreign suppliers to extend credit to domestic farms, the Moldovan government has attempted to provide insurance against policy changes that would undermine the performance of the loan (World Bank, 1996).

While explicit, planned supplier credit can be very desirable, an insidious strain of unplanned, unsupervised inter-firm credit has tended to proliferate rapidly in the CEECs and NISs. As we saw in Chapter 5, the bilateral monopoly structure inherited from the command system locked firms into a small number of buyers. Faced with a buyer that cannot or will not pay on time, a seller may perceive little choice but to extend credit *ex post*. Legal action is costly, the outcome is uncertain.

If legal proceedings lead the buyer to declare bankruptcy, the supplier's own financial position may ultimately be compromised. Due to transaction costs, it is difficult to diversify among buyers and seek out more solvent buyers. As a result, inter-firm credit builds. Meanwhile, the seller itself may miss payments to its own suppliers because it has been forced to extend credit to its buyers. Thus, it is necessary to take measures that will curtail the spread of inter-firm credit and allow one firm to fail without producing drastic contagion effects. For example, turnover taxes could be collected at the time of payment or no later than some pre-set time limit such as sixty days after invoicing even if payment has not yet been received.

## 17.6. Hardening Budget Constraints: Financial Versus Non-financial Firms

A determined effort to harden the budget constraint of firms in the agri-food sector and non-financial firms throughout the economy is essential to the establishment of an effective market system. As a recent World Bank study states:

> [u]nder no circumstances should the financial system be used as a convenient mechanism for rescuing insolvent enterprises, public or private, in agriculture or elsewhere.
>
> (World Bank, 1995, p. 90)

Hardening budget constraints in the financial sector is somewhat more problematic to say the least. Despite some expert advice to the contrary, widespread bank failure cannot be permitted by governments in the CEECs and the NISs any more than it could be in the modern market economies. Due to the massive structural changes underway and the inexperience with financial regulations, it would be fanciful to suppose that no bail-outs would be necessary. Manifestly insolvent banks should certainly be permitted to fail, but other linked financial institutions that would otherwise be solvent should be supported. Thus, every bank should not be bailed out every time. Since current bank bail-outs can produce the expectation of future bail-outs, allowing the failure of mismanaged banks would send a useful signal without crippling the entire financial system. Meanwhile, requiring deposit insurance would minimize the impact on savers.

## 17.7. Conclusions

This chapter has shown that numerous financial issues and obstacles arise on the road to a market economy. Respecting the independence of the central bank and building and maintaining an effective two-tier financial system is crucial. While it is necessary to establish monetary control and to stabilize the price level, an inflationary burst at the outset of the transition is probably inevitable. Further, such inflation has the benefit of reducing the relative importance of old monolithic financial institutions relative to new entrants and reducing the relative importance of bad loans in the portfolios of old institutions. At non-farm stages of agrifood production, there are important opportunities for financing by means of venture capital funds. Since the rural segments of the agrifood sector are likely to face banks with dominant local positions, governments should encourage the development of grassroots institutions such as credit unions. Systems of planned supplier credit can also be helpful at the farm level. In the agrifood sector and throughout the economy it is necessary to reduce unplanned *ex post* inter-firm credit that arises when bills cannot be paid. While governments can and should permit limited bank failures in the cases of clear insolvency and mismanagement, widespread bank failures cannot be permitted to disrupt or derail the transformation process.

## Notes

1. Stiglitz (1992) suggests that it is an open question as to whether deposit insurance is useful in modern market economies where savers have access to safe assets such as treasury bills.
2. See the World Bank (1996) for inflation rates for each CEEC and NIS over the period 1990–1995. A one-year snapshot for 1993 is provided by Table 1.1 in Chapter 1 of this volume which reports figures from the UN Economic Commission for Europe. By 1993, inflation rates tended to be on the way down in most of the CEECs and Baltic States, but near their peak in the other NISs.
3. During the initial bursts of inflation in the CEECs and the NISs, the real interest rates for both savers and borrowers tended to be negative. Since there was still a margin between the rates offered to the two groups, the opportunity for profitable intermediation remained.
4. In many of the CEECs and the NISs, food prices, whether subject to regulation or not, have risen more slowly than consumer prices in general.

## References

Adam, J. (1995) The transition to a market economy in Hungary. *Europe–Asia Studies* 47, 989–1006.

Frydman, R. and Rapaczynski, R.F. (1994) *Privatization in Eastern Europe: Is the State Withering Away?* Central European University Press, Budapest.

Stiglitz, J. (1992) The design of financial systems for the newly emerging democracies in Eastern Europe. In: Clague, C. and Rausser, G.C. (eds) *The Emergence of Market Economies in Eastern Europe.* Blackwell, Cambridge, Massachusetts, pp. 161–184.

Sturgess, I. (1994) Credit and marketing in Bulgarian agriculture during the transition period. Paper presented to the symposium on The Economic Transition in Eastern Europe: the Case of Bulgaria, Exeter, April.

World Bank (1994) *Ukraine; the Agricultural Sector in Transition*. A World Bank Country Study. International Bank for Reconstruction and Development, Washington DC, 197 pp.

World Bank (1995) *Farm Restructuring and Land Tenure in Reforming Socialist Economies: a Comparative Analysis of Eastern and Central Europe*. World Bank Discussion Papers 268. International Bank for Reconstruction and Development, Washington DC, 148 pp.

World Bank (1996) *From Plan to Market*. World Development Report 1996. Oxford University Press, Oxford, 241 pp.

# 18 A System of Commercial Law

The full legal system in modern market economies consists of a number of key layers or functions. The first of these is the legislative or rule-making function itself. Next, is the investigative or fact-finding function which is fulfilled by the police and various bureaus. Many bodies are involved in the evaluative or decision-making function including the courts, regulatory tribunals and other quasi-judicial bodies, and various departments of the government bureaucracy. Finally, the enforcement function is fulfilled primarily by the police and the penal system. There is little doubt that reform at each of these layers is critical in the transition from a command to a market-based economy. Indeed at the top of Clague's list of 11 tasks of the transition one finds:

> 1. Setting up the legal infrastructure for the private sector: commercial and contract law, anti-trust and labour law, environmental and health regulations; rules regarding foreign partnerships and wholly-owned foreign companies; courts to settle disputes and enforce the laws.
>
> (Clague, 1992, p. 5)

In order to provide a stable business environment, there must be a separation between political and judicial matters. While politicians and political parties are heavily involved in the legislative process, impartiality is crucial at all other levels. The police, courts, regulatory tribunals and prisons should ideally operate at arms length from the political process. Judicial and quasi-judicial decisions should be based on the rule of law rather than power politics. By contrast, capricious political actions by politicians and/or bureaucrats severely weakens the incentives to engage in investment and economic activity. The planning system intrinsically involved a high degree of direct intervention and

313

regulation in all economic matters. In practice both bureaucrats and party operatives wielded enormous economic power under the command system. Since political and bureaucratic interference in the economy was formerly the norm, establishing a fair and effective legal system in the former command economies represents a considerable challenge.[1,2] Further, even at the legislative level, there is a need for reorientation away from direct control and towards facilitating the operation of markets and mitigating instances of market failure. In this latter regard, it should be remembered that an important economic role for government remains at the end of the transition. Even in the modern market economies, governments routinely provide public goods such as national security, alleviate externalities such as pollution, provide various forms of social assurance not supplied by markets such as unemployment insurance, rectify perceived income inequities through means such as progressive taxation and public education, and intervene where markets themselves are not competitive. Of course, given their starting points the Central and Eastern European Countries (CEECs) and New Independent States (NISs) of the former Soviet Union should be establishing effective independent legal systems at the same time that they move decisively in the direction of deregulation, privatization and withdrawal from direct economic intervention.

The legal system itself plays at least four key roles in market economies. First and foremost, the legal system must define and enforce property rights. Such property rights must be defined relative to intellectual as well as physical assets. Second, corporate law must address a wide variety of issues pertaining to the definition, structure and continuity of firms. On the one hand, it is necessary to define the legitimate forms of ownership and governance for firms, and on the other hand, bankruptcy, merger and takeover procedures must be defined. Third, contract law should promote coordination among firms in fields where inter-firm cooperation is in the social interest. Finally, competition law should prevent collusion among firms where inter-firm cooperation is not in the social interest. These four facets of market-oriented commercial law were either absent or underdeveloped under the command system.[3,4] In this chapter, it is necessary to consider the need for development in each of these four areas.

## 18.1. Property Law

In order to promote efficiency, it is extremely important that property rights be clearly assigned and easily transferable at low cost. Since state ownership was central to both the ideology and practice of the

Communist system, re-assigning property to individuals and firms and establishing mechanisms for the exchange of property rights are absolutely fundamental aspects of the transition process. Progress on both fronts has been slow and arduous, particularly in the NISs. Matters are complicated further because the assignment of property rights is often intrinsically linked to the question of whether enterprises are to be reorganized. As the analysis in Chapter 7 indicated, the question of assigning property rights over land cannot be separated from the question of whether state and collective farms are to be taken apart. Adam has observed that Hungary's land restitution policy caused 'extreme fragmentation of the land' that, in turn,

> ... created impediments to large-scale production and introduced a feeling of uncertainty into reorganized collective farms which has manifest itself in a huge decline in arable land, a dramatic decline in the animal population ... and lower efficiency.
>
> (Adam, 1995, p. 996)

The point is not that restitution is an inherently bad (or good) way of assigning property rights. Whatever the mode of assigning property rights, however, it is essential to provide support for a land market where transaction costs are low, and land titles can be expeditiously exchanged and registered. Market-driven adjustments in farm size can then take place smoothly. Part III, especially Chapters 6 and 8, discussed the issues that have arisen with various modes of privatization such as voucher systems that have been applied at the non-farm stages of agrifood production.

Virtually all of the former command economies have moved to prohibit foreign ownership of agricultural land. On the other hand, as we saw in Chapter 15, many of the CEECs have remained open to various forms of foreign investment and foreign participation in other areas of the agrifood sector.

Property rights over intellectual property as well as physical property are important. For example, patents encourage the development of new products and techniques by conferring a temporary monopoly, while trademarks facilitate the establishment of a firms' reputation.[5] In many areas such as genetics and genetic engineering, and pharmaceuticals, technological development is extremely important to the agrifood sector. The World Trade Organization (WTO) requires uniform world standards of intellectual property protection in its Trade-Related Intellectual Property-Rights (TRIPs) code. For example, a 20 year patent length is now required. Since most of the CEECs are already members and most of the NISs have applied, they will be bound by these world standards.

World standards for intellectual property protection are a mixed blessing from the standpoint of the CEECs and NISs. On the one hand, Diwan and Rodrik (1991) argue that enhanced intellectual property protection in developing countries, and, by extension, the liberalizing former command economies, will provide the incentive for the development of appropriate local technologies by domestic and foreign firms. Taylor (1993) and others have argued that enhanced intellectual property protection will also encourage technology transfer via foreign investment, licensing, etc. On the other hand, several authors have argued that the uniform world standards such as those in the WTO TRIPs code seriously disadvantage developing countries, and, by extension, the CEECs and NISs, *vis-à-vis* developed countries (e.g. Chin and Grossman, 1990; Deardorff, 1992; Richardson and Gaisford, 1996). While there is a useful role for international coordination because otherwise countries would tend to under-protect intellectual property and free-ride on the protective efforts of their neighbours, countries are not symmetric. Since technology tends to be developed in the advanced market economies, the temporary monopoly profits associated with intellectual property tend to accrue primarily in those countries. Thus, while the protection of intellectual property rights are uniform across countries, the benefits are highly asymmetric. Given that the world standards of the TRIPs code are now a fact of life, it seems that the CEECs and NISs might be wise to concentrate the preponderance of their scarce enforcement resources in other areas of commercial law.[6] Unfortunately, flawed world trade law, which is certainly not the fault of the CEECs and NISs, yields paradoxical enforcement incentives. As suggested in Chapter 15, however, poor enforcement of intellectual property rights may lead to higher levels of direct entry by agribusiness transnationals than may be desired.

## 18.2. Corporate Law

The permissible types of organization for firms that have been recognized in the CEECs are similar to those which are standard in western Europe and North America. While there is some variation across countries in the laws relating to corporate governance, Gray (1993) notes that most follow the German model of 'hands-on management by an administrative board ... and oversight functions by an independent supervisory board (p. 9).' Gray also goes on to note that:

> [t]he activities of directors should in principle be limited not only by shareholder oversight, but also laws on fiduciary responsibility, conflicts of interest, insider trading and fraud. Yet legal principles such as these are underdeveloped in the CEE [Central and Eastern European] countries.
>
> (Gray, 1993, p. 9)

Since the transition from a command to a market-based system is a shake-down period, laws governing exit or bankruptcy are extremely important.

> The typical order of priority [for bankruptcy proceedings] in advanced market economies gives secured creditors first priority with regard to secured property, and then gives priority to salaries, tax and other public liens, supplier credits and general unsecured creditors, with shareholders being the residual claimant ... In the new Hungarian law, these rules of priority are shifted to place the claims of workers for salaries and severance pay (considered here as liquidation costs) above that of secured creditors. This is likely to dampen the incentives of secured creditors to initiate bankruptcy, reduce the role of banks in enterprise restructuring and ultimately constrain the development of secured credit as a financial instrument.
>
> (Gray, 1993, p. 12)

In formulating bankruptcy law in general, there is a trade-off between providing good incentives for banks to extend credit and for banks to take on a supervisory role.

On the one hand, in virtually all the CEECs and NISs there is a disincentive to extend credit because of legal barriers that make it difficult for creditors to precipitate bankruptcy proceedings.[7] It is also true that rules of priority that increase the risk to secured creditors would have the same effect. Further, if only one country such as Hungary goes the route of lowering the priority of banks, it would find itself at a competitive disadvantage *vis-à-vis* other countries on international credit markets. On the other hand, a reduced incentive for secured creditors to precipitate bankruptcy proceedings is not altogether a bad thing in the CEECs and NISs. Moreover, there is likely to be an increased rather than a reduced incentive for banks to assume a supervisory role over enterprise management and, where necessary, precipitate restructuring. Given the acute need to oversee management discussed in the previous chapter (see Frydman and Rapaczynski, 1994), the Hungarian order of priority in bankruptcy may in fact be a reasonable balance at least as a transition measure.

There is also a case for giving priority to suppliers over banks as a temporary transition measure. Inter-firm credit has expanded rapidly in the transition period, so that the bankruptcy of one firm threatens the viability of its suppliers and customers. Giving priority to suppliers who have extended credit would help de-couple one firm from other firms with which it has business dealings in the event of bankruptcy. This is particularly important in agrifood chains because the failure of many links may be more difficult to repair than a single link and food security will be improved. If the priority of suppliers in bankruptcy proceedings is raised, however, it becomes important for other measures to be implemented to reduce the incentive of firms to extend inter-firm credit. Although bank loans would be more vulnerable, the discussion in Chapter 17 suggested that it was desirable and perhaps inevitable that enterprise budget constraints should be hardened more rapidly than those of banks and other financial intermediaries. Thus, lower priority in bankruptcy might be seen as a part of a quid pro quo for a somewhat slower hardening of banks' budget constraints.

## 18.3. Contract Law

In some situations it is beneficial to encourage cooperation between firms and in other circumstances it is desirable to discourage such cooperation. As rule of thumb, horizontal collusion amongst groups of sellers or groups of buyers restricts trade and impairs efficiency, while vertical coordination across a market between a buyer and a seller improves efficiency. Of course, there is in practice a considerable grey area between these extremes. For example, buyer–seller contracts that have exclusive dealings provisions may serve to limit horizontal competition as well as promote vertical coordination. In this section we discuss attempts to encourage buyer–seller coordination through contract law and in the next section we consider attempts to curtail collusion amongst sellers or buyers via competition law.

Generally, contracts between a buyer and a seller promote efficiency and limit opportunism. Both of these features of contracts are particularly important in the CEECs and NISs where bilateral monopoly situations abound (recall Chapter 5). If the two parties can commit themselves to act in accordance with their contractual obligations, they will be able to share the maximum total surplus. The distribution of this surplus, however, will depend on the bargaining strengths of the buyer and seller and will not normally be the competitive solution (see Gravelle and Rees, 1992). Asymmetric information (see Myerson and Satterthwaite, 1983) and asset specificity (see Klein *et al.*, 1978) compound the difficulties involved in designing and executing contracts, but in theory efficiency would still be attainable (see Tirole, 1988) if it were not for the transaction costs associated with the contracts themselves.

Contracts must either be self-enforcing or enforceable through the judicial system at low cost. When future cooperation is important to both parties, contracts may be self-enforcing in the sense that both parties would fully meet their obligations even if there were no outside authority committed to enforcement. Clearly, future cooperation is potentially important if both the buyer and seller anticipate future dealings. Indeed, in such situations the two parties may enter into a long-term contract. Even if there will not be regular on-going dealings between the two parties to a current contract, each party may place a sufficiently high value on similar future arrangements with other parties to make it worthwhile to invest in a good reputation by fully honouring their present commitments. Self-enforcing contracts, however, become more problematic as the degree of uncertainty rises. A high probability of bankruptcy for either party is particularly damaging to the prospects for self-enforcement. Unfortunately, economic and political uncertainty is rife in the CEECs and NISs because of the huge economic adjustments that are presupposed by the transition to a market-based economy and the associated political upheaval. Thus, self-enforcing contracts are a highly dubious proposition at best.

Contract law is intended to fill the breach when contracts would not be self-enforcing. In the CEECs and NISs it would be highly desirable to have an effective, inexpensive enforcement authority. Nevertheless, there is very limited experience with contract law in these countries so that recourse to the courts as a means of enforcing contracts, even where this is possible, is at present a lengthy, highly uncertain and costly process. Surprisingly enough, the fact that the court systems of the CEECs and NISs are ill-equipped to directly interpret and enforce contracts leaves open some desirable options that would pose greater difficulties in the developed market economies. Before turning to those options, it is important to consider why contract enforcement is a complicated issue.

There is much more to contract enforcement than simply punishing parties that deliberately renege on their commitments. A credible threat to bring the coercive power of the state to bear on delinquent parties is insufficient to prevent all commercial disputes. Modern business contracts have become increasingly complex and as a result there are many legitimate disputes which arise over the interpretation of contract provisions. Further, as contracts become more complex, transaction costs rise and eventually make it uneconomic to attempt to cover every possible contingency. Thus, commercial disputes between firms are also natural in the event of unanticipated circumstances. The inevitability of legitimate commercial conflicts based on differing understanding of contract provisions or unanticipated contingencies opens the door to opportunism where firms strategically renege on contracts (see Chapter 4).

The discussion in Chapter 16 suggested that private commercial arbitration was a possible alternative to the court system for settling commercial disputes.[8] We saw that arbitration may be the only reasonable alternative during the transition period while formal judicial machinery is being revamped. On the grounds of continuity, it is also constructive to actively foster arbitration as a transition measure because it was a common dispute settlement procedure for various types of international contracts even under the command system.

> On the international front, arbitration has long been relied upon in CEE joint venture contracts, and most CEE countries are party to at least some of the important international conventions. On the domestic front, however, CEE countries are just beginning to encourage the development of arbitration as an alternative to litigation. Their Chambers of Commerce, which were often involved in international arbitration during socialist economies, are leading this effort. While still very new, these efforts deserve the active support and encouragement of the international community.
>
> (Gray, 1993, p. 16)

Thus, when a firm seeks another domestic firm or a foreign firm as a potential trade and/or investment partner, both parties are likely to have confidence in the arbitration process and perhaps direct experience.

While arbitration may be the only reasonable mode of dispute settlement during the transition period, the question of the appropriate roles for arbitration versus the courts remains for the longer term. The answers to this longer term question have important immediate consequences for the types of development to be pursued within the formal legal system. Even in the modern market economies there is growing evidence that court systems have become increasingly less effective and efficient in settling commercial disputes. Recourse to the courts has become increasingly costly. As well as the direct transaction costs such as legal fees and court costs there are a number of indirect costs. The time, inclusive of appeals, to obtain a decision can be very long and the rules of evidence may require a firm to publicly reveal strategic information that is harmful to its competitive position. Further, the adversarial nature of the court system, coupled with the fact that decisions are often bound by precedent, tends to work against Pareto efficiency. Alternative judgments where both parties could be better off are often precluded. Even when a commercial dispute results in formal legal action, out-of-court settlements are an increasingly common means of avoiding some of these costs.

Arbitration is often a less costly means of commercial dispute resolution. While arbitration does sacrifice the due process of the formal legal system it offers gains in flexibility and more timely closure on disputes.

> In light of problems of the courts, arbitration may be an alternative and efficient form of resolving domestic commercial disputes. Moving towards arbitration and away from litigation in effect 'privatizes' dispute resolution itself, which can be highly desirable when government capacity is severely stretched.
>
> (Gray, 1993, pp. 15–16)

As mentioned in Chapter 16, arbitration also promotes a degree of sector-specific specialization that is very important in the agrifood industry where production is highly uncertain and subject to annual irreversibilities, and when products are perishable and heterogeneous. Arbitration also generates competitive benefits. In addition to the obvious pressure to keep transaction costs as low as possible, competition provides a strong incentive for arbitrators to develop reputations for efficiency and impartiality. Since arbitration proceedings are typically held in private, the parties can also avoid exposing corporate secrets. Further, expert witnesses can participate directly in the arbitration process without the intermediation of lawyers and, indeed, some members of arbitration panels may themselves be experts in a field rather than lawyers.

Since there are long-term as well as transitional advantages to arbitration or privatized dispute settlement relative to the courts, the CEECs and NISs should attempt to promote arbitration on an on-going basis. This means that the formal legal apparatus that is developed ought to be primarily complementary with arbitration rather than a substitute for it. One of the major shortcomings of the arbitration process is uncertainty whether the courts will enforce arbitration awards in the event that either party reneges on its initial commitment to accept the awards as binding. This possibility for opportunistic behaviour at the end of the arbitration process reduces the incentive to enter into the process in the first place. Kerr and Perdikis suggest that there are three important reasons why national courts might overturn arbitration decisions.

> 1. They [the arbitration decisions] are seen to contravene in a fundamental way the intent of domestic legislation or the principles of justice that the courts apply.
> 2. Third parties might be adversely affected by the arbitration decision.
> 3. The arbitration decision was not in the national interest.
>
> (Kerr and Perdikis, 1995, p. 206)

Domestic legislation in the CEECs and NISs could do much to reduce the uncertainty over the enforcement of arbitration awards. While the first and second reasons for overturning arbitration awards place legitimate boundaries on those awards, these boundaries should be drawn explicitly and allow for a wide range of awards. In order to reduce uncertainty, unnecessary obstacles to arbitration and restrictive limits on arbitration awards should be avoided. Generally, appeal to courts should only be permissible on the grounds that the compliance with the award would lead one of the parties to violate domestic law. Although the third reason for overturning decisions pertains more to international commercial disputes than to domestic ones, it is extremely open ended and should be entirely ruled out in domestic legislation.[9]

Thus, the primary role of the courts should be to stand ready to enforce arbitration awards. While the parties to a commercial dispute could be free to opt for formal court procedures rather than arbitration, it would not generally be economic to do so. At least in so far as domestic commercial disputes are concerned, the private sector dispute-settlement procedures being advocated for the CEECs and NISs are rather different from those that are currently followed in the modern market economies where established vested interests favour the continued dominance of the formal legal system. Although many disputes are resolved out of court, the modern market economies to a large extent remain shackled to the higher cost court system. Thus, the area of contract law is one in which starting fresh is an opportunity rather than a liability for the CEECs and NISs.

## 18.4. Competition Law

Market power is a pervasive legacy of the command system. The discussion in Chapters 4 and 5 suggested that the limited business channels established by the planning process during the communist era coupled with high transaction costs have tended to severely limit the development of competition. As a result, markets continue to be thin; bilateral monopolies and oligopolies abound, often on a regional basis. Further, ex-command firms, whether privatized or not, are typically much larger than their counterparts in the modern market economies. Within the agrifood sector, this discrepancy in firm size is particularly striking in the case of former collective and state farms. As we saw in Chapter 7, any movement towards smaller farms has tended to be weak except in Romania and Albania where land redistribution has been mandatory.

In the current circumstances, it is tempting to believe that aggressive competition law should be a priority in the CEECs and NISs. Ideally, competition among firms could be promoted, and, at the same time, the internal governance of firms could be improved by reducing firm size. Indeed, some form of anti-monopoly law is now standard at least among the CEECs. As in western Europe and North America, competition law is administered by a specialized body or tribunal and it typically addresses both issues of structure and conduct. Generally, on the market structure front, firms in dominant positions can be dismantled and mergers and take-overs are monitored to limit the extent of industry concentration. Meanwhile, on the conduct front, activities such as collusion and predatory pricing are typically prohibited (Gray, 1993).

Unfortunately, even in the modern market economies it is not always clear what should be prohibited. As we have seen, it is often difficult to determine whether or not a particular form of business relationship serves the public interest.

> *Even industrial countries have found it notoriously difficult to differentiate a restraint of trade that reduces efficiency from a legitimate business deal that raises efficiency in the short-run or long-run.* Sophisticated economic analysis . . . shows that many vertical restraints (such as tying of sales, resale price maintenance, refusals to deal, discriminatory pricing) may enhance efficiency under some circumstances . . .
>
> (Gray, 1993, p. 14; Gray's emphasis)

Thus, Willig argues for strict anti-monopoly measures in areas such as horizontal restraints where the public interest is clear.

> In view of the critical importance of competitive conduct to an emerging free market economy, there is a strong case for anti-monopoly law that includes *bright-line* rules against cartel behaviour along with criminal sanctions. Only in this way can the government deliver the clear and powerful statement of what business conduct is expected and what is forbidden, and, thereby, make plain the linkage between the drive for free markets and the requisite new business code of conduct.
>
> (Willig, 1992, p. 191)

He goes on, however, to warn against overly-expansive anti-monopoly laws.

> Anti-monopoly laws with broad provisions permitting intervention against dominant-firm behaviour and 'price-gouging' pose the danger of chilling the very investment and entrepreneurship that emerging economies sorely need.
>
> (Willig, 1992, p. 195)

Further, there are some industries in which competition is not desirable or even feasible. For example, regulation or public ownership is often a preferable alternative, and sometimes the only possibility, in the case of public utilities and other natural monopolies.

Chapter 5 contained a further important warning that the transition to a market-based economic system poses additional problems for anti-monopoly law pertaining to market structure. Evidence from the CEECs and NISs seems to suggest that it is easiest to reorganize and privatize large state enterprises on functional lines. For example, it may be easier to split major functions such as the entire dairy operation or equipment servicing component from a state farm than it is to create competing smaller farm units. Unfortunately, the increased vertical segmentation that arises when firms are dismantled along functional lines tends to create new bilateral monopoly situations and, by so doing, may impair rather than enhance efficiency.

In cases where state enterprises can be subdivided into a number of competing horizontal units, the prospects for the successful application of competition policy are brighter. In the agrifood sector, for example, there appears to be a significant opportunity for increased competition at the retail and distribution stages as well as the farm level itself. Small-scale independent trucking would be particularly advantageous since it would enhance competition amongst producers on parallel food chains. There are still significant short-term hurdles, however. Newly created competitive firms that find themselves sandwiched between monopolistic (or oligopolistic) sellers and monopsonistic (or oligopsonistic) buyers are likely to encounter acute transitional difficulties. While such competitive firms would be expected to earn zero profits in a long-run general equilibrium, this is of little solace for operators that are caught in an extreme short-run squeeze. Of course, such a squeeze is a real practical concern for family farms and small-scale private truckers. Not surprisingly, experience in the CEECs and NISs shows that the voluntary entry of new small-scale firms has rarely been rapid (see Csaki (1995) and Brooks and Lerman (1993) in reference to family farms). The mandatory dissolution of large state enterprises in no way eases the very real adjustment problems. To avoid large output declines and widespread bankruptcy among the newly created agribusinesses, temporary support measures may be necessary.

Competition law also presents a procedural issue that is particularly important during the transition to a market economy.

> In most cases the approach [undertaken by the CEECs] appears to be 'rule of reason' (meaning the illegality of a particular activity is judged on a case-by-case basis) rather than 'per se' (meaning the activity is illegal under any circumstances), giving the authorities almost unlimited discretion to choose which cases to prosecute or whom to grant exemptions in particular cases.
>
> (Gray, 1993, p. 14)

Given the general ambiguities that arise with competition law as well as the peculiarities inherent in the transition from a command to a market-based system, rule-of-reason procedures definitely make a great deal of sense in many areas in the long run. Nonetheless, as the quotation suggests, such procedures not only add to the general economic uncertainty that inhibits business, but also may undermine long-term institutional support if, on inspection, the judgements appear to be capricious, open to corruption or politically motivated.

It would be difficult to avoid the conclusion that competition law should be implemented and applied with caution during the transition. Fortunately, as shown in Chapter 13, a gradual liberalization of international trade would provide additional external pressure that would promote competition in the CEECs and NISs.

## 18.5. Enforcement

While developing the legal institutions which underlie an efficient market system is extremely important to the process of transition, without effective enforcement the effort will be largely wasted. Even the best designed commercial legal system will lose credibility if it is not enforced. Problems with enforcement will arise from two sources: (i) if the enforcement bodies such as the police or various investigative bureaus do not have the authority or resources to accomplish their responsibilities; and (ii) if the process is open to corruption or political interference. Both of these problems are widespread in the CEECs and the NISs.

In Communist states the police, judicial and investigative apparatus was used to keep the Party in power and assist in achieving Party objectives. Its hand was heavy and oppressive. Hence, there is widespread distrust of these institutions both among the wider populace and politicians. As a result, their authority has been significantly reduced during liberalization. Politicians are wary of increasing these institutions' powers. This is particularly the case for business conduct. Further, the police and other investigative arms of governments have been starved of resources. As a result, they are badly paid and demoralized. There have been almost no resources for training. As the command system had no need for commercial crime units, these skills are completely lacking in the CEECs and the NISs. Police and bureaucrats in rural areas are particularly poorly trained for these new responsibilities. Additional resources for training need to be found if enforcement is to be effective.

Poorly paid and demoralized police may be corrupt. While much of this corruption is small scale, commercial regulation holds out the possibility of very lucrative opportunities to extract bribes. In rural areas, agribusinesses may be the only source of significant corruption income. The police may even be used as part of a firm's competitive strategy – bribing police to harass one's competitors. Corruption in the police cannot be eliminated through edicts. Only if the police are reasonably well paid and receive political support when they use the authority granted to them will they gain sufficient pride in their work so that an *esprit de corps* can be re-kindled. Then the police will themselves take the measures necessary to eliminate their corrupt members. Specialized training in areas such as corporate, computer or tax evasion crimes can go a long way to creating an *esprit de corps* . Having modern skills will allow police to escape their previous stereotypes and improve their self-worth.

The judiciary and police must also be free from political interference. Only then can agribusiness firms have faith in the law . Given the long history of politicized law enforcement, it may be difficult for politicians to withdraw their influence on the legal system. Until politicians realize the damage that political interference in legal systems can cause, there is little chance that agribusinesses will have sufficient faith in the law to invest for the long term.

## 18.6. Conclusions

An effective system of commercial law is a necessary prerequisite for full transition to a modern market economy. Opaque and capriciously enforced laws increase information and monitoring costs for firms. Trust based on personal relationships becomes the only secure way to organize transactions. High transaction costs reduce competitiveness and the increased risks will reduce investment levels.

Depoliticization of all aspects of the legal system is crucial. Given the long history of political involvement, this process is likely to be extremely difficult. There also has to be a political commitment to providing sufficient resources for the legal system to function at an effective level, to reduce the incentives to be corrupt and for widespread retraining. Considerable foreign expertise can be drawn upon for the design of commercial legal systems and for training. Depoliticization of the legal system must be accomplished domestically.

## Notes

1. To a degree, the initial fall from power of the communist party tended to promote less interventionism and the move to a fairer legal system. Similarly, the resurgence of former communists in many countries such as Poland and Russia may ultimately slow down the pace of reform on both of these fronts.
2. The well-documented rise of organized crime in Russia is a particularly disturbing departure from the effective rule of law.
3. Some countries have been able to build on their pre-communist experience at least to a limited extent. This is particularly true in the CEECs where the command system was implemented after the end of World War II.
4. Of course, many other aspects of the legal system are pertinent to the development of a market system such as law pertaining to: labour relations, the environment, and banking and the financial sector. For example, the development of environmental law will be important both for the economy in general and for the agrifood sector in particular because of the extent of environmental degradation that occurred under the command system. Further, privatization has often left open the question of liability for past environmental damage. This reduces the incentive to invest because it adds a new dimension of uncertainty in the assessment of the value of the firms in question (see Gray, 1993).
5. Further, copyright protection which encourages creative effort has been expanded to cover computer software in addition to books, music, etc.
6. In a recent series of disputes with the US over the copyright protection of computer software, China, which is not yet a WTO member, appears to have been testing for the minimum threshold of enforcement effort that would just avoid US trade retaliation. In view of the preceding discussion, China's stance is hardly surprising.

7. Attempts to foreclose on collateralized loans often results in legal difficulties as well.
8. Conciliation is another private-sector means to help resolve disputes. Whereas both parties agree in advance to be bound by the decisions of an arbitrator, a conciliator merely offers advice to the parties to a dispute. Arbitration is sometimes used as a second step if conciliation has not resulted in a successful resolution of the dispute.
9. While there are obvious *ex post* strategic benefits from promoting the interests of domestic as opposed to foreign firms anywhere in the legal system, the *ex ante* effect is to reduce the number of mutually beneficial undertakings between foreign and domestic firms. Given all of the other obstacles to trade that were enumerated in Chapters 13 and 14, the legal system should preclude such national opportunism.

# References

Adam, J. (1995) The transition to a market economy in Hungary. *Europe–Asia Studies* 47, 989–1006.

Brooks, K. and Lerman, Z. (1993) Land reform and farm restructuring in Russia: 1992 status. *American Journal of Agricultural Economics* 75, 1254–1259.

Chin, J. and Grossman, G. (1990) Intellectual property rights and north–south trade. In: Jones, R.W. and Krueger, A.O. (eds) *The Political Economy of International Law*. Blackwell, Cambridge, Massachusetts, pp. 90–107.

Clague, C. (1992) Introduction: the journey to a market economy. In: Clague, C. and Rausser, G.C. (eds) *The Emergence of Market Economies in Eastern Europe*. Blackwell, Cambridge. Massachusetts, pp. 1–22.

Csaki, C. (1995) Presidential address: where is agriculture heading in Central and Eastern Europe? Emerging markets and the new role for the government. In: Peters, G.H. and Hedley, D.D. (eds) *Agricultural Competitiveness and Policy Choice*. Dartmouth, Aldershot, pp. 22–41.

Deardorff, A.V. (1992) Welfare effects of global patent protection. *Economica* 59, 35–51.

Diwan, I. and Rodrik, D. (1991) Patents, appropriate technology, and north–south trade. *Journal of International Economics* 30, 27–47.

Frydman, R. and Rapaczynski, R.F. (1994) *Privatization in Eastern Europe: Is the State Withering Away?* Central European University Press, Budapest.

Gravelle, H. and Rees, R. (1992) *Microeconomics*. 2nd edn. Longman, London, 752 pp.

Gray, C.W. (1993) *Evolving Legal Frameworks for Private Sector Development in Central and Eastern Europe*. World Bank Discussion Papers, 209. International Bank for Reconstruction and Development, Washington DC, 153 pp.

Kerr, W.A. and Perdikis, N. (1995) *The Economics of International Business*. Chapman and Hall, London, 274 pp.

Klein, B., Crawford, R.G. and Alchian, A.A. (1978) Vertical integration, appropriable rents, and the competitive contracting process. *Journal of Law and Economics* 28, 297–326.

Myerson, R.B. and Satterthwaite, M.A. (1983) Efficient mechanisms for bilateral trading. *Journal of Economic Theory* 29, 265–281.

Richardson, R.S. and Gaisford J.D. (1996) North–south disputes over intellectual property. *Canadian Journal of Economics* 19, Special Issue Part 2, S376–S381.

Taylor, M.S. (1993) TRIPs, trade, and technology transfer. *Canadian Journal of Economics* 26, 625–637.

Tirole, J. (1988) *The Theory of Industrial Organization*. MIT Press, Cambridge. Massachusetts, 479 pp.

Willig, R. (1992) Anti-monopoly policies and institutions. In: Clague, C. and Rausser, G.C. (eds) *The Emergence of Market Economies in Eastern Europe*. Blackwell, Cambridge, Massachusetts, pp. 187–196.

# 19 Health, Food Safety and Consumer Protection Regulations

Governments in most modern market economies play a key role in ensuring that the food and beverages which consumers purchase and consume are safe. Furthermore, they ensure that any potential negative side-effects which are not prohibited by law are clearly communicated to consumers (in the case of tobacco products, for example). Health issues relate to communicable diseases which can infect humans or animals and are dealt with through government veterinary inspection and monitoring services. To achieve this end, a series of health, food safety and consumer protection regulations have evolved as the agrifood systems in modern market economies have evolved. In transition economies, with the gradual privatization of food production, processing and distribution, responsibility for food safety and for food production methods which could affect consumer health moves from a concern solely of the state's to a combination of state and private enterprise responsibility. This chapter discusses the reasons why public provision of food safety, health and consumer protection regulations is necessary in a market economy, identifies some of the key issues concerning food safety and consumer protection in economies in transition and suggests steps which should be taken to establish effective control measures.

## 19.1. Why Do We Need Health, Food Safety and Consumer Protection Regulations?

Food handling and production practices at all levels of the agri-food chain, from input supplier (i.e. the effects on human health of

329

agricultural chemicals, pesticides, etc.) through to retailer can influence the quality and safety of the final product which the consumer purchases. Some animal diseases, for example, anthrax in cattle, can be passed to humans. Other diseases can impair the economic efficiency of animal production even if food safety or human health are not threatened. The personal hygiene practices of food handlers affect the probability of disease being transmitted to others through food products. The use of chemical food production processes can impair food quality and safety if the wrong types or levels of chemicals come into contact with food. Perishable products must be processed and stored under strict temperature controls to avoid deterioration and spoilage which in certain instances can make the product toxic. Cross-contamination between different foodstuffs can be a problem if raw and cooked food products are stored in close proximity. The list could go on; suffice it to say that there are an almost endless number of ways in which food safety and quality can be compromised, thereby threatening the health of consumers.

Government regulations are required because information asymmetry exists between upstream and downstream firms in the agrifood chain and between all firms in the chain and consumers. Information asymmetry arises when parties to an exchange have unequal access to information and manifests itself in the problems of moral hazard and adverse selection (see Chapter 4). Moral hazard arises after a transaction has been concluded because the actions of an economic agent are not fully observable to others, while adverse selection arises where hidden information exists prior to a transaction. Both types of information asymmetry can lead to the conditions whereby food safety and consumer health can be compromised.

Adverse selection (*ex ante* opportunism) can arise because the safety and health characteristics of food products cannot be detected visually, hence, a consumer (or downstream food processor or distributor) may unwittingly purchase a substandard product or 'lemon' (see Chapter 4). As was discussed in Chapter 11, three types of goods can be distinguished: search goods are those whose quality characteristics can easily be verified by visual inspection prior to purchase; experience goods are those whose quality characteristics can only be verified after purchase and consumption has taken place; and credence goods are those whose quality cannot be verified by non-experts even after purchase and consumption (Nelson, 1970; Darby and Karni, 1973). When consumers purchase food, two characteristics of importance are often food safety and nutrition. These can be regarded as characteristics which make food items experience or credence goods. A food product which has immediate negative impacts on consumers' health would be classed as an experience good; a repeat purchase would be unlikely following a negative experience such as food poisoning (assuming a repeat purchase was even possible and that the severity of the subsequent food poisoning was not fatal). In some circumstances, however, the negative side-effects of consuming contaminated food may not be known for some time or it may be difficult to determine which food item caused a health problem. When purchasing credence goods, consumers must rely on the reputation of the seller in making a purchase decision. If the negative side-effects do not occur for a prolonged period, however, relying on a seller's reputation alone is an inefficient way of ensuring food safety. The presence of imperfect information therefore leads to market failure (the market mechanism fails to provide adequate information about a product). Market failure is a well accepted justification for government intervention in a market economy.

Moral hazard can arise between firms along an agrifood chain after a transaction has been concluded.[1] In an ongoing contractual arrangement between two firms, the principal may be unable to observe fully the actions of the agent. Suppose a food processor has a contract to supply a retailer with sausages of a pre-agreed specification. After the contract has been agreed to and the specifications determined, the processor may discover that it is possible to substitute a different, lower quality (and even potentially harmful) dye or powder to produce a sausage of the correct colour. The principal (the retailer) is unable to observe this change without constantly monitoring the production practices of the agent (the processor). Sufficient levels of monitoring would often be extremely costly. Alternatively, frequent monitoring of incoming supplies would be necessary to detect the change in ingredients. Again, this is costly for the retailer, particularly a small-scale retailer without its own laboratory facilities.

Consequently, the presence of information asymmetry which creates risks of adverse selection and moral hazard can result in market failure and higher monitoring costs for those downstream in the agrifood chain, ultimately including consumers. From market failure and transaction cost perspectives, therefore, it is more efficient to have the government monitor and enforce food safety and food health regulations for the protection of consumers. Although large-scale retail firms could invest in food-testing laboratories, as is the case for the supermarket multiples (or chains) in many modern market economies, if they were required to monitor the food production, processing and handling practices of all upstream firms in all food chains, the monitoring and enforcement task would be monumental. Considerable duplicate monitoring would occur between retailers and between firms at different stages of the food chain. For the multitude of small private food retailers emerging in several transition economies, this task would be impossible. Similarly, it is not possible for individual consumers to adequately monitor the safety of the food which they purchase.

A similar argument can be made for the provision of consumer protection regulations (i.e. trading standards laws) to prevent food firms from falsely labelling and advertising food products. The monitoring costs involved in verifying food manufacturers' claims would be simply too high; market failure occurs if this task is left to the private sector.

Publicly-run food safety, health and consumer protection services are also in the interests of agrifood firms because consumers often rely on the reputation of sellers when purchasing food products. A public endorsement of the quality and safety of a firm's products serves to enhance this reputation. In the extreme, with no public monitoring of food safety and quality, consumers can only ensure food safety by incurring prohibitive monitoring costs and would probably seek to internalize these costs through vertical integration – in other words, through the consumption of home-grown food. The importance of garden plots as a source of food in many former command economies has increased, partly due to food security problems but perhaps also because of concerns over food safety.

## 19.2. The Situation in Transition Economies

As with many elements of the transition process, food safety standards differ widely across the former command economies. With the removal or scaling back of government control of the food chain, health, safety and hygiene standards have fallen. This has been aggravated by the dire financial position in which many state or former state owned enterprises now find themselves. Where laws stipulating food safety and health standards do exist, they are inconsistently enforced. Some of the worst conditions exist in the New Independent States (NISs). Reports suggest that, not only have the volumes of food production fallen drastically, the standards under which that food is produced have also slipped. Less than 20% of Russian food manufacturing plants may fulfil the hygiene and quality standards required by Russian law. The OECD paints a bleak picture of the situation in Russia:

> Concerns about food safety continued to grow in 1994. The collapse of government controls over agriculture and the creation of new individual farms outside the state procurement network raised general food safety concerns. Reports of counterfeit or adulterated imported and domestic food products were a staple theme in the press, perhaps most spectacularly in stories of large-scale poisoning as a result of drinking

industrial alcohol labelled and sold as vodka. Almost one-fifth of all livestock products sold to the state procurement agencies in the first half of 1994 (19 per cent of cattle and poultry, 21 per cent of fluid milk) had come from areas of the country which had experienced substantial radiation contamination from the Chernobyl and other disasters.

(OECD, 1995a, p.147)

Clearly, the problem of radiation contamination from the Chernobyl disaster does not only affect the Russian food chain, the accident occurred in what is now the Ukraine, with substantial radiation fall-out occurring over what is now Belarus. While the western European countries affected by radioactive fall-out from Chernobyl put into place measures aimed at preventing contaminated agricultural produce from entering the food chain (Kerr *et al.*, 1989), it is not at all clear that similar policies were adopted and rigorously enforced in the NISs. Hence, longer-term repercussions of the disaster for food safety and consumer health are likely in these countries.

The widespread poisoning that resulted from the deliberate mislabelling of industrial alcohol as vodka referred to above highlights the danger of false labelling and packaging of products to deliberately mislead consumers into purchasing a food or beverage. In the extreme, this may involve packaging a harmful substance (industrial alcohol) and selling it as a food or beverage. Accurate labelling is also essential for those consumers with allergies to certain ingredients. The opening of large numbers of private restaurants has overwhelmed inspection services. Training in the safe handling of food for both management personnel and employees often simply does not exist.

Some CEECs and NISs have begun to reform their food safety and consumer protection regulations. The Hungarian Food Law of 1994 introduced a broad set of rules modelled closely on European Union (EU) regulations. Issues covered by the legislation include food quality regulations but also animal breeding, feeding and health protection, national recognition of new plant varieties, trade in plant and seed material and wine production as well as tariff and subsidy regulations related to agricultural trade in general (OECD, 1995a). The wide-ranging nature of the law suggests that it may be too general to have a significant impact on food safety standards, although this will depend on the efficacy with which it is enforced.

Responsibility for food quality and sanitary control in the Polish agrifood chain is divided among several institutions. Sanitary inspection is carried out by National Sanitary Inspection, a division of the Ministry of Health and Social Welfare. Inspection of agricultural deliveries (to processing plants) and food processing, as well as veterinary (sanitary) and phytosanitary[2] inspection of agrifood products is the responsibility of the Ministry of Agriculture and Food Economy. Food inspectors from the Quality Inspection Office of the Ministry of Foreign Economic Relations certify that imported and exported agrifood products conform with quality standards. A number of research institutes (e.g. the Veterinary Institute, the Plant Protection Institute and Voivoidship (regional) Quarantine and Plant Protection Stations are all involved in pest and disease control (OECD, 1995b). These institutions all operate in the public domain.

Considerable duplication of responsibility for food safety standards clearly exists in Poland, with at least three government ministries partly responsible for inspection at different stages of the agrifood chain, plus a number of separate research institutes involved in monitoring plant and animal disease. A system which divides responsibilities in this way makes it very difficult to provide a comprehensive and cohesive food safety protection service. The Polish government, recognizing the need for rationalization of this system, has proposed a restructuring programme which would allow import and export certification procedures to be carried out in the private sector. Furthermore, the transfer of quality and sanitary control functions to the Subsector of Interprofessional Organizations (a conglomerate of farmers, traders, processors and consumers) is being considered (OECD, 1995b). While these moves would reduce the duplication and inefficiency of the current multi-government departmental system, private sector operation re-introduces the danger of market failure in the provision of adequate food safety information to consumers. Private sector control of food quality and safety inspection and certification services will require government supervision to ensure that the interests of consumers are represented. The provision of food safety by the private sector relies on building reputations; a long and complicated process at best.

Responsibility for food safety and quality in the Czech Republic is also divided among a number of public institutions:

> Sanitary and quality control activities are . . . divided among many institutions and central authorities, with no clear differentiation in some areas of responsibilities. This lack of clarity should be removed as part of the harmonization process that is closely linked to the application of the Czech Republic for EU membership.
>
> (OECD, 1995c, p. 139)

The drive for EU membership by some CEECs may provide a strong impetus for reform of food safety regulations and standards. European Union countries have harmonized their food safety standards and inspection procedures; any new member would have to conform to these European-wide standards. Furthermore, an improvement in food quality and credible assurances as to food safety will be essential if former command economies are to develop viable export markets for their agrifood products in the future.

## 19.3. Developing Consumer Protection Institutions

In modern market economies, a detailed set of food safety, animal and plant disease control and consumer protection laws exist. As discussed above, as part of the move towards a Single European Market within the EU, many of these regulations have been harmonized across member countries. This has provided member countries with an impetus to revamp and modernize their own standards, as was the case when the UK government introduced its 1990 Food Safety Act.[3] In addition, inspection procedures for plant and animal diseases which could be transmittable to humans through the food chain have been harmonized. Similar moves to harmonize food quality inspection procedures have occurred between other trading partners, for example, between the US and Canada under the North American Free Trade Agreement. Common food safety, sanitary and phytosanitary health standards reduce transaction costs for businesses because they avoid the need for duplicate testing (whereby exports must not only meet the home country's standards but also the importing country's standards) and they reduce the potential for opportunistic use of inspection procedures and food quality/safety standards as non-tariff barriers.

In most modern market economies, food and animal/plant health inspections are the responsibility of some level of government (usually national). An effective inspection service is necessary for enforcement purposes. The first step for economies in transition, therefore, is to develop an enforceable set of food safety and animal and plant health (sanitary and phytosanitary) regulations. The second step is to ensure that these laws are effectively enforced. Rather than divide responsibility for enforcement under a number of government ministries, it is usually more effective if one ministry holds that responsibility. Enforcement must also be seen to be credible. If consumers do not have confidence in the safety and general healthiness of the food which they obtain through the agrifood system, then the public inspection service has failed to correct the initial market failure. Whether a ministry of agriculture or ministry of industry should be responsible for ensuring food safety, with the possible concern of a conflict of interest within the ministry between producer/processor and consumer interests, or whether a separate Ministry of Food should be created to monitor standards in the agrifood chain is a point for further debate within the CEECs and the NISs.

Consumer protection regulations to prevent false labelling, packaging and advertising of food are also of paramount importance. In modern market economies, consumers follow information on food labels to determine the content, nutritional value and pecuniary value of food products. The requirement that firms label accurately the contents, ingredients, nutritional value and, often, price per kilogram, of food products is stipulated by law in most modern market economies. Trading standards laws ensure that products are not misrepresented, i.e. if consumers think they are buying vodka, then they should be buying vodka, not industrial alcohol labelled as vodka. Trading standards laws also prevent false or misleading advertising by food manufacturers or distributors. Food labelling laws ensure the accuracy of ingredients and nutritional information. The absence of effective consumer protection regulations in former command economies opens the door for practices which endanger the health of the general population. Corruption among inspection personnel will allow some firms to act opportunistically. As a result, the credibility of the entire agrifood system is debased. These problems should be of great concern to policy makers as the provision of a safe food supply is an essential element of food security – there would be little point in having abundant supplies of food if people believed that food to be unsafe to eat.

Enforcement of food safety, health and consumer protection regulations requires that stiff penalties be handed down to those found breaking these laws and endangering public health. For example, the UK government increased the penalties under its 1990 Food Safety Act, introducing a substantially higher maximum fine per offence and made it possible for courts to close businesses, prohibit individuals from engaging in food-related activities in the future and imprison those convicted of serious offences under the Act. The powers of enforcement held by environmental health protection officers were also strengthened under this Act. While the UK Food Safety Act (1990) may not be the appropriate model for former command economies, it demonstrates how a modern market economy, with comparatively strict regulations regarding food safety and quality, has recently acted to strengthen these regulations (Hobbs and Kerr, 1991). Those CEECs and NISs seeking to improve food safety and health standards should also introduce penalties sufficient to deter would-be offenders within their agrifood systems.

Finally, although the more widely publicized cases of food poisoning and food-related illnesses in CEECs and the NISs may have been caused by deliberate acts on the part of unscrupulous agents in the agrifood chain, the majority of problems arise unintentionally as a result of poor food handling and preparation practices. Many of these problems could be avoided through proper education and training of food handlers. In some modern market economies, food handlers (i.e.

those employed in food manufacturing plants, cooks, even those involved in the distribution of food) are required to pass a basic course in food handling, hygiene and food safety. Greater attention to personal hygiene prior to coming into contact with food is one simple and inexpensive method of reducing the incidence of diseases being transmitted through the food chain. Proper temperature control and storage facilities for perishable products, particularly meat products, can reduce problems of bacterial contamination. Often this can be achieved through education of food manufacturing plant managers and workers. There may be a role for governments in providing these basic education and training courses, at least until general standards of practice in food industries have improved. Alternatively, industry-wide training could be provided through cooperation between firms within an industry, thus achieving economies of scale in the provision of training courses. By way of example, the Danish meat processing industry (which is dominated by four voluntary farmer-owned cooperatives) operates a Meat Training College through which all meat processing workers receive training. This ranges from short in-plant food hygiene and preparation courses to four year apprenticeships. The cost of the training is borne by the state. This enables the Danish industry not only to ensure high standards of food safety but also to produce products of a consistent quality and to consistent specifications (Hobbs, 1996). Of course, Denmark is a major world exporter of pork and pork products.

## 19.4. Conclusions

Food safety, health and consumer protection regulations are essential foundations of a food supply which is both safe and nutritious. These regulations are either inappropriate for countries wishing to become market economies or ineffectively enforced in most CEECs and NISs. Market failure arises because of information asymmetry between consumers and agrifood firms; this leads to high transaction costs for consumers and for downstream firms in the agrifood chain. Consequently, governments in most modern market economies recognize the need for publicly provided food safety and consumer protection regulations to correct the market failure and lower the subsequent monitoring costs.

Access to a reliable, safe supply of food is a fundamental requirement of any society. If the agrifood system cannot be relied upon to provide this, citizens will be forced to resort to outside sources of food, such as home-produced food which they know to be safe and available. While this may be an option for those in rural areas, it is not feasible for most urban residents. The political consequences of this in terms of civil unrest are clear. Food security concerns are discussed in the next chapter; however, beyond ensuring that supplies of food are available, governments also have a responsibility to ensure that this food is safe to consume.

## Notes

1. Note that the problem of moral hazard probably only applies between firms in an agri-food chain because it is *ex post* opportunism, occurring after the contractual agreement has been reached (although not necessarily after the product or service has been delivered). It therefore applies to long-term supply relationships between two parties. Consumers tend to purchase food prducts in the spot market rather than under contractual arrangement, hence, they are directly subject to the risk of adverse selection (*ex ante* opportunism) rather than moral hazard.
2. The term phytosanitary refers to plant-related health issues, while the term sanitary in this context refers to animal-related health issues.
3. This replaced the 1984 Food Act in England and Wales and the 1956 Food and Drugs Act in Scotland and was introduced partly in response to increasing consumer concerns over food safety after a series of widely publicized cases of food poisoning and partly in response to the EU Official Controls of Foodstuffs Directive requiring harmonization of food handling and safety standards among member states (see Hobbs and Kerr, 1992).

## References

Darby, M.R. and Karni, E. (1973) Free competition and the optimal amount of fraud. *Journal of Law and Economics* 16, 67–88.

Hobbs, J.E. (1996) *Danish Pork in Asia-Pacific Rim Markets: a Culture of Excellence.* EPRI Study No. 96-01, Excellence in the Pacific Research Institute, University of Lethbridge, Canada, 74 pp.

Hobbs, J.E. and Kerr, W.A. (1991) Implications of the Food Safety Act 1990 for Scottish agribusiness. *Scottish Agricultural Economics Review* 6, 51–59.

Hobbs, J.E. and Kerr, W.A. (1992) Costs of monitoring food safety and vertical coordination in agribusiness: what can be learned from the British Food Safety Act, 1990? *Agribusiness, An International Journal* 8, 575–584.

Kerr, W.A., Kwaczek, A.S. and Mooney, S. (1989) Disaster policy and nuclear liability: insights from post-Chernobyl agriculture in the United Kingdom. *Environmental Management.* 13, 521–527.

Nelson, P. (1970) Information and consumer research. *Journal of Political Economy* 78, 311–329.

OECD (1995a) *Agricultural Policies, Markets and Trade in Central and Eastern European Countries, Selected New Independent States, Mongolia and China: Monitoring and Outlook 1995.* Centre for Co-operation with the Economies in Transition, Organization for Economic Co-operation and Development, Paris, 232 pp.

OECD (1995b) *Review of Agricultural Policies: Poland.* Centre for Co-operation with the Economies in Transition, Organization for Economic Co-operation and Development, Paris, 285 pp.

OECD (1995c) *Review of Agricultural Policies: Czech Republic.* Centre for Co-operation with the Economies in Transition, Organization for Economic Co-operation and Development, Paris, 298 pp.

# 20 Ensuring Food Security

At the most basic level, the transformation of the agrifood systems in the former command economies should provide a degree of food security which is acceptable to the citizenry. This is true both throughout the entire process of transition and at its eventual conclusion. Conceptually, food security has two important characteristics. First, it is concerned with minimums – an acceptable level of food security can be considered the lower bound constraint on the performance of an agrifood system. It is a level where trade-offs cannot be envisioned. This does not mean, however, that it is an absolute standard. Each society must define its acceptable minimum.

The second fundamental characteristic of food security is that it be all encompassing. Whatever minimum acceptable level is decided upon must be extended to all citizens. Again, this suggests that there can be no trade-offs. A system which provides food security for some but not for others does not provide food security. Famines, malnutrition and the distribution of unsafe food can all be considered breakdowns in food security.

While the conceptual characteristics of food security are relatively easy to identify, the level of food security which is considered acceptable by a society is not. The minimum acceptable standard will vary considerably from society to society and is probably highly correlated to levels of economic development. The apparent levels of concern with food safety among British consumers relating to bovine spongiform encephalopathy (BSE) – commonly called mad cow disease – for example, would probably not arise in any of the former command economies.[1] Equally, the long queues and short-run shortages which characterized the food systems in command economies would not be acceptable in the United Kingdom. A British government would probably not survive.

Food security is a relative concept. This makes it difficult to establish standards by which the performance of the agrifood systems in transition can be judged. One approach would be to compare them with the pre-reform system. This has the advantage of using expectations from the recent past as a benchmark. If one accepts the concept of food security as being an acceptable minimum without trade-offs, then the citizens of the former command economies should be no worse off during and after the transition process than before. Hopefully, they would be better off, but they may be willing to accept the same level of food security as existed before transition, provided that other improvements are manifest in the wider economy.

One should not confuse food security with levels of food consumption. Increased or decreased average levels of food consumption are possible while the level of food security remains constant. If some individuals fall below the minimum level of food security, even if the average level of consumption increases, food security will not be attained. Falling levels of consumption can be acceptable so long as the minimum is not breached.

## 20.1. Food Security in the Command Era

Did the command system provide food security? In the post-Stalin era, the answer is, by and large, yes. When a command economy did not succeed in providing food security, it was not for lack of trying. It took as its first principle that everyone should be fed. It may be, however, that the methods chosen to achieve the goal of food security often led to conditions whereby the agrifood system produced little in excess of the minimum in the short run and inhibited rising standards over the long run. Providing food security took precedence over ideological orthodoxy in many instances.

Food prices were kept low so that all consumers could afford food. Even with the huge distances in the Soviet Union, prices in areas at the furthest point from food producing regions were kept low enough to ensure that no one was excluded from the market. Of course, low prices led to excess demand for food. In some cases, the excess demand was handled through formal rationing. Rationing is an official policy response to food security problems. By not allowing prices to rise to clear the market, and thereby shutting some individuals out of the market, rationing is introduced specifically to ensure that all receive a portion. While rationing of any individual commodity may not provide a minimum standard of food security, food rationing systems, taken as a package which includes both rationed items and those available for purchase at reasonable prices, are generally designed to produce an acceptable level of food security. This package approach was followed in command economies when direct rationing was put in place.

More commonly, however, the excess demand problem created by low fixed prices was handled informally via food queues. Queuing gives advantages to those such as pensioners who, while they have low fixed incomes, also have a low opportunity cost in terms of their time. Queuing, with its first come, first served ordering of purchases, allows those with more time to increase the likelihood of acquiring the desired item. When shortages were particularly severe, it was common for employers in command economies to allow selected workers time off to queue. Selection was often based on perceived need, e.g. single mothers with large families. By itself, queuing will not solve the excess demand problem. When combined with black markets, however, queuing allows the food security problem to be solved largely on the basis of opportunity cost. Those with a perceived high opportunity cost to their time will volunteer to leave the queue and acquire the desired products in higher priced, but lower opportunity cost markets. The persistence with which authorities were willing to turn a blind eye to black markets in food suggests they understood the role these markets played in providing food security. Given the power of the security forces in communist states, these black markets would only have operated with the Party's acquiescence. Hence, the long queues and black markets which characterized agrifood systems in command economies should be considered a cumbersome method of providing food security given the system's inability to provide adequate quantities of low priced food.

Food security in command economies was also provided through crude food stock management policies – particularly the available pool of livestock. When there were good harvests, livestock numbers were allowed to increase. In times of poor harvest and, hence, low availability of grain for animal feed, livestock were slaughtered both to free up grain and to increase the quantity of meat available. While this stock management policy provided a measure of short run food security, it inhibited the long term development of the livestock industries. Excessive culling in periods of poor harvests made the process of genetic improvement more difficult because good breeding stock was sent to slaughter, while in times of herd expansion, poor quality breeding animals were retained in the herds. The phases of rapid herd expansion and contraction meant that the quantity of meat available fluctuated considerably.

In the last decades of the command era, in an attempt to improve food security and to boost meat consumption, large scale purchases of foreign grain were made. This was a major departure from the policy of self-sufficiency which was followed in most other sectors of the economy. Typically, imports were prioritized with defence requirements and advanced technology receiving the majority of foreign exchange allocations. Other sectors were forced into more cumbersome countertrade[2] arrangements if imports were desired. The willingness to import large quantities of grain when needed is very strong evidence of the Party's commitment to providing food security.

Probably the strongest evidence of the concern for food security in the former command economies was the acceptance of private plots. Ideologically, private plots were an anathema to party purists. Wadekin suggests:

> Politically and ideologically it is alien because it involves private ownership of cattle as well as of small agricultural implements and buildings, thus contradicting the Marxist–Leninist position on socialization of the means of production.
>
> (Wadekin, 1973, p. 2)

Foods security was improved by the existence of private plots in a number of ways. The official system for food distribution was notoriously inefficient and wasteful when it came to perishable fruits and vegetables. Private plots and unofficial markets were the means used to reduce waste caused by spoilage (Hedlund, 1984).

The very success of the private plots tended to emphasize the inefficiencies associated with collective and state farms. As a result, attempts were made at various times to clamp down on private plots[3] and eliminate them as a source of official embarrassment. Concerns over food security would inevitably prevail.

> Thus, although the basic Soviet ideological position on private farming has remained essentially the same ever since collectivization, practical economic and social policies have shifted to cope with the problem of food supply. Soviet leaders have had to utilize the contributions of small private producers to solve the food problem.
>
> (Wadekin, 1973, p. 3)

Hence, it seems that communist regimes went to considerable lengths to provide food security. They were willing to relax some of their most closely held beliefs in support of improving food security, e.g. private means of production, national self-sufficiency, uncontrolled market prices. All of the attempts at improving food security had to be carried out within the broader constraints of the centrally planned economy. Thus, while the communist regimes were willing to go to extraordinary lengths to provide food security within the system, attacking the fundamental problem was not contemplated. This remained the case right through to the end of the command era (van Atta, 1990).

The concerns of the Communist party regarding food security were simply the most basic of politics. Food security is one of the keys to political longevity.

> The importance in eastern bloc countries of secure food supplies at low prices takes on an importance that, for a Westerner, might be hard to comprehend. By providing food and other basic necessities at low prices, the state to some extent compensates its population for the loss of a broad range of political rights. This then is one side of the crisis (in agriculture) – the future ability of the Soviet regime to buy legitimacy by fulfilling popular expectations for improvements in food supply.
>
> (Hedlund, 1984, p. 1)

For all the waste and inefficiencies which characterize the command economies' food sectors, a reasonable degree of food security was provided for the populous.

The market system has been touted as the means to eliminate the waste which is endemic in command economies and as the route to improve production efficiency. For those observing and experiencing market reforms, there is probably no more obvious indicator of the ability of economic systems to deliver on their promises than the ability to provide food security. Hence, food security is an extremely important political issue in the process of transformation. The provision of food security has been a major concern of politicians in the former command economies. If there is widespread deterioration in food security, the political fortunes of those who have a track record in providing it are likely to be in the ascendence. The deterioration in the provision of a broad range of social security items – health care, pension income, crime prevention – as well as food security lie at the heart of the return to power of reformed communist parties in a number of countries and their strengthening in electoral popularity in most other countries. Clearly, food security during transformation has become an important political issue. A discussion of the ability of the emerging market system to provide improved food security is therefore warranted.

## 20.2. Food Security During Transformation

The question of food security during transition is complex. In part, this relates to the varied paths to reform of the agrifood system instituted in the individual CEECs and NISs. Some general trends, however, are fairly clear. Food consumption has fallen. The decline is particularly pronounced for high protein food sources such as meat and milk products. Even in the relatively prosperous Czech Republic, meat consumption declined by about 16% between 1989 and 1994 (OECD, 1995). In

Armenia, by contrast, dairy product consumption declined by 75.5%, meat consumption by 71.8% and egg consumption by 47.3%. On the other hand, potato consumption increased by almost 30% and bread consumption by over 3% (World Bank, 1995). Similar shifts in food consumption patterns are found in almost all former command economies. In part, this change in the mix of food products consumed reflects changing relative prices. As meat and dairy products were the most heavily subsidized in the command era, the liberalization of prices – even partial liberalization – has meant that the relative prices of these commodities has increased. The changes in diet, however, also reflect falling real incomes. One of the longest standing and universal empirical observations by agricultural economists is that meat and other animal protein sources have relatively large positive income elasticities. As incomes fall, one would expect consumption of these products to be adversely impacted more heavily than food staples. According to the OECD, in the Czech Republic:

> the expenditure pattern between food and non-food items did not change significantly in the process, although consumer demand for some food products decreased between 1989 and 1994. The average share of household income spent on food remained almost unchanged through the period, varying from 31.6 to 32.6 percent.
>
> (OECD, 1995, p. 66)

Over the same period, the consumption of subtropical and tropical fruits increased by over 50%, from 16.9 to 26.5 kg per capita (OECD, 1995). These changing patterns of consumption do not suggest a widespread deterioration in average food security. At the very least, one would expect consumers worried about food security to increase their budgetary expenditures on food, not decrease them as incomes fall (as implied by the constant proportion of falling income spent on food), and not to increase expenditures on luxury goods such as tropical fruit.

In parts of Asian Russia and in some of the central Asian former Soviet Republics, the decline in food consumption has been more dramatic. However, nowhere does there appear to be a widespread decline in food consumption below acceptable minimums. The severe decline in food consumption in some of the eastern sections of the old Soviet Union reflect, in part, the very considerable effort which the command system expended on providing food security. With independence, the withdrawal of the former massive Russian subsidies has led to considerable disarray.

Averages, particularly in relation to food security, can cover up large deficiencies in the food systems' operation. To some extent, the concern with food security is a concern with who falls through the cracks of the food system. The food security issue surrounding the movement to a market driven agrifood system has two aspects: (i) discovering who might fall through the cracks on the road to a market system; and (ii) the policies put in place to alleviate problems with food security.

Whose food security has deteriorated as a result of the process of transition? There are three important factors which will lead to a deterioration in an individual's level of food security during the transition to a market economy: (i) a relatively fixed income; (ii) distance from food surplus areas; and (iii) difficulty in accessing imported foodstuffs.

There seems little doubt that those who are dependent on the state for their entire income, particularly those who are unable to work, have suffered a deterioration in their food security. The groups particularly affected have been older pensioners and the disabled. The problem does not necessarily extend to the officially unemployed. The freeing of the labour market has opened opportunities for a variety of informal jobs and grey market activities. Entrepreneurship has become the new way of life. Participation in grey market activities to provide second, or even third, incomes was well developed even in the command era. Freed from the limitations on such activities imposed by the state, there has been a rapid expansion of such activities. Participation in emerging market activities is open to many whose state jobs no longer provide sufficient income for adequate food security. When those employed by the state go without pay for long periods, it must be concluded that they are generating income from the unofficial sector on a large scale. Chernina (1995) suggests that those who are really poor are those who cannot take advantage of second job opportunities. Many pensioners, to offset their falling pensions, have joined the labour market as independent providers of goods or services.

Those who are unable to work due to disability, including those pensioners truly too old to work, have suffered a decrease in their food security. Their pension incomes have simply not kept up with food price increases. The common sight of very old pensioners selling their meagre possessions in the street is stark proof of their impoverishment. The World Bank suggested that in Armenia:

> there is the risk that households most negatively affected by the change to market economy (i.e. single, aged individuals, the landless and/or those previously heavily dependent on wage labour) will be financially unable to provide a sufficient diet for their members.
>
> (World Bank, 1995, p. 13)

It would appear that pensioners who do not have the ability to enter into income generating activities arising from the liberalization of markets, no longer receive adequate food security through the income provided by the state. Their lack of food security, however, has less to do with how well the agrifood system functions than it does with the general decline in almost all aspects of social security. A more efficient (and, hence, lower priced) agrifood system could, in some cases, restore an acceptable level of food security to at least some of those currently disadvantaged. A more efficient agrifood system is, however, a long term solution. Direct government intervention is required in the short run. It is probably not surprising that some of the reformed Communist parties most fervent supporters are pensioners. Unfortunately, the types of direct intervention proposed by reformed communists are likely to be government control of prices and other forms of interference in markets. These policies may improve short term food security but will inhibit the long term improvements to food security which will arise from increased efficiency.

A second group whose food security has declined significantly are those who live in food deficit areas. Food security levels decline as the distance from centres of agricultural production increases. This is because neither the nascent markets nor the remnants of the former command system can consistently coordinate the movement of food over long distances. It is over large distances that the true inadequacies of the agrifood system become apparent – vertical coordination breaks down. Most of these problems arise in non-European Russia and the Central Asian republics of the NISs. For the latter, the problem has been exacerbated by the barriers to trade and international commercial relations arising from their political independence and the absence of trans-border infrastructure which characterized the closed Soviet society. The lack of transportation infrastructure to the former Soviet Union's neighbours in central Asia inhibits the flow of imports which could be used to counteract the current inability of the agrifood system to vertically coordinate the movement of food from surplus to deficit markets.

The major problem with the vertical coordination of food over long distances is the high level of transaction costs.[4] Without adequate market institutions, coordination tends to break down. Over long distances there are very high costs for potential buyers attempting to locate alternative suppliers. Even if alternative suppliers can be identified, the costs of assessing their reliability are prohibitively high. Without either the ability to obtain information on a potential supplier's reliability *ex ante* to a transaction or a functioning commercial legal system to enforce contracts *ex post*, the risks associated with long distance transactions are enormous. As pointed out in Chapter 16, the banking system has still not developed to the point where cheques written in one area can be cashed and cleared in another area, except between a few major cities. As a result, transactions must be undertaken on a cash basis. Even in modern market economies, the risks associated with international transactions undertaken over long distances require extremely complex market institutions to induce firms to enter into transactions –

documentary letters of credit, forfaiting, international commercial arbitration, etc. (Kerr and Perdikis, 1995).

The inability to organize transactions at great distances suggests that
if food moves through the new market system between two centres a
great distance apart, it will do so through a large number of transactions
undertaken over short distances. This reduces the efficiency of the
system and adds greatly to its cost. Costs increase as the number of
middlemen (all of whom require a return for their services) increases.
Multiple transactions will also take a great deal of time to organize.
Cash-based transactions without grading or other measures of quality
control require visual inspection. This means that foodstuffs must be
unloaded for inspection by buyers. This adds time and losses due to
increased handling.

The first effect of this system of multiple transactions is that the
movement of perishables becomes problematic. Clearly, this system of
vertical coordination cannot provide a viable cold chain for products
such as meat and fresh vegetables. Even staples such as potatoes will
suffer from multiple handlings. Grain and some pre-packaged foods
such as tinned goods will be less affected. The net result is that the
range of products available at the end of long food distribution chains
are likely to be quite limited.

Food security, however, is more likely to suffer from the increased
probability of breakdown which comes from a system consisting of multiple transactions and constrained by severely limited information. High
information costs make it difficult to match excess supply with excess
demand. This will lead to situations where suppliers do not receive
accurate price signals and, hence, inadequate quantities will be dispatched from food producing areas. Intermediate buyers will have
imperfect information regarding the prices they can expect in the
market where they expect to sell, causing them to underbid. As a result,
those at great distances from food surplus areas are likely to suffer disruptions in their food supplies. With only intermittent supplies of
staples, food security declines. Further, if supplies are intermittent,
hoarding and speculation will drive up the price of any existing stocks
of food, exacerbating the food security problems of the poor.

The remnants of the food chains which existed in the command
era are no longer able to provide food security to those in areas where
insufficient food is produced. The official sector of the food supply
system, while in many cases now privatized, is still heavily dependent
on the state due to its need for subsidies. As a result, the state can still
exert considerable influence on this sector regarding food security
issues. The effectiveness of this influence, however, has been considerably reduced.

Direct government intervention to supply food to areas of the country where unacceptable levels of food security are observed means that supplies must be diverted from more profitable alternatives. The primary means by which attempts at improving food security are made is through ministerial orders, not through the expenditure of state funds. Given that enforcement of ministerial orders is virtually impossible, managers, or their employees acting independently, can divert food out of the official channels for profit. As a result, official supplies to food deficit areas are disrupted.

The privatization of individual components of the food distribution system has, of course, led to the problems associated with bilateral monopolies. The more bilateral transactions, the greater the likelihood of a breakdown in negotiations. As the number of bilateral transactions increases with the length of the distribution chain, the probability of disruptions to supplies increases. Wengren cites reports in Russian newspapers of shortages in state stores:

> In late December 1992, a time when stores often have more food in preparation for New Years and Russian Christmas, beef could not be found in 12 cities, vegetable oil in 37, sugar in 22, rye bread in 59, potatoes in 3, milk in 3 and white bread in 35 cities.
>
> (Wegren, 1994, p. 296, fn. 60)[5]

Clearly, the combination of inadequately developed markets and the loss of control in the formal state distribution system has reduced food security in Siberia and other peripheral regions of the NISs.

On the other hand, food security has probably increased in urban areas near to food producing areas. This is particularly the case in most of the CEECs where food production is well dispersed and distances are short. This is where the market system has worked most effectively. Movements of products from the farm to urban markets is simple, often involving loading the truck, driving to the local market, and then using the truck as a market stall. Only one transaction is involved. Even if a middleman is used, cash transactions will suffice and trust based on personal relationships can develop. Consumers have relatively low cost alternatives to former state stores and can normally choose among a number of suppliers. Losses to perishable products are reduced by direct-to-consumer transactions. Price signals are received directly by the farms, allowing them to make better informed production decisions. Of course, markets developed in this fashion in the market economies. While one may wish to create more sophisticated food marketing systems, in the absence of a clear blueprint as to how that may be accomplished, traditional marketplaces provide a tried alternative when distances between on-farm production and consumers are short.[6]

The final determinant of food security is the ability to import. While imported food products can be found in all areas of the CEECs and the

NISs, in most cases the products available do little to contribute to food security. For the most part, the products available tend to be high value packaged foods or speciality items such as fresh fruit. In general, they are expensive.

All former command economies, however, have suffered short term disruptions in their food systems during the process of transition. These disruptions may be attributed to: (i) inadequate financial resources to undertake production at the farm level; (ii) disequilibrium in the processing and distribution sector as a result of underdeveloped market institutions or chains of bilateral monopolies; and (iii) the collapse of the international trade system of the Council for Mutual Economic Assistance (CMEA). Those countries with good transportation links with western Europe (Poland, the Czech Republic, Hungary) were able to mitigate disruptions by importing food staples. In some cases, food imports could be obtained at concessionary prices from surplus EU stocks.

Foreign suppliers were able to extend their existing distribution systems into nearby countries. As long as the distribution channel could remain under their direct control, then their relative efficiency ensured that food stuffs were consistently delivered into the emerging markets. When distances became too great or the bureaucratic encumberances too onerous, products had to be sold into the local distribution system and the reliability of shipments declined. In cases where parts of the existing state distribution system were acquired by foreign food companies as extensions to their existing networks, the distribution of imported products was more widespread. Foreign suppliers were often better able to respond to price signals indicating shortages than were local suppliers. Food security increased as a result.

Food security also encompasses food safety. While little information exists regarding food safety levels as discussed in Chapter 19, it seems likely that they have deteriorated. The existing food inspection bureaucracies are not flexible enough nor sufficiently well funded to adequately monitor the plethora of small food handling firms which have sprung up since liberalization. The absence of funds for reinvestment in the large former state enterprises means that food safety-related equipment, including the equipment in monitoring laboratories, will not be replaced as it wears out. Transportation equipment is also wearing out, accelerating the problems associated with distance. According to the World Bank (1995, p. 142) 'Transport for perishable products is inadequate to meet health and sanitation requirements. This lack poses serious health problems'. Further, there is no reason to believe that inspection personnel are immune to the corruption that affects all other elements of the civil service. The veterinary system is also less able to cope with health problems which has led, for example, to increased instances of Brucellosis in humans (World Bank, 1995).

The main reason that food safety problems have not become a critical issue is that the food safety system was always somewhat problematic in the Communist era, plus a large percentage of food went through the system of uninspected private plots and unofficial markets. This meant that consumers had to take a high degree of individual responsibility for the safety of their food in the past.

## 20.3. Policy Responses to Declining Food Security

In some cases the state has been relatively passive in its response to declining food security. For example, with the liberalization of labour markets and declining employment, in Russia there has been some re-migration from cities to rural areas: 'urban dwellers began to filter back to the countryside to find reliable sources of food, to become peasant farmers or simply to privatize a plot' (Wegren, 1994, p. 297). The Russian government did, however, act to facilitate the process of re-migration in search of food security.

> In 1991–92 urban dwellers received 16 million of the 19 million plots of land that were distributed throughout Russia. Urban dwellers found free land and easy credit attractive incentives as urban conditions deteriorated. Access to even small plots of land, which could be privatized for a minimal application fee, gave urban residents a source of security and access to a reliable food supply.
>
> (Wegren, 1994, p. 297)

Governments, and to a certain degree private employers, have been willing to allow their employees to exploit the food security aspects of private plots to their fullest. Time off work is often given during the important labour periods in gardening.[7] These absences from work, if nothing else, have an opportunity cost in work foregone. Clearly, one policy response to difficulties with food security has been the tacit condonement of a move to subsistence agriculture by large numbers of the population. In the economic development literature, subsistence agriculture is put forward as clear evidence of the failure of the market system to develop. It arises when food security risks are so great that they offset the efficiencies which arise from labour specialization, monoculture cash cropping and economies of scale. Small plot subsistence agriculture can provide a modicum of food security, but the opportunity cost in relatively modern economies is probably very high.

The major policy response to food security concerns, however, has been to continue the command era policy of keeping food prices low. At its heart, this is an attempt to increase the effective income of those unable to secure additional income in the liberalized economy. As Peters suggests, in command economies:

> it can be said that prices are used to distribute income, while government directives determine the allocation of resources. In market systems, prices allocate resources and government directives, operating through taxation, subsidies and regulations, adjust the resulting income distribution to meet societies priorities. In transition economies, there has first been a tendency to liberalize food prices more slowly than other prices; that is to attempt to use the price system still as a means of determining the pattern of income distribution.
>
> (Peters, 1995, p. 194)[8]

The problem with continuing to use the traditional price control mechanisms combined with queuing to address problems with food security is that much of the ability to allocate resources has been given up in the process of liberalization. Opportunities for diversion of food out of official channels abound. Liberalization of some prices and privatization of monopoly elements along the food production and distribution chain has meant that very large subsidies are necessary at some point in the food chain if prices are to be kept low for consumers. These subsidies are difficult to finance as governments' ability to collect taxes has declined. The result has been deficit financing through monetary expansion. General inflation has reduced the real incomes of those the pricing policy was designed to assist. Further, as large scale food diversion has taken place, the queuing aspect of the old food security system can no longer ensure that those with low opportunity costs for their time can obtain food security. As more food ends up in physical marketplaces, income rather than time gains in importance as a means to acquire food.

The failure to deal directly with the core problem of temporary short-falls in food security – inadequate incomes – has inhibited a solution to the longer run efficiency problem. Only by allowing markets to fully develop can the price declines required to provide food security to the widest possible number of individuals be realized. Otherwise, a large percentage of the population will continue to fall through the cracks and suffer the consequences of inadequate food security. Unless the income problem is tackled directly, similar mechanisms to those used in modern market economies to assist those who fall through the cracks due to market failure and inadequate income policies will be all that remain as policy options in transition economies. It is striking that in the case of Armenia, the World Bank was reduced to recommending that:

> there have been long traditions of giving food to relatives and neighbours in need in the villages and towns. These traditions need to be encouraged

and perhaps institutionalized to ensure adequate nutrition across the entire rural population through, for example, interventions by local churches and establishment of community food banks.

(World Bank, 1995, p. 13)

Charity, in various forms, is a laudable endeavour. In modern market economies, however, reliance on charity to provide for individuals' food security is generally identified as evidence of market and/or policy failure.

## 20.4. Policies to Ensure Food Security in Transition and Beyond

There needs to be a fundamental reorientation of food policy to ensure food security in the former command economies. Incomes rather than prices should become the focus of food security policy. Direct income subsidies to groups whose food security is at risk will prove to be a superior long run policy (Peters, 1995) rather than direct intervention in food markets. One alternative to direct financial transfers targeted at the needy might be to institute a food stamp programme similar to that in the USA. This would not lead to price distortions. The best way to ensure food security, however, is to allow consumers to spend their incomes as they wish. The choice of how to spend one's income should be spread over both food and non-food items. In this way, the relative prices necessary for markets to allocate goods can be established. Food stamps may appear to act more directly to ensure food security but preventing the development of secondary markets in food stamps may take considerable enforcement resources and, in any case, it would not be possible to prevent people from selling food obtained with food stamps. Food security will not be universal, and the secondary markets will produce inefficiencies.

Some individuals may be willing to trade off their individual food security for other goods and services. The objective of a government's food security policy should be that a minimum acceptable level of food security should be available to everyone, not that all individuals must take advantage of that level of food security. Providing adequate incomes can accomplish this. If individuals wish to jeopardize their food security through their own choice, that is not the concern of government policy. It may be that certain individuals whose food security depends on the actions of others could be provided with food directly – schoolchildren come immediately to mind. School lunch programmes, with all their failings in modern market economies, may provide a means of ensuring children's food security during transition.

Once incomes are sufficient to provide food security in the short run, policies should be put in place to ensure that a full set of market institutions develop as quickly as possible. The most fundamental of these policies is the creation and enforcement of property rights throughout the food chain (Cheung, 1982). Property rights allow individuals to capture the rents from the use, lease or sale of property. Once these rents are secure, individuals will tend to use those market mechanisms which earn rents in the manner which incurs the lowest transaction costs. The competitive pressure to lower transaction costs will increase the efficiency of the food system and eventually lower food prices, thereby improving food security.

Institutional arrangements to lower the costs of information for existing and potential businesses will help eliminate the monopoly position created by the privatization of large state enterprises. This market broadening will reduce monopoly rents, again lowering food costs.

Investments in technology to reduce information costs will be particularly important for improving the performance of the food distribution system over long distances. Combined with improved property rights enforcement through an effective system of commercial law, reduced information costs will increase the distances over which transactions can be conducted, eliminating layers of costly middlemen. The development of a banking system which will allow cheques (or more modern forms of electronic payment) to clear over long distances is a final key element in improving food security. Food security in areas which are not self-sufficient in food production should improve. Until these changes can be made, maximum use should be made of the potential of garden plots. As food prices fall with the decrease in transaction costs, individuals with high opportunity costs will stop working their plots for the convenience of purchasing food supplies. Their plots could be sold to individuals with lower opportunity costs, hence, improving their food security. The consolidation of plots could lead to small commercial market gardens.

Finally, additional resources need to be found for the enforcement of food safety standards. Given that the number of food handling establishments has increased many-fold as a result of liberalization, many of those involved in food handling are ignorant of even rudimentary food safety procedures. As replacement investment in food safety equipment has been curtailed, food inspection becomes critical. Without adequate inspection, food security will be at risk.

# 20.5. Conclusions

Issues of food security are at the heart of the problems associated with transformation of the agrifood systems of former command economies. Shortages (or high prices) can be tolerated for most non-food products. If skirts or televisions are not available or are so highly priced that only the rich can afford them, one can make the old skirt last a little longer or do without a television until the market begins to function. At least over the intermediate run, too few skirts or television sets are not likely to lead to political instability.

Food (and energy), on the other hand, require continuous flows of products at prices (or in terms of opportunity costs in the case of queuing) that all citizens can afford. If food security falls below accepted minimums, political unrest will be the result. Fear of a political backlash has made it difficult, if not impossible, for politicians to abandon the mechanisms used to provide food security in the command era and to turn to untried market mechanisms. The rapid price rises that accompanied the liberalization of prices in other sectors only confirmed politicians' worst fears. While advice from those with expertise in modern market economies was forthcoming for large numbers of issues, little was given regarding food security. Even if macroeconomic stability could be achieved, it was perceived as too dangerous to take a hands-off approach and allow markets to develop naturally.

The result was a continuation of partial price controls and attempts at allocation through government directives. This half-hearted attempt at liberalization produced the worst of all worlds. Prices, and hence resource allocation signals, remain distorted. This inhibited the full development of markets and the inevitable decline in food prices which would arise from reductions in transaction costs. On the other hand, price controls and direct allocation can no longer provide equivalent levels of food security as in the past. Liberalization has meant that opportunities for profitable diversion of food into the private sector were created. Less and less food has been available at controlled prices.

The problems of food security have been particularly acute in markets which are long distances from sources of food production. Food security declined and the distortions which inhibited improvements to production and distribution inefficiency have remained. Improvements to food security have been slow to materialize except where imports were easily accessible. The strong showing of reformed communist parties in most of the former command economies relates to concerns over declining food security.

There can be no going back. Unless the command system is restored in full, with severe punishments for food diversion, continuing attempts to improve food security by controlling prices is doomed to failure. They can only lead to further deterioration in food security. Treating food security as an income problem, on the other hand, will provide the incentives for long-term solutions to food security problems. By targeting for direct income support, those whose incomes are too low to attain acceptable levels of food security, the root of the problem will be addressed. Identifying those in need will not be easy, and some people will still fall through the cracks but the basis by which markets can assume the major responsibility for food security will have been created.

# Notes

1. This does not mean that the disease itself would not be of concern to the population, only that the food safety measures currently in place in the UK would probably be acceptable.
2. Countertrade involves the direct exchange of goods for goods rather than prices and currency exchanges.
3. A good description of the changing political fortunes of private plots can be found in Hedlund (1989).
4. One account of the difficulties associated with long distance transactions in Russia can be found in Kerr *et al.* (1994).
5. It is interesting to note that shortages were greatest for those products where the possibilities for diversion are largest; vegetable oil, sugar, rye and wheat flour are all relatively non-perishable staples. Beef, which is often frozen, and milk require cold chain facilities which do not exist outside the former state system.
6. While farmers' markets can improve food security, the high transaction costs associated with farmers' markets discussed in Chapter 8 mean that efficiency of the agrifood system is reduced.
7. In a case familiar to the authors, senior executives from an oilfield east of the Urals – an area which is not considered to have agricultural potential and is at the end of a very tenuous food distribution chain – spent most of their summers tending their garden plots with the blessing of their employers. As these executives were well paid by Russian standards (some worked for foreign oil companies) it is unlikely that they faced serious income constraints. Food deliveries had simply become too erratic, particularly in the winter. Clearly, the food security provided by the combination of the reformed command institutions and new private market traders is not considered adequate by either the individuals or their employers. In another case, attempts were being made to organize visits of three weeks duration to Canada for managers of former collective farms and other agricultural officials. Scheduling proved difficult as the officials, who were all very keen to visit the West, did not feel that they could leave during the gardening season.
8. The quote is actually from Peters' summary of a paper by Csaba Csaki of the World Bank and Stanley Johnson of Iowa State University presented to a symposium organized by the International Association of Agricultural Economists in Kiev, Ukraine, October 11–16, 1993.

# References

Chernina, N.V. (1995) Poverty as a social phenomenon in Russian society. *Russian Social Science Review* 36 (4), 3–15.

Cheung, S.N.S. (1982) *Will China Go Capitalist?* Hobart Paper No. 94, Institute of Economic Affairs, London.

Hedlund, S. (1984) *Crisis in Soviet Agriculture*. Croom Helm, London, 228 pp.

Hedlund, S. (1989) *Private Agriculture in the Soviet Union*. Routledge, London, 208 pp.

Kerr, W.A. and Perdikis, N. (1995) *The Economics of International Business*. Chapman and Hall, London, 274 pp.

Kerr, W.A., Hobbs, J.E. and Gaisford, J.D. (1994) Privatization of the Russian agrifood chain: management constraints, under-investment and declining food security. In: Hagelarr, G. (ed.) *Managment Studies and the Agribusiness: Management of Agri-Chains*. Wageningen Agricultural University, The Netherlands, pp. 118–128.

OECD (1995) *Review of Agricultural Policies – Czech Republic*. Centre for Co-operation with the Economies in Transition, Organization for Economic Co-operation and Development, Paris.

Peters, G.H. (1995) Agricultural economics: an educational and research agenda for nations in transition. *Agricultural Economics* 12, 193–240.

van Atta, D. (1990) Full-scale, like collectivization but without collectivization excesses. In: Mishkoff, W. (ed.) *Peristroika in the Countryside*. M.E. Sharp Inc., London, pp. 81–106.

Wadekin, K. (1973) *The Private Sector in Soviet Agriculture*. University of California Press, Berkeley, 407 pp.

Wegren, S.K. (1994) Weapons of the weak: rural responses to urban bias and consequences for land reform in Russia. *The Soviet and Post-Soviet Review* 21, 283–318.

World Bank (1995) *Armenia – The Challenge of Reform in the Agricultural Sector*. World Bank, Washington DC, 198 pp.

# 21 Agrifood Systems on Different Paths

From the evidence presented in this book, it should be clear that no single model for the transformation of the agribusiness sector has been consistently followed by governments in the former command economies. Probably, no such model even exists. As with the transformation of the wider economy, governments and, for that matter, entrepreneurs (both enthusiastic volunteers and reticent members of the existing managerial cohort) have been feeling their way awkwardly through transition. While there has been considerable foreign effort aimed at providing advice at the macroeconomics level as well as more limited amounts of microeconomic/managerial advice for individual enterprises, sectoral level advice has been largely absent. For the agrifood sector, what was needed was operational advice regarding how to replace food supply chains based on command with those based on market institutions. The result has been that, in the absence of a model to follow, and with the pressure of events, a plethora of observable transition initiatives have emerged. This diversity, however, may provide some indications of what the model should be. Certainly, there is no shortage of prototypes. No two countries have used the same policies. Further, it is probably safe to say that the process of transition is being undertaken in different ways among the various agribusiness supply chains within any country. The transition of the agrifood system has many paths.

## 21.1. The Reasons for Many Paths

The central question of the transition process in the agrifood sector is how to replace supply chains based on command with those based on market institutions. This does not mean that macro-economic stability is unimportant. High levels of inflation makes the business planning process exceedingly complex and inhibits the development of market institutions. Providing managerial advice is also of fundamental importance to the process of transition. While entrepreneurs have proliferated, they are not necessarily business people. A good idea and a will to succeed need to be supplemented with good business skills. The decades where the possession of business skills was considered one of the least acceptable characteristics a person could have meant that, over time, these skills largely disappeared. They now must be taught. Further, as the existing managerial cohort has been able to protect its vested interests and remain in decision-making roles, it needs extensive training in business skills if transition to a market economy is to be accomplished. While some expertise is available in the former command economies to provide advice for the training of both entrepreneurs and command era managers, the shortage of business skills remains acute (Kerr, 1996).

Reform of entire food supply chains, however, remains the key element for the transition of agribusiness. In most other sectors, except energy, greater emphasis can be given to the problems faced at individual technically separable stages of production and distribution because breakdowns in the supply chain cause much less hardship. As discussed in Chapter 20, if skirts do not move efficiently through all of the stages of their supply chain, there is no real urgency to making them do so. The volumes, perishability, and food security aspects of the agribusiness sector make reform of the supply chain as a unit a priority.

Privatization, one of the central themes of the transition process, is enterprise specific. Privatizing individual units in the supply chain provides no insights as to how the individual unit will function as part of the supply chain.

The liberalization of prices is supposed to provide the means by which the supply chain is coordinated. However, without well functioning market institutions, freeing prices simply allows firms which are positioned in supply chains as monopolies to set monopoly prices. Of course, an inefficiently functioning supply chain results, particularly if the firm is one partner to a non-cooperative bilateral monopoly. Those shepherding the transition process in the agribusiness sector in many of the former command economies have intuitively understood that these difficulties would be the outcome of privatization and the freeing of prices. As a result, there has been less deregulating of food supply chains than other sectors. Where prices were freed and privatization carried out, high transaction costs and problems with bilateral monopolies have been the result.

Ideology, culture and geography have all had their part to play in determining the paths followed by individual governments and the way in which governments have treated individual industries. Where renamed but unreformed communists have remained in power, or in some cases have been restored to power, the hand of the state has remained heavy. Where reformers have formed the government, privatization and the freeing of prices has progressed furthest.

In countries with an entrepreneurial and capitalist history and cultural heritage such as the Czech Republic, a viable model of a market system had previously existed and skills could be remembered – skills that could be passed on to younger members of society. Russia had only a small cohort of capitalists superimposed on a largely traditional economy in 1917. Those with capitalist skills had largely disappeared over 65 years of communist rule and, in any case, little that was applicable to the modern era would have been available to pass on. Individuals with business skills were even less in evidence prior to communists regimes in very traditional societies such as those in the republics of the old Soviet Central Asia.

Countries close to western Europe could easily learn from first hand examples of the workings of market economies across their borders. The benefits of market economies were also much more obvious for the citizens of those countries. These examples increased the pressure for reform. Further east, the examples from market economies are less accessible and the benefits of a market economy more imagined then experienced. The information costs associated with identifying alternative economic arrangements are far higher both for policy makers and the general populace in Russia, the Ukraine, and Kyrgyzstan than they are in the Czech Republic, Hungary or Albania. If the costs of acquiring information regarding alternative forms of economic organization are high, it will be more difficult to initiate change in social and economic institutions (Cheung, 1982). Of course, the ability to tightly control information on alternative market institutions was one of the reasons why communist regimes were able to forestall change for so long against mounting evidence of the relative failure of economic organizations based on command. Media information from the modern market economies was prohibited, jammed and otherwise difficult to acquire. Travel was strictly controlled. Open discussions of alternative economic systems was punishable by, at the very least, banishment to the nether regions of the Soviet empire. While the Berlin Wall and other border restrictions were often portrayed in the press as having been put in place to prevent a mass exodus to the modern market economies, their primary function was to prevent people from communist regimes from visiting western Europe and other market economies and returning with alternative ideas which could be shared with others. When free access was allowed, the overwhelming majority of people did not become emigrants, but rather chose to remain living in the same place.[1] To understand the fear that ideas can bring to regimes which do not allow free economic and political debate, one has only to observe the difficulty China had in dealing with information faxed to the country in the wake of the Tiananmen Square riots and its current attempts to deal with access to the internet (Anon., 1996).

The liberalization initiated by the Gorbachev regime was primarily a liberalization of information. Alternatives could be discussed, proposed and debated. The end result was, of course, not what had been envisioned at the outset. The lowering of information costs set in motion great social, economic, and political changes. As suggested above, the lowering of information costs has not been everywhere equal – in large part due to geography.

The uptake and translation of market principles has not been accomplished at the same rate or in a homogeneous fashion in all former command economies. Hence, the existing command structure remains in different degrees. One central aspect of the command era, however, has been universally purged in the reform process. Central planning is no more. This does not mean that governments no longer direct activities in the sectors where they still have the power to make decisions or that they do not attempt to influence private enterprises. There is, however, no longer a master plan. In the sectors still directly controlled by government, the various interests vie with each other for resources. The asking for resources is now from the bottom up rather than the top down allocation procedures of the past.

Agribusiness enterprises which remain within the control of the state compete with other sectors for resources. The entire sector, and particularly farms, have not faired well in this competition. Other sectors which, for example, conjure up images of modernity or which have the trappings of westernization have tended to fire the imagination of politicians. New private sector enterprises suffer from severe capital shortages as well. The privatized former state enterprises suffer the worst of both worlds, absence of commercial capital and the need to beg for subsidies from the government. The result is more chaos than order. The chaos arises from a combination of uncoordinated government decision-making and private sector decision-making in the absence of smoothly functioning markets. The degree to which decision makers' rationality is bounded is very high. While each decision may be rational given the constraints decision makers operate within, the invisible hand of Adam Smith cannot yet be relied upon to produce the result envisioned from a market system.

Of course, given the degree of experimentation which is taking place in the reorganization of food supply chains, some supply chains are performing better than others. All chains are evolving. The pace of evolution differs considerably. They represent different paths. The real question becomes then, do all the paths lead to the same destination?

Does it matter which transitional path a supply chain follows as long as the end result is the same? As long as the concerns regarding food security discussed in the previous chapter are satisfactorily resolved, does it matter how long or how convoluted the process of transformation is? In the absence of a well defined blueprint for transformation of the agrifood system, all evolutionary processes are likely to be somewhat wasteful. No one can identify the least wasteful path. Economics, as a social science, has at best a tenuous understanding of how economies in disequilibrium function. If not all paths lead to the same destination, however, then there may be genuine cause for concern.

## 21.2. Different Paths to Where?

There has been an implicit assumption by those in modern market economies and many in the former command economies that the process of transition has a clearly definable end point. The expected end point of the divergent paths is well understood. It is an economy where market forces provide the signals for economic decisions. The right to capture the gains from private property creates the ability to accumulate; avarice and self-interest motivate the desire to accumulate currently, and in the future. In short, a market economy based on a capitalist hypothesis. Government is perceived as having three roles: (i) providing institutional arrangements that facilitate the functioning of the market system with minimal transaction costs; (ii) intervening when market failure occurs; and (iii) acting to rectify income distribution problems in ways which minimize the distortion of markets. The degree to which governments intervene in relation to points (ii) and (iii) is expected to vary among countries and does not represent a different end point to the transition process. The essence of the stylized end point is a situation in which market forces direct the allocation of the vast majority of resources and the benefits of economic effort accrue largely to the owners of private property including an individual's labour and human capital.

Those involved in the transition process have had only two models of economic organization in mind. One was based on command, the other a market economy described in stylized fashion above. With the abandonment of the command system, the only alternative model and, hence, the only logical end point for transition was a market economy. Why did this bi-polar view of the world arise? In part, it reflects the simplistic 'us' versus 'them' view fostered by the Cold War. In its economic dimension, the conflict was characterized as a contest between command and market systems of organizing economic activity.

Couching the rhetoric in terms of contests may have naturally led to a perception that the victor should supplant or replace the vanquished in the contest. The end of the communist era and governments' subsequent rejections of command as a method of organizing economic activity was certainly widely perceived as a victory for the market system.

Of course, those who came to power in the former command economies and, to a considerable degree the wider population, could observe the benefits, in terms of material standards of living, which modern market-based economies could provide. There was a desire to acquire those benefits. Adopting that form of economic organization was seen as the means by which those desires could be fulfilled. Simply desiring a market economy, however, does not mean that it will arise. Over time there has come to be a realization that achieving a modern market economy will be a long, complicated process. From the simple advice provided at the beginning of the liberalization process – primarily privatize and free prices and the rest will take care of itself – the advice has become more complex, and greater and greater emphasis is being placed on institution building. The end point, however, is still expected to be an economy which resembles those in modern market economies. Advice is given in hope of speeding up the transition to that desired end point.

Transition is the process of moving from one equilibrium to another.[2] Command and market economies are, however, only two possible equilibria. The question is whether there are other equilibria which are possible end points for the transition process and are the former command economies evolving towards them?

What is a stable equilibrium? It is a state of the economy where the forces leading to change can be at least counterbalanced by the forces which work to prevent change. In other words, when the costs of initiating change exceed the costs required to prevent change.

What are the costs of initiating change? First, as discussed above, there are the costs associated with acquiring information on alternative models of economic organization. Second, there are the costs associated with overcoming the objections of those who do not wish change to occur. Those who do not wish change to occur will have a vested interest in preventing change. For change to occur, those who stand to benefit must be able to provide sufficient compensation to induce those who stand to lose to willingly accept the change. The higher these costs, the more difficult it will be to move from an equilibrium.

Which vested interests who do not wish further change to occur might be created in the process of transition? In answering this question, we will find that the path chosen may well determine the type of equilibrium eventually reached. The answer is tied to the first role of government in a modern market economy – providing institutional arrangements that facilitate the functioning of the market system with a minimum of transaction costs. According to Cheung (1982), well specified and enforced property rights are a necessary condition for a well functioning market system. Well specified property rights allow the individuals who own the rights to property to capture the benefits from that property either by selling the rights to the property or themselves putting it to productive use.

To a greater or lesser degree, property rights remain inadequately defined in the former command economies. Even where property rights are relatively well defined, enforcement is problematic at best. While the question of poorly defined property rights to agricultural land has been discussed in Chapter 7, the problem is endemic to the entire food chain. The property rights of the owners of privatized state businesses are particularly weak. There are restrictions on to whom they can sell their firms. There are constraints on the quantity of labour they must use. A portion of their output may have to be delivered to the official food distribution system at fixed prices. Former owners or other interested parties are often able to put forward counter-claims to ownership.

Property is also subject to confiscation by organized crime because law enforcement is weak. The myriad of permissions which must be obtained from government means that there is no right to create a business.

Poorly defined property rights have two effects. First, they reduce the probability that those who invest in productive assets will reap the rewards from that investment. This reduces the incentive to invest, hence, leading to underinvestment. In the former command economies, which require massive reinvestment, this underinvestment could greatly inhibit the modernization process.

Second, poorly defined property rights opens the door for others to capture part of the rewards generated by property. The primary means of capturing these rewards is corruption. Without the right to freely sell or buy property, permission must be sought. Those who have the right to grant permission can charge for doing so, hence, capturing a portion of the rewards. This is a movement away from allocating of resources according to market forces to allocation through the sale of permission (termed licensing by Cheung, 1982). Paying a bribe to receive the licence raises the transaction costs of firms. When neither property rights nor the licences granted by bureaucrats can be enforced, organized crime is able to extract protection moneys. The effect on firms is the same. In one case the granter of a licence is a bureaucrat, in the other a crime boss.

Corruption is now endemic in many former command economies. Vaksberg describes two distinctive features of corruption in Russia.

First, it has spread across the nation with lightening speed like an epidemic (pandemic is perhaps a more accurate term) that blights every sphere of activity and penetrates every pore of society to the point where it becomes the rule rather than the exception in business relations.

Second, the main carriers of the virus are the former apparatchiks, members of the nomenclature at various levels, who for years had instructed us in right living and had exacted compliance with the rules of party ethics . . . the former functionaries once appointed by the party to be 'captains of industry' hold all the key positions in the central and regional administrations and in the economy generally.

(Vaksberg, 1995, p. 210)

If a former state farm wishes to purchase fertilizer but the privatized fertilizer company is not allowed to freely sell its output because the government has not been willing to liberalize the market, then a licence must be obtained to acquire fertilizer. This allows the official who has the right to license fertilizer sales to extract a fee for granting the licence. While this adds to a firm's costs, the transaction takes place once the licence is acquired. However, in the poorly organized state apparatus of the former command economies, paying the bribe does not necessarily lead to the expected service being delivered. A new official requiring an additional licence may well be encountered the next day. The problem of multiple licences has been further exacerbated by the existence of multiple layers of government in countries such as Russia. Of course, the ability to extract rents from licensing creates an incentive for inventing licences. New licences have the effect of reducing property rights. According to Vaksberg, this licensing system is evolving in some of the former command economies:

It is . . . former functionaries who have full control of real estate; they sell, allocate, and lease land and buildings for bribes that may be in money, assets, or reciprocal favours . . . those who have something to offer in return for bribes – for example, managers of assets that have the right to sell – are for the most part the middle ranks of the former party nomenclature who have learned the modern market jargon . . .

(Vaksberg, 1995, p. 211)

If the income of officials with the ability to grant licences comes largely from the ability to create and sell licences, this leads to a vested interest in retaining the system of licences. As a result, they will work to ensure that property rights remain poorly defined and enforced. A new equilibrium with high transaction costs and allocation by licence may be created (MacKay and Kerr, 1995).

Cheung (1982) has defined this equilibrium as a licensing economy. He suggests this may be the model which best fits economies such as India or Indonesia. The results are low levels of investment in productive activities and extremely high transaction costs for firms. Endemic corruption is extremely difficult to overcome due to the sheer size of the vested interests involved. As a result, the transformation to a modern market economy may be thwarted (Sandbrook, 1986).

Of course, allowing only two possible stylized end points for the transition process is a gross simplification. A continuum of stable equilibriums can be envisioned with trade-offs between the degree to which property rights are well defined and enforced and the level of transaction costs (MacKay and Kerr, 1995). In part, this was the crux of the early debate over the sequencing of reforms and the rate at which reforms should take place (Gelb and Gray, 1991). As we have seen, there has been little consistency across countries or supply chains in the sequencing of reforms in the agrifood sector. Operationally, it is not prudent to put one's trust in Coase's (1960) conclusion that in the absence of transaction costs, it does not matter who receives the property rights. First, property rights have not been well defined. Second, and as a result of the first, transaction costs are very much in evidence. Control of the use and disposition of resources has been distributed to bureaucrats rather than property rights having been created. In many cases, control of resources has passed from the centre to former local party officials or managers from the communist era. While, in part, this reflects the shortage of managerial skills, it may also be seen as part of the compensation required to overcome resistance to any degree of reform. Turning the control of property over to local functionaries has created opportunities to profit from licensing activities and, as a result, considerably increased the cost of continuing the process of reform. The existing cohort of party functionaries and enterprise managers had to be well skilled in the politics of allocation prior to liberalization. Giving control of resources to this group may have created a situation where they are able to maximize the returns to their peculiar form of human capital.

The agrifood sector is particularly vulnerable to the incomplete allocation of property rights. In part, this is due to the problems associated with redistribution of land discussed in Chapter 7, but is primarily due to the concerns over food security discussed in Chapter 20. Continued control of prices means that the owners of property are not allowed to dispose of their property as they wish. Restrictions on the sale of agribusiness firms – to foreign firms which would bid the highest price, for example – constrains the development of markets. Refusing to allow the closure of inefficient and unprofitable enterprises due to their strategic position in a supply chain prevents efficient enterprises from entering the market. Subsidies to former state farm input supply firms reduce the profitability of the private enterprises which might grow to replace them. The creation of property rights, however, is better advanced in some facets of the food supply chain, e.g. retailing and services. Failure to effectively enforce those property rights which do exist is widespread.

The slower pace of reform in the agrifood sector means that opportunities to benefit from licensing activities by officials abound. As a result, the cost of initiating further reforms may be high. It is also easy for those with a vested interest in the *status quo* to make arguments against further reforms on the basis of food security concerns. As a result, agribusiness may remain a lagging sector in the process of transformation.

## 21.3. Conclusions

The key to the transformation of the agrifood sector in the former command economies is reducing the costs associated with initiating change. The creation and enforcement of property rights is a necessary condition for the fostering of any modern market economy. Distribution of the control of resources without the creation and enforcement of property rights leads to a vested interest in preventing the further extension of property rights.

Competition appears to be one of the keys to reducing the costs associated with altering the *status quo*. If the benefits from corruption arise from the ability to license economic activity, competition can reduce those benefits (Schleifer and Vishny, 1993). A Czech agribusiness firm faced with paying a bribe to receive an allocation of essential inputs – a chemical food preservative for example – may simply be able to obtain the input from Germany. The option of purchasing in Germany reduces the rent which the official can extract from his licence. The cheaper the alternative source of supply, the lower the rent. Low cost communications, good transportation infrastructure and open borders reduce the cost of acquiring products from Germany. The lower the rent which can be extracted from the licence, the less incentive the official has to block reforms.

Internal competition works in much the same way. As decision-making regarding resource allocation is decentralized, firms may be able to seek out alternative holders of licences for the allocation of the goods or government services required. Introducing competition into the licensing system reduces monopoly rents. The key to increasing competition is reducing the cost of acquiring information on alternative sources for those goods and services not allocated by the market. Of course, lowering information costs through improved communication systems will have the positive externality of also reducing the costs of market information for agribusiness firms and the cost of learning about alternative economic systems and institutional arrangements. The latter will add to the pressure for change.

The costs of acquiring information, however, are likely to remain high in some areas of the CEECs and the NISs. The distance which agribusiness firms are from market economies will be an important determinant of these costs. Internal distances and population density will also play a part. The emphasis that governments place on improving the ability to communicate will also be of crucial importance.

The essence of globalization is communications. In modern market economies, governments complain that the electronic communications revolution is making it difficult for them to control and regulate events. This will be true in the former command economies. Hence, the degree to which the agribusiness sectors in the former command economies are allowed to join the international system is likely to be a crucial factor in determining where the transitional paths ultimately lead.

## Notes

1. Even in East Germany, where a common language existed and where the cultural transition would be relatively easy, both during the early years of the communist regime and after its fall, most people chose to stay. Only when the option to visit the west was closed off did people, particularly skilled people, choose to flee in large numbers.
2. It may, of course, be possible that transition leads only to more transition – to a state of permanent disequilibrium or equilibriums which are not stable.

## References

Anon. (1996) *Far Eastern Economic Review*, July 4, p. 12.

Cheung, S.N.S. (1982) *Will China Go Capitalist?* Hobart Paper, No. 94, Institute of Economic Affairs, London.

Coase, R. (1960) The problem of social cost. *Journal of Law and Economics* 3, 1–44.

Gelb, A. and Gray, R. (1991) *The Transformation of Economies in Central and Eastern Europe: Issues, Progress, and Prospects*. Policy and Research Series 17, World Bank, Washington DC.

Kerr, W.A. (1996) Marketing education for Russian marketing educators. *Journal of Marketing Education* 18, 39–49.

MacKay, E. and Kerr, W.A. (1995) *Competition, Excellence and Economic Transition: China Transforming into What?* EPRI Report No. 95-03, Excellence in the Pacific Research Institute, University of Lethbridge, Lethbridge, Canada, 30 pp.

Sandbrook, R. (1986) The state and economic stagnation in tropical Africa. *World Development* 14, 319–332.

Schleifer, A. and Vishny, R. (1993) Corruption. *Quarterly Journal of Economics* 18, 599–617.

Vaksberg, A. (1995) Fitting the punishment to the crime, and politics too. In: Isham, H. (ed.) *Remaking Russia*. M.E. Sharp, New York, pp. 203–215.

# 22 Joining the International System

The difficulties of planning and executing trade between planned economies, which were discussed in Chapter 13, dictated that the economies and agrifood systems of the Soviet bloc countries remained relatively self-sufficient in the communist era. After the collapse of communism, a further erosion of trade between the Central and Eastern European Countries (CEECs) and the New Independent States (NISs) occurred because of the dissolution of the Council for Mutual Economic Assistance (CMEA), the move to world prices for oil from Russia and the Caspian states and, no doubt, the underlying political suspicions of Russia.

For the NISs themselves, additional forces of fragmentation have been at work. Under communism, the Soviet Union functioned as a single planned economy. The dissolution of the Soviet Union brought with it difficulties in dividing assets and administrative bodies and tended to divert some energy and attention from market reform. In very large measure, the dissolution of the Soviet Union initially resulted in the creation of mini-command economies at what had been the republic level. The Commonwealth of Independent States (CIS) which was created to minimize the effects of economic fragmentation has been, at best, only partially successful. Much of the international trade of the NISs is still orchestrated by state trading monopolies or their successor organizations, often on a barter basis (OECD, 1995).

Even within the Soviet Union, economic planning in the agrifood sector emphasized a fairly high degree of regional self-sufficiency to try to reduce transport costs, spoilage, etc. Consequently, the degree of regional specialization and intra-regional integration within the food system of the Soviet Union was lower than would otherwise be expected on the basis of the immense and highly varied geography. In some respects, this unusually high degree of food autonomy has been a short term blessing for the NISs. As suggested in Chapter 21, the discipline imposed by foreign competition may be one of the major determinants in reducing vested interests in the *status quo*. Reducing vested interests is fundamental to continued transition to a market-based economy. While the new difficulties of international as opposed to intra-union trade between the NISs have aggravated the problems of food distribution and contributed to the overall economic decline, major food crises have not yet been manifest. Nevertheless, the initial erosion of trading links has resulted in increased isolation of the NISs that has become a handicap to growth and development.

Integration into the international system is one of several necessary conditions for a successful transition for the CEECs and NISs. In Chapters 13 and 14 we discussed the import and export sides of international trade and trade policy from the perspective of an individual CEEC or NIS. Nevertheless, in addition to national governments, the international trading system consists of many important supra-national arrangement and organizations. For example, at the level of global trade, there is the General Agreement on Tariffs and Trade (GATT) and now the broader World Trade Organization (WTO).[1] At the regional level, there are economic blocs such as the European Union (EU) and the North American Free Trade Agreement (NAFTA). There are also separate institutions and arrangements that govern international trade in specific commodities.[2]

The very existence of multilateral organizations and arrangements stands as stark evidence that free trade does not constitute a spontaneous equilibrium for the international economic system even though such a state would be efficient in the absence of market failure. As mentioned in Chapter 13, it is individually rational for national and even sub-national governments to directly and indirectly intervene in international trade in the course of pursuing a multitude of domestic political goals such as re-election and related economic goals such as supporting the incomes of various sectors or groups, stimulating regional development and promoting overall growth. For example, in the modern market economies where farm sectors have been subject to prolonged pressure from sectoral technological progress and overall economic growth coupled with low income elasticities for most food products, governments have implemented many support policies that have had a pronounced impact on international trade.

In equilibrium, the governments of all countries will intervene in international trade.[3] There are many countries trading many goods, the goals and motivations of governments are highly complex and not necessarily consistent, and the range of policy instruments goes well beyond tariffs. Free trade will not arise as a spontaneous outcome in the real world with all its complexity. Thus, international trade relations can be characterized as a prisoners' dilemma. Each country's own distortionary interventions in the world markets further its national interests, or at least the interests of its government, but the distortionary actions of other countries usually work in the opposite direction. Consequently, all countries typically end up worse off.[4] The incompletely resolved conflict over agricultural export subsidies between the EU and the US that developed in the 1980s is just one such case. As a result of the prisoners' dilemma, the maintenance of liberal trade relations at the global and regional levels through cooperative multi-lateral efforts is the interest of virtually every country. It is in this context that the evolution of a supra-national institutional structure can be best understood.

This chapter will explore the multilateral dimension of the international system from the standpoint of the CEECs and NISs. The CEECs and NISs individually and jointly are already participating in a plethora of trade arrangements and organizations. The evolving multi-lateral arrangements involving the CEECs and NISs will be considered in conjunction with some crucial questions for the future. How important to the CEECs and NISs is membership in the GATT and the WTO? Are trade organizations and arrangements among the NISs and/or the CEECs themselves important? Should these countries seek membership in the EU and, if so, under what terms? What does international participation mean for the agrifood systems in the CEECs and NISs?

## 22.1. Membership in the GATT and WTO

Most of the CEECs had joined the GATT prior to the end of 1995 (OECD, 1995) and, thus, were eligible for membership in the WTO from its inception on 1 January 1995. The memberships of the Czech Republic, Hungary, Romania, and the Slovak Republic were effective as of 1 January 1995; Poland's membership was effective as of 1 July 1995; and the membership of Slovenia was effective as of 30 July 1995. By mid-1996, working groups were considering membership for Albania, Bulgaria, Croatia, Macedonia and all of the NISs except Turkmenistan and Azerbaijan (WTO web-site, September 1996).

There are at least three important reasons why membership in the GATT and WTO is crucial to the CEECs and NISs. To begin with, GATT forces some desirable disciplines on the CEECs that have become members and would do the same for the NISs. In other words, the GATT prevents member countries from taking measures that are counterproductive from the standpoint of their own broadly-defined national welfare, but for which there are some domestic clientele with vested interests. For the CEECs that have become members, the Uruguay Round Agreement requires that Most-Favoured Nation (MFN) status be granted immediately to all other members and that non-tariff measures be converted to tariffs. These requirements are very desirable in that they will add transparency to trading relationships and prevent discrimination against any trade partner.[5] Further, in Chapter 13 we argued that tariffication itself has desirable pro-competitive effects which are important in diminishing the problems of bilateral market power that are a legacy of the command system. When import quantities are not set directly, domestic producers are constrained by the inflow of additional imports if they attempt to raise prices (Helpman and Krugman, 1989). Members are also bound by upper limits on their tariffs. As we saw in Chapter 13, however, the Uruguay Round does not require much, if any, tariff reduction from the new members in Central and Eastern Europe. Due to the particular nature of their circumstances, the CEECs that have become GATT members were not required to bind their tariff ceilings at the levels prevailing prior to the Uruguay Round. Since the CEECs bound most of their tariffs at rates higher than their actual rates, self-discipline will be required for them to maintain, let alone gradually reduce, their tariff rates (OECD, 1995). Similar self-discipline would be required to move towards a uniform tariff structure that would allow world prices to dictate a more efficient use of resources across import-competing products and reduce the opportunities for rent-seeking by domestic firms and industries. Membership in the GATT will, however, limit the use of export subsidies.

The second key reason why GATT membership is important to the CEECs and NISs is that it provides guaranteed access to foreign markets on an MFN basis. By joining GATT, a CEEC or NIS assures that other members will not discriminate against its exports. This benefit from access is of a one-time nature that is not directly related to the benefits of further trade liberalization by foreign countries. Uruguay Round trade liberalization may, however, prove to be very beneficial for many of the CEECs and NISs. Whether they are GATT members or not, the Uruguay Round trade-liberalization commitments made by the modern market economies are likely to have broadly, but not exclusively, favourable effects for the CEECs and NISs. On agricultural and non-agricultural markets alike, Uruguay Round tariff reductions in the modern market economies will be beneficial to those CEECs and NISs that are exporters because the world price will rise. Likewise, the upward price change on those agricultural markets such as grains where the modern market economies are required to reduce export subsidies will be beneficial to those CEECs and NISs that are also exporters. Of course, on both the above types of markets, the higher world prices occasioned by the

Uruguay Round will reduce overall (i.e. consumer plus producer) welfare for any CEEC or NIS that is an importer despite the advantage to import-competing domestic producers. Thus, any given country will be adversely affected by foreign liberalization in some markets even though the overall impact is likely to be favourable.

The third reason why acceding to the GATT and WTO is desirable is the least obvious but perhaps the most important. Membership in the GATT is valuable as a commitment and signalling device. By accepting the disciplines of the GATT, it becomes more difficult for any country to back-slide into protecting or promoting individual domestic vested interests. Rent-seeking interests can be resisted more readily. For the CEECs who are members and the Baltic countries that are completing the process of joining, the GATT commitments will help consolidate the market-reform process and provide a more certain environment for both domestic and foreign investment. Increased trade broadens markets and lowers information costs.

## 22.2. Trade Among the CEECs and NISs

Trade relations between the CEECs is in a state of flux pending the deepening of connections with the EU. In 1993, the division of Czechoslovakia into the Czech Republic and Slovakia was accompanied by the formation of a customs union between the two republics, but a recent study notes a growing number of trade frictions in this relationship (OECD, 1995). A Central European Free Trade Agreement was also concluded between the Czech Republic, Hungary, Poland and Slovakia in 1993. The Baltic Free Trade Agreement between Estonia, Latvia and Lithuania has been in effect since mid-1994 but agriculture was initially omitted from this deal.[6] Estonia is also negotiating free trade agreements with the Czech Republic, Hungary, Poland and Slovakia (OECD, 1995). Lithuania has concluded an agreement easing the transshipment of Russian products in return for access to the Russian market on an MFN basis (OECD, 1995).

There has been a near complete collapse of trade relations between the new republics of the former Yugoslavia because of armed hostilities, first in Croatia and then in Bosnia Hertzegovina, and sanctions against Serbia. It remains to be seen whether the US-brokered peace agreement for Bosnia and the lifting of sanctions on Serbia will be successful in stabilizing political, let alone economic, relations. Slovenia, the northernmost of the new republics has managed to avoid major hostilities and has begun establishing ties with the CEECs and western Europe. Although Slovenia has attempted to maintain liberal trade with the former Yugoslav Republic of Macedonia through a free trade agreement, there have been a number of snags in the implementation process. Slovenia has also established more successful trade agreements with Hungary, the Czech Republic and Slovakia (OECD, 1995).

The dismantling of the Soviet Union was accompanied by the birth of the Commonwealth of Independent States. While the CIS remains in existence, it has not even begun to provide a structure for market-based trade among the NISs. Indeed, the CIS functions in a manner that is reminiscent of the Council for Mutual Economic Assistance (CMEA) that governed trade between the countries of the former Soviet bloc.

> The New Independent States lack the infrastructure for international trade. Slow and uncertain clearing of payments, arbitrary and unpredictable customs duties and transportation problems further complicate international trade among them.
>
> Attempts to re-establish trade ties based on intergovernmental agreements rather than market exchanges have repeatedly been made among the successor states of the former Soviet Union, both within and outside the loose framework provided by the Commonwealth of Independent States . . .
>
> (OECD, 1995, p. 116)

State trading monopolies and old style barter trade remain prominent.

There is no doubt that during the communist era, the proportion of intra-CMEA trade to total trade for each of the CEECs and the Soviet Union itself was kept artificially high for political reasons. In effect, imports were diverted from low cost external sources to higher cost CMEA sources. On the other hand, there is little evidence that the intra-CMEA trade volumes themselves were artificially high. Rather, total trade was too low. The high transaction costs associated with the conduct of trade by command economies served as a brake to all types of trade including intra-CMEA trade. Given the general stagnation, it is not surprising that trade volumes among the CEECs and NISs have declined markedly. Nonetheless, permanent or long term declines in such regional trade are neither natural nor desirable.

In many respects, the CEECs and NISs are natural trading partners for one another. Of course, geography favours strong regional trade linkages. Transportation costs for intra-regional trade tend to be relatively low. Further, transshipment through adjacent countries is necessary to access third-country markets in western Europe and the rest of the world. This is especially true for countries such as Belarus and Kazakhstan that are landlocked and do not border directly on western Europe. Given the comparable income levels, low-product quality will also tend to be less of a problem for intra-regional trade than for trade with the modern market economies. Incidentally, this same argument suggests that it may be worthwhile for the CEECs and NISs to explore trade opportunities with Less Developed Countries and especially the Newly Industrializing Countries of Southeast Asia.

Trade policy measures that serve further to reduce regional trade among the CEECs and NISs, such as withholding MFN status or complicating transshipment procedures, should definitely be avoided. While supra-national arrangements that promote regional trade such as the Central European Free Trade Agreement and the Baltic Free Trade Agreement are desirable, consolidation of the many arrangements would serve to reduce transaction costs. For the NISs, the resuscitation of the CIS and its transformation into an effective market-based trade agreement should be a priority in addition to joining the GATT and WTO. While the initial reluctance or inability to conduct intra-NIS trade on a market basis is understandable, continued reliance on barter trade conducted by state trading monopolies will have a very deleterious effect on the further development of markets and the reduction of transaction costs. A revamped CIS is especially important for the central Asian NISs. For most, if not all, of these countries, Russia will remain the dominant potential market regardless of the underlying politics.

## 22.3. The European Union: To Join or Not to Join?

Linkages with western Europe have been complicated by developments in the west as well as the east. By early 1995, Bulgaria, the Czech Republic, Hungary, Lithuania, Poland, Romania and Slovakia had each concluded Association Agreements or Europe Agreements (EAs) with the European Union. Bilateral trade agreements with the EU have also been negotiated by Estonia, Latvia and Slovenia. These trade treaties were expected to be upgraded to Association Agreements. While Albania has a less far-reaching agreement on commercial and economic cooperation with the EU, at time of writing, it had not yet sought an Association Agreement (OECD, 1995).

The Czech Republic, Hungary, Poland and Slovakia had each concluded free trade agreements with the European Free Trade Area (EFTA), which then consisted of Austria, Finland, Iceland, Norway, Sweden and Switzerland (plus Liechtenstein), by late 1993. Agricultural trade, however, was governed on the basis of separate bilateral agreements with each EFTA country. During 1994, Estonia and Finland also had an operational free trade agreement that included many agricultural products. The supra-national structure of east–west trade relations within Europe was altered when Austria, Finland and Sweden joined the EU at the beginning of 1995 and, as a result, were forced to abrogate all other free trade agreements. Consequently, there has been a move to re-open and broaden the EAs to reflect the extension of the EU (OECD, 1995).

The EAs will provide for free trade in manufactures and many services after a ten year phase-in period. This adjustment period is asymmetric in the sense that the EU will provide access to its markets more rapidly than the CEECs and Baltic Countries (Köves, 1992). While the EAs do include reciprocal market access provisions in agricultural and food products, the end-state does not include free trade in such products. In fact, even the agricultural exports to which the CEECs are entitled under the EAs have often failed to materialize.

> An issue of considerable importance for agricultural trade between the
> CEECs and the EC (European Community) is that of the apparent
> under-utilization of the quotas under which concessionary tariff rates
> apply as a result of the Europe Agreements. While some quotas were fully
> utilized in 1992 and 1993, some were very little used. Utilization was
> greater in 1994, helped by the further reduction in duties to 40 per cent.
> Among the reasons advanced to explain the under-filling of quotas are
> shortages of supplies, lack of experience on the part of the CEECs,
> problems of meeting quality and health standards, and unfamiliar
> administrative procedures.
>
> (OECD, 1995, pp. 15–16)

Thus, transaction costs coupled with product-quality problems help explain the poor performance of the CEEC agricultural sector in yet another area. As well as definite upper limits to the liberalization of agricultural trade, the EAs fail to provide for any significant labour mobility. While the EAs do mention the possibility of EU membership at some future date in their preambles, such membership is by no means a certainty (Köves, 1992, p. 96).

The possible membership of the CEECs and Baltic countries presents a conundrum for the EU.

> Following the enlargement of the EC to 15 countries on 1 January 1995,
> further enlargement to include some of the Central and Eastern European
> Countries (CEECs) is a priority issue. Decisions have taken place at the
> highest political levels by EC leaders that such an enlargement should
> take place to heal the divisions that have separated European countries
> since 1945.
>
> (OECD, 1995, p. 17)

Despite these protestations of political desire to expand EU membership eastward, the harmonization of economic activity and policy between the would-be members and the EU countries is considered to be a difficult problem. Agriculture is one of the most controversial sectors. The OECD (1995) cites disparities in 'economic levels' as a factor that 'may delay accession longer than the CEECs would wish' (p.17). For several years, however, it has been clear that the on-going slump in the CEECs in combination with the deepening of the economic linkages within the existing EU under the 1992 Single European Market Initiative was actually widening the economic disparities between the EU and the CEECs making it even more difficult for the latter countries to meet conditions of membership (Köves, 1992). In reality, economic disparities are a reason for a lengthy phase-in period rather than for postponing membership. As in the case of Greece, Spain and Portugal, limited convergence of the CEECs and the EU may be more likely with membership than without it (Kwieciński, 1994).

From the EU perspective, the fiscal problems of incorporating the CEECs fully into the Common Agricultural Policy (CAP) in its current form, even with its 1992 reforms, are so enormous as to be virtually unthinkable. These problems may be alleviated to a certain extent as the EU moves from highly distortionary price supports to less distortionary income supports to fulfil its Uruguay Round GATT commitments. Indeed, it seems likely that to accommodate the membership of the CEECs, the move to income-based support may be made more rapidly. Further, there seems to be a very strong possibility that new CEEC members will not be eligible for full income support from the EU and may not be eligible for any such support. In particular, it has been suggested that future EU income support be tied to past price support (see Munk, 1992). While this result may be politically expedient from both sides, the economic hypocrisy is evident. Reasonably well-off farmers in western Europe will receive ongoing income support from the EU in partial recompense for their loss of price supports. Meanwhile, their much poorer counterparts in the acceding CEECs and Baltic countries who are going through a much more wrenching transformation to a market-based system will receive little or nothing.

The Baltic States and the CEECs, with the possible exception of Albania, seem determined to pursue membership in the EU. Hungary and Poland made official applications in April 1994 (OECD, 1995). No doubt, this is natural in view of the proximity of western Europe, revulsion with past Soviet domination, and possibly the fear of future domination by Russia. Further, especially for the Czech Republic, Hungary and Slovakia which are landlocked, many of the natural transportation corridors to the rest of the world run through western Europe. The maximum eastward extension of the EU seems clear. Beyond the Baltic Countries and, possibly at some date in the much more distant future, the Ukraine, there is unlikely to be sufficient mutual interest for any of the other NISs to become full members in the EU.

Membership in the EU would be a mixed blessing at best for individual countries and especially for the former CMEA region considered as a whole. This is especially true in agriculture, where EU markets remain highly distorted by the CAP and new members are unlikely to receive full EU price or income support. Trade diversion has the potential to become a major problem both for former CMEA countries that become members of the EU and especially for those that remain outside. For those that join, high priced exports from within the EU will displace lower cost exports from other former CMEA countries and the rest of the world. Particularly for highly protected agricultural commodities, trade diversion could become a serious problem. The static and dynamic trade creation gains[7] from preferential access to western European markets would easily dominate the trade diversion losses for front-line countries such as the Czech Republic, Hungary, Poland, Slovakia and Slovenia. The case is less clear for the Baltic countries and for Bulgaria and Romania.

For those that remain outside the EU as others join, trade diversion is a much more serious problem. As the new joiners become more integrated with the rest of the EU, their trade would be diverted away from the countries remaining outside. Meanwhile, the offsetting benefits of trade creation would be negligible for the outsiders.[8] Thus, there are important external effects when one country joins. On the one hand, a country is hurt as a trading partner joins, but, on the other hand, its own net benefits from joining rise. For example, the benefits of membership for Hungary are higher now that Austria is acceding to the EU and, similarly, the benefits for Slovakia would rise dramatically if the Czech Republic became a member. This type of cascading effect seems to have already been a factor in the extension of EAs to Romania, Bulgaria and now the Baltic countries.

Fortunately, the possibilities are not limited to full EU membership or nothing; various forms of affiliate membership are also possible. In fact, the existing EAs provide a model for affiliate status that is practical and desirable. In such a model each affiliate member would be entitled to unrestricted trade with all full members in non-agricultural goods and services. As in the existing EAs, less restricted access to selected agricultural markets might be arranged on a bilateral basis between all full members as a group and each affiliate member. While the partial exclusion of agriculture would contravene the spirit of the GATT provisions which require that preferential trading arrangements allow for free trade in virtually all goods, the special treatment of agriculture has precedents in the existing EAs and other agreements including the NAFTA. From the EU perspective, this type of affiliation would avoid the highly contentious problems of participation in the CAP and labour mobility.[9] Free trade in manufactured goods and services between various affiliate members could be handled in a number of ways. The closest affiliates might have customs union status with one another and the full members. For such countries, tariffs would have to be harmonized with the

EU, but cumbersome rules of origin would not act as a barrier to internal trade. More loosely affiliated countries could preserve their autonomy concerning external trade policy, but they would be subject to rules of origin in their trade with full members and close affiliates. Thus, the existing EAs can be viewed as a form of the latter type of loose affiliation.

The choice between full membership and close affiliation or loose affiliation depends on the individual circumstances of a particular CEEC or NIS and the terms that can be negotiated with the EU. There is a real possibility that one or other form of affiliation would be more desirable than full membership in the EU for all countries. Here the terms of participation in the CAP are likely to be a decisive factor. In the likely event that the transfers received from western Europe under the CAP will be negligible, the very large domestic expenditures required to implement CAP support policies in individual CEECs in conjunction with the policy distortions themselves may more than negate the benefits of freer access to EU markets. In such a case, close affiliation with the EU could well prove to be superior to full membership. Meanwhile, as an affiliate member a country would still be free to push for the liberalization of agricultural trade on a world basis through the WTO.

It would seem that the Czech Republic, Hungary, Poland and Slovenia should attempt to negotiate full membership or close(r) affiliate status and make their choice depending on the terms that can be negotiated. Slovakia's decision is likely to depend significantly on that of the Czech Republic as well as its own terms. The large agriculture sectors of Hungary and Poland in particular, will make negotiating full membership extremely difficult. Anything from full membership to loose affiliation might be reasonable for the Baltic countries, Albania, Bulgaria and Romania. Here again, the decision of each country should depend on its own terms of membership and the actions of important proximate trade partners. From a European and world geopolitical standpoint, it is extremely important not to shut Belarus, Russia and the Ukraine out of European trade. Given the mutual economic interests, at least loose affiliation of these countries with the EU does seem possible as well as desirable. From a broader world economic perspective, an argument might be made for extending affiliate membership in the EU to the Central Asian NISs as well. Here, however, the high transaction costs coupled with high transport costs reduce the potential trade creation gains relative to trade diversion losses and weaken the mutual interests in affiliation. Thus, in terms of joining the international system, as well as in the national domain, the prospects for the Central Asian NISs are particularly grim. As a first step, negotiating transshipment agreements and upgrading transportation links to non-NIS markets would seem a priority.

## 22.4. Conclusions[10]

Trade in commodities and products is only one aspect of the much discussed globalization – one of the defining features of the last decade of the twentieth century. Globalization, however, is probably a misnomer. The communications, financial, trade and intellectual integration which comprises globalization is manifest most strongly in modern market economies. It is also very evident in the rapidly developing economies of the Asia Pacific Rim and, to a lesser extent, in Latin America. There are pockets of global connections in most other countries but they are islands in a sea of isolated (largely rural) societies whose economic organization is closer to that of the late 19th (or possibly 18th or even 14th) century in the now modern market economies. The technology which underlies globalization has reduced transaction costs sufficiently so that location has little meaning both in space and time. While moving goods still takes time and resources, the costs of arranging international transactions has fallen dramatically. Agribusinesses in modern market economies are fully integrated into, and sometimes leaders in, the globalization process. The economic integration made possible by globalization has increased the relative income of those who participate relative to those who do not.

Will the agrifood sectors of the CEECs and the NISs be able to participate in globalization or will they become isolated (and poor) backwaters? The first prerequisite of participation is being able to engage in international commercial activities. This is why joining the WTO and other trade organizations is vitally important. If trade is heavily restricted, access to globalization technology will be wasted. The CEECs and the NISs need to push for as much market access as possible for agrifood products in their trade negotiations and be willing to open their markets to foreign competition. Having to compete on an international basis is the only way that these agrifood systems can participate in the long run benefits of globalization. Learning how to compete is difficult, and the changes it may bring, painful. One only has to look at the protection demanded by the farm sectors of modern market economies to understand the desire to forestall the changes expected from competition. Unless the rest of society is willing to support the incomes of farmers through their taxes, over time isolation will mean poverty for farmers and rural communities as it does in much of the developing world.

Isolation is caused by high transaction costs. In a world where fresh New Zealand lamb can be delivered to Canadian tables and average Italian families can enjoy fresh flowers grown in Latin America, transportation need no longer be a formidable barrier to agricultural trade. The evidence in this book suggests that transaction costs in the agrifood systems of former command economies are extremely high, both when international trade is attempted and within the domestic market. There can be few stronger images of isolation than the manager of a privatized collective farm in the eastern Ukraine with no effective means to communicate with the outside world on a daily basis, much less minute by minute which is the lifeblood of modern agribusiness. The manager is in charge of a large enterprise with considerable resources, a lot more resources than are available to many agribusiness managers in modern market economies. Unless the manager can transact with other firms, however, the resources are virtually useless and those who have a stake in the farm will not be well served. Globalization will pass the manager and the farm's stakeholders by.

The process of transition is clearly proceeding at different paces in the various countries that comprise the CEECs and the NISs. In some of the countries which have greatest agricultural potential, such as the Ukraine, the process has barely begun. Even in those countries, such as the Czech Republic, which are making good progress, transition has been far more difficult than was ever imagined at the outset. Market institutions have been slow to develop to replace the planner's visible hand. Without these institutions, prices can not fulfil their coordination role efficiently and the transaction costs faced by agribusinesses are high. As a result, nowhere in the CEECs and the NISs has the agrifood sector approached its potential. Only by joining the international system and taking steps domestically to foster the evolution of market institutions can the potential that exists in the agrifood sector and the wider economy be realized.

# Notes

1. The WTO arose from the Uruguay Round GATT negotiations. The WTO supervises the new General Agreement on Trade in Services (GATS) and the Agreement on Trade-Related Intellectual Property Rights (TRIPs) as well as the GATT. The broader world economic system also includes many other institutions such as the International Monetary Fund, the World Bank, and the United Nations Conference on Trade and Development.

2. For example, world oil markets are strongly influenced by the Organization of Petroleum Exporting Countries (OPEC) and much of the trade in textiles and apparel has come under the so-called Multi-Fibre Arrangements (MFAs). Of course, OPEC acts explicitly in the interests of member exporting countries. While the MFAs have coordinated trade between exporters (mainly developing countries) and importers (mainly developed countries), these arrangements have primarily served the protectionist requirements of the importing countries. When the Uruguay Round is fully phased in, trade in textiles and apparel will be brought back under GATT disciplines.

3. Johnson (1953) and Dixit (1987) have examined the formal Nash equilibrium in tariffs for a stylized two-country, two-good world in which each country pursues a simple national welfare maximizing goal. As explained in Chapter 5, a Nash equilibrium is

a state in which none of the players (in the current case, national governments) regrets the strategy that it has chosen.

4. Johnson (1953) and Kennan and Riezman (1988) have shown that a country with sufficient market power might be better off even after retaliation. It is unlikely, however, that even economic powerhouses such as the US, the EU and Japan possesses such market power. Nevertheless, the impact of the Nash equilibrium relative to free trade is unlikely to be symmetric across countries. While large countries are typically at a relative advantage, this rule of thumb need not always hold (see Gaisford and Kerr, 1992).

5. Further, certain incentives for rent-seeking by foreign firms desiring preferential access are eliminated by extending MFN status to all trade partners and requiring tariffication. For example, if Russia were to remain outside the WTO Canadian beef producers, or indeed, the Canadian government might lobby for tariff concessions or a more generous import quota. Such rent-seeking is, of course, inefficient on a global basis.

6. The proposed extension to agriculture has involved extensive rules of origin to deal with transshipment issues because, to date, Estonia has maintained a much less restrictive regime for food products than the other Baltic States (OECD, 1995).

7. Static trade creation gains arise from new trade that is created when low-cost exports are available within the EU, while dynamic trade creation gains materialize over time as faster income growth brought about by membership generates additional trade.

8. If the new joiners grow more rapidly as a result of membership, dynamic trade creation can occur as more imports are demanded from insiders and from outsiders alike.

9. Coordination of social and economic policy is also a major issue, but the possible move to a common currency no longer seems to be a significant stumbling block. It is now virtually certain that a significant subset of existing EU countries will either be unwilling or unable to meet the criteria for participation in a common currency.

10. This section acts as both the conclusion to the chapter and to the book.

# References

Dixit, A.K. (1987) Strategic aspects of trade policy. In: Bewley, T. (ed.) *Advances in Economic Theory, Proceedings of the 5th World Congress of the Econometrics Society.* Cambridge University Press, Cambridge, UK, pp. 329–362.

Gaisford, J.D. and Kerr, W.A. (1992) Which country loses the least in trade wars. *Australian Journal of Agricultural Economics* 36, 249–274.

Helpman E. and Krugman, P.R. (1989) *Trade Policy and Market Structure.* MIT Press, Cambridge, Massachusetts, 191 pp.

Johnson, H.G. (1953) Optimum tariffs and retaliation. *Review of Economic Studies* 21, 142–153.

Kennan, J. and Riezman, R. (1988) Do big countries win tariff wars? *International Economic Review* 29, 81–85.

Köves, A. (1992) *Central and East European Economies in Transition.* Westview Press, Boulder, Colorado, 150 pp.

Kwieciński, A. (1994) Poland's membership in the European Union by the year 2000 – a realistic objective? Agricultural aspects. Paper presented to a conference on Agribusiness and the Food Industry in Central and Eastern Europe, Budapest, Hungary, March.

Munk, K.J. (1992) The development of agricultural policies and trade relations in response to the transformation in Central and Eastern Europe. Paper presented to the International Agricultural Trade Research Consortium, St Petersburg, Florida, December.

OECD (1995) *Agricultural Policies, Markets and Trade in the Central and Eastern European Countries, Selected New Independent States, Mongolia and China: Monitoring and Outlook 1995*. Centre for Co-operation with the Economies in Transition, Organization for Economic Co-operation and Development, 232 pp.

# Index